Nanoscale Science and Engineering Forum

Presentations at the 2009 AIChE Annual Meeting

Nashville, Tennessee, USA
8-13 November 2009

ISBN: 978-1-61567-943-0

Printed from e-media with permission by:

Curran Associates, Inc.
57 Morehouse Lane
Red Hook, NY 12571

Some format issues inherent in the e-media version may also appear in this print version.

Printed by Curran Associates, Inc. (2010)

For permission requests, please contact AIChE
at the address below.

AIChE
3 Park Avenue
New York, NY 10016-5991

Phone: (203) 702-7660
Fax: (203) 775-5177

www.aiche.org

Additional copies of this publication are available from:

Curran Associates, Inc.
57 Morehouse Lane
Red Hook, NY 12571 USA
Phone: 845-758-0400
Fax: 845-758-2634
Email: curran@proceedings.com
Web: www.proceedings.com

TABLE OF CONTENTS

Development of An Undergraduate Nanotechnology Certificate Program at Georgia Tech

Sankar Nair, School of Chemical & Biomolecular Engineering, Georgia Institute of Technology, Atlanta, GA

This talk describes recent and ongoing initiatives in nanotechnology education at Georgia Tech, focusing primarily on the development of a Nanotechnology Certificate Program for undergraduates (supported by NSF-NUE). The talk will present an overview of this collaborative program, which is targeted at teaching the fundamentals of nanoscience and nanoengineering to senior-level undergraduates and graduates through a combination of traditional classroom learning and inquiry-based team learning. In doing so, we fuse together the roles of nanofabrication techniques, molecular assembly of nanomaterials, nanoscale characterization techniques, and fundamentals of nanophysics/nanochemistry towards the production of nanoscale devices. Our proposed program - with faculty participation from six disciplines - is based on a 15-credit curriculum of collaborative instruction, which includes a hands-on laboratory course, multidisciplinary lecture courses, an invited speaker program, and a 3-6 credit research project in any nanoscience/nanotechnology research group at GT. Students in these courses will have ample opportunities for collaborative learning and development of technical communication skills via conceptual design projects and live demonstrations of state-of-the-art "top-down" fabrication of nanodevices. In the Nanosystems Laboratory course based on both collaborative learning and guided-inquiry instructional techniques, we are developing four Modules that allow students to synthesize nanomaterials, computationally design a process for their assembly into individually addressable devices, implement the nanodevice design experimentally, and analyze its properties and functionality. Concurrently, they will be exposed to talks and discussions involving invited speakers who are recognized experts in the economic, industrial, environmental, and other societal impacts of nanotechnology practice.

See more of this Session: Education Issues in Nanotechnology
See more of this Group/Topical: Nanoscale Science and Engineering Forum

NUE - Teaching Undergraduates Nanomanufacturing Engineering (TUNE)

James D. Palmer, Chemical Engineering Department, Louisiana Tech University, Ruston, LA

Hisham Hegab, Mechanical Engineering, Louisiana Tech University, Ruston, LA

The goal of this project is to produce nanosystems engineering graduates with the critical thinking and manufacturing skills necessary to meet the nation's workforce needs in nanotechnology. As indicated by recent reports (National Nanotechnology Advisory Panel review of the National Nanotechnology Initiative) and industrial surveys of nanotechnology workforce needs, one of the barriers to future commercialization of nanotechnology is workers with abilities to scale-up nanomanufacturing from the research lab to pilot/small scale production levels. While there are a number of nanotechnology laboratory and lecture classes nationwide, the majority of the experiences demonstrate 'recipes' to produce nanostructures but do very little to prepare students with the critical thinking skills required to assess and improve nanomanufacturing processes.

This project will address these issues through the following activities. First, creating a new junior level nanomanufacturing laboratory course that focuses beyond fundamental nanofabrication techniques to consider the production engineering aspects of nanomanufacturing (i.e., quality, uniformity, production costs of the product) through effective characterization of the manufacturing process. Second, integrating a focus on nanomanufacturing into capstone senior projects that will build upon the students knowledge and skills obtained in their junior year to tackle "open-ended" process design problems. Third, connecting industry to Louisiana Tech's Nanosystems Engineering degree program to provide assessment and input on specific knowledge and skills inventories that are required for workforce needs. Fourth, a faculty workshop to provide project assessment and training of Nanosystems Engineering faculty to ensure the proper balance of nanoscience versus nanoengineering in capstone projects. Over 20% of the faculty identified to participate in the workshops are either minority or women. The increased visibility of these faculty will allow them to become role models to a larger portion of our student population.

See more of this Session: Education Issues in Nanotechnology
See more of this Group/Topical: Nanoscale Science and Engineering Forum

Hands-On Teaching of Nanoscale Fundamentals with the Light Microscope Equivalent of This Century - Scanning Probe Microscopy

Daniel B. Knorr Jr., Chemical Engineering, University of Washington, Seattle, WA

Mehmet Sarikaya, Materials Science and Engineering, University of Washington, Seattle, WA

Rene M. Overney, Chemical Engineering, University of Washington, Seattle, WA

One of the most effective ways of introducing nanotechnology in science and engineering undergraduate curricula is hands-on teaching with the tools that precipitated nanoscience and nanotechnology, namely, scanning probe methods (SPM) such as scanning tunneling microscopy (STM) and scanning force microscopy (SFM). Currently, key factors that hamper the use of SPM based educational modules in laboratory settings of 2-year and 4-year higher educational programs are

(1) the high capital costs to acquire multiple instruments that give every student access,

(2) limited support for maintaining laboratories, and

(3) the lack of time and experience to develop appropriate effective teaching modules that incorporate this new technology.

We have responded to the above needs by introducing a new paradigm for operating and maintaining SPM instrumentations to make resources more readily available. Further, we have developed and tested a variety of hands-on SPM undergraduate teaching modules, which we have made widely available electronically (http://depts.washington.edu/nanolab/NUE UNIQUE/NUE UNIQUE.htm).

Over the past three years, we have developed an effective teaching laboratory with leased top-notch equipment (scanning probe methods) providing true hands-on experience in authentic undergraduate laboratory settings. Thereby, we have been working with a SPM manufacturer with the intent to jump start a new paradigm for acquiring SPM teaching products for educational institutions.

Thus, the focus has been to develop teaching material and opportunities that are transferable and adaptable by most educational institutions. To date, we have run three successful undergraduate laboratory workshops, where students were able to have extensive hands-on experience on six SPM modes of operation including:

(i) electrostatic force microscopy involving photovoltaic polymeric materials,

(ii) tunneling microscopy and the determination of the workfunction,

(iii) modulation force microscopy with a glass transition analysis,

(iv) force-displacement analysis under controlled humidity conditions to distinguish capillary forces from Van der Waals interactions,

(v) non-contact force microscopy for protein adsorption studies in air and liquid environment,

(vi) SPM dip-pin nanolithography.

Student response surveys were conducted to ascertain perceived benefit and to determine whether or not program objectives were met. We found that these modules provided a truly hands-on experience in a classroom laboratory setting with a small student to instrument ratio of 4:1.

Further information and laboratory modules themselves are available at:http://depts.washington.edu/nanolab/NUE UNIQUE/NUE UNIQUE.htm.

*NUE UNIQUE was initiated by the National Science Foundation under Grant No. 0634088, and is supported by GEMSEC a NSF Material Science and Engineering Center at the University of Washington

See more of this Session: Education Issues in Nanotechnology
See more of this Group/Topical: Nanoscale Science and Engineering Forum

Integrating Nanofabrication Experiments into the Undergraduate Teaching Laboratory

Erin L. Jablonski, Chemical Engineering, Bucknell University, Lewisburg, PA

Recently, through the support of an NSF-NUE grant, Bucknell has established a nanofabrication laboratory appropriate to the undergraduate setting. With the goal of advancing the Nanotechnology Educational Initiative in the College of Engineering, nanofabrication experiments have been designed for several courses in the undergraduate engineering curriculum at Bucknell, including "Science of Materials" and "Colloid, Surface & Nanoscience". This talk will present the how the available nanofabrication facilities are incorporated across the engineering curricula, how faculty workshops and student internships have been utilized to integrate nanofabrication experiments in several disciplines, and how individual courses have been impacted. Using a project-based design approach, students are introduced to the methods and materials of fabrication simultaneously in the classroom and laboratory. Experiments designed for implementation in numerous teaching laboratories, appropriate to the undergraduate level, will be discussed. Specifically, an experiment in photolithography that uses inexpensive materials (UV-curable adhesive, glass slides, broad spectrum UV lamp) to demonstrate how a photomask is used to image a pattern on a substrate will be detailed.

See more of this Session: Education Issues in Nanotechnology
See more of this Group/Topical: Nanoscale Science and Engineering Forum

Nanoparticle Delivery Vehicle to Mimic Bacterial Invasion of Lung Epithelium

Timothy Brenza, Division of Pharmaceutics, The University of Iowa, Iowa City, IA

M. A. Apicella, Microbiology, The University of Iowa, Iowa City, IA

Jennifer Fiegel, Chemical and Biochemical Engineering and Pharmaceutics, The University of Iowa, Iowa City, IA

A variety of phylogenetically distinct bacterial pathogens, such as non-typeable *Haemophilus influenzae* (NTHi), invade host cells in the upper airways by binding the platelet-activating factor (PAF) receptor. Lipooligosaccharide (LOS) glycoforms naturally expressed on the bacterial cell surface facilitate bacteria-epithelial interactions. We are utilizing LOS ligands isolated from the bacteria surface to coat the surface of nanosized delivery vehicles to mimic the invasion characteristics of NTHi, enabling targeting of drug molecules to the respiratory epithelium. We have observed that modification of the surface of nanoparticles with absorbed LOS glycoforms from the 2019 NTHi strain increased cellular adherence and penetration when applied to bronchial epithelial cell cultures *in vitro*. However, when LOS modified nanoparticles were applied to a more physiologically relevant model of the respiratory system their diffusion through the respiratory fluid barrier is hindered. Therefore, the goal of our research is to minimize nanoparticle interactions with respiratory fluids while retaining the improved cellular adherence and penetration afforded by the LOS coating.

Commercially available fluorescent nanoparticles were modified with various NTHi LOS glycoforms, polyethylene glycol (PEG; 5-40 kDa), and combinations thereof. Upon coating, the modified nanoparticles exhibited an increase in size proportional to the molecular weight of the coating material and more neutral surface charge. Calu-3 bronchial epithelial cells were grown to confluent monolayers under air-interfaced culture conditions on semi-permeable Transwell membranes. The adherence and uptake of nanoparticles into the epithelial cells was determined under three apical fluid conditions: no apical fluid, natural secretions produced by the cells and synthetically prepared mucus mimetic. Laser scanning confocal microscopy was used to differentiate the nanoparticles associated with the cell monolayer from those that remained in the apical fluid. Our results show that transport through mucosal fluids and uptake into lung epithelium is improved through optimization of: PEG molecular weight, PEG surface coverage and choice of LOS bacterial strain.

See more of this Session: Micro- and Nanodevices for Targeted Therapeutics
See more of this Group/Topical: Nanoscale Science and Engineering Forum

PEGylated Composite Nanoparticles for Photodynamic Therapy

Stephanie J. Budijono, Chemical Engineering, Princeton University, Princeton, NJ

Jingning Shan, Mechanical and Aerospace Engineering, Princeton University, Princeton, NJ

Yiguang Ju, Mechanical and Aerospace Engineering, Princeton University, Princeton, NJ

Robert K. Prud'homme, Chemical Engineering, Princeton University, Princeton, NJ

Photodynamic Therapy (PDT) is an approved treatment for non-small cell lung cancers. The therapy relies on activation of a photosensitizer by visible light to produce cytotoxic singlet oxygen. The current method has significant limitations. Most photosensitizers (PS) have limited water solubility and thus non-deliverable to target tumors, the visible light used to excite these photosensitizers has a limited penetration distance in tissue and the mode of singlet oxygen delivery is not targeted specifically to cancer cells.

This paper presents a composite nanoparticle design where upconverting nanophosphors (UCNP) and a photosensitizer (PS) are simultaneously packaged in a biodegradable nanoparticle compartment. The UCNP, composed of NaY_4:Yb^{3+}, Er^{3+}, emits visible light, when it is exposed to near infra-red light (λ=980 nm). The visible light emitted can be used to excite the PS. The near infra red light, with a greater tissue penetration distance compared to visible light, will answer the tissue penetration distance problem, hence making our formulation accessible to deep-seated tumors.

The UCNP and the photosensitizer (meso-tetraphenyl porphine (mTPP)) are packaged in a single compartment using biocompatible, FDA-approved polymeric materials. Candidate block-copolymers include PEG-PCL (poly(ethylene glycol)-block-poly(caprolactone), PEG-PLGA (poly(ethylene glycol)-block-poly(lactide-co-glycolide) acid) and PEG-PLA (poly(ethylene glycol)-block-poly(lactide)) of varying molecular weights. The CNPs are manufactured using Flash NanoPrecipitation technology which provides homogenous mixing, resulting in good control over nanoparticle size with high reproducibility.

We found that PEG-PLGA and PEG-PLA provides better particle stabilization in water and in physiological conditions compared to PEG-PCL. We further found that PEG-PLA with molecular weights 5000-b-5000 Da and 5000-b-10,000 Da formed superior nanoparticle vehicle, while PEG-PLA with molecular weights of 5000-b-20,000 Da did not form stable nanoparticles. With a high ratio of photosensitizer to drug (1:3, by weight) in the composite, these composite nanoparticles are capable of producing cytotoxic singlet oxygen. Functionalization of PEG for active targeting ligand attachment on the surface of nanoparticles is underway.

See more of this Session: Micro- and Nanodevices for Targeted Therapeutics
See more of this Group/Topical: Nanoscale Science and Engineering Forum

Shape-Shifting Particles for Drug Delivery

Jinwook Yoo, Chemical Engineering, University of Californica - Santa Barbara, Santa Barbara, CA

Samir Mitragotri, Chemical Engineering, University of California, Santa Barbara, Santa Barbara, CA

Size and surface chemistry have traditionally served as two dominating design parameters of drug delivery carriers. In the last few years, however, there has been growing recognition that shape is a key parameter that dictates applications of polymeric carriers for drug delivery. Nevertheless, the field of shape engineering is still in its infancy and comprehensive role of particle shape in drug delivery is yet to be understood.

Herein we investigated the role of shape in achieving two important functions in drug delivery: avoiding phagocytosis in macrophages and promoting endocytosis in target cells. Particles were synthesized from PLGA with two shapes, spheres and elliptical disks, and their internalization was studied in murine macrophages and endothelial cells (human umbilical vein endothelial cells). The results showed that macrophages as well as endothelial cells favored elliptical disks over spheres for internalization by a factor of about four. While reduced phagocytosis of elliptical disks by macrophages is highly desired, reduction of endocytosis due to the same morphological features is undesired.

To address this challenge, shape-shifting particles were developed. These particles were initially elliptical disk-shaped and changed their shape to spheres over time. The degree and the rate of the shape change depend on several factors including glass transition temperature, intrinsic viscosity, particle size and environmental temperature. By controlling these parameters, particles were made to change their shape over a time ranging from a few minutes to a few days. The ability to change the shape of particles in real time upon administration into the body provides an additional parameter to engineer polymeric carriers for applications in drug delivery.

See more of this Session: Micro- and Nanodevices for Targeted Therapeutics
See more of this Group/Topical: Nanoscale Science and Engineering Forum

Engineering Particle Geometry for Target Specific Adhesion Using Simplified Synthetic Microvascular Networks

Nishit Doshi, Chemical Engineering, University of California, Santa barbara, Santa Barbara, CA

Balabhaskar Prabhakarpandian, Biomedical Technology, CFD Research Corporation, Huntsville, AL

Angela Ramsey, Biomedical Technology, CFD Research Corporation, Huntsville, AL

Kapil Pant, Biomedical Technology, CFD Research Corporation, Huntsville, AL

Samir Mitragotri, Chemical Engineering, University of California, Santa Barbara, CA

Designing novel carriers by optimizing important parameters such as size, shape, composition and surface chemistry, and mechanical properties has resulted in remarkable accomplishments in the field of drug delivery and diagnostic imaging. The next step is to elucidate the biological interactions of these carriers. In this study, the effect of particle geometry, which has recently received significant attention, on the transport of particles through the vasculature and adhesion propensity at the target site has been illustrated. Experiments were performed in simplified synthetic microvascular networks that mimic the in vivo environment.

A synthetic system that simulates the vasculature offers great advantage since it allows variation of a single design parameter to study the particle interactions. Hence, a simplified synthetic microvascular network in the form of a bifurcating channel was synthesized by soft lithographic technique for this study. Particles of different shapes (spheres, rods, elliptical disks and oblate ellipsoids) and sizes (1μm, 3μm and 6μm) were fabricated by the film stretching technique. The surface of the channels and the particles were modified using complementary biomolecules to determine the effect of particle shape on receptor mediated adhesion.

The attachment propensity of spheres, elliptical disks, rods and oblate ellipsoids of different sizes was compared in a bifurcating microchannel. It was observed that elongated particles (rods and elliptical disks) exhibit higher adhesion propensity than particles of other shapes and the difference is more pronounced at larger sizes. Increasing the extent of stretching of 3μm elliptical disks from 4 to 6.5 increased the attachment two fold in both the inlet (straight) section of the channel and at the junction. Moreover, particle adhesion at the junction was found to be 3-6 fold higher than the straight sections. Larger particles, due to higher settling velocities tend to attach more (~2 fold) at the inlet compared to smaller particles. Static experiments were also performed where no significant difference in the attachment of particles of different geometry was observed, thus illustrating the importance of studying particles in flow. Further experiments with more complex networks and with cells grown in the microchannels will help understand the interactions better.

Particle geometry plays an important role in transport properties and targeting efficiency of particles. Such information will be useful in optimizing the design of carriers for specific biomedical applications.

See more of this Session: Micro- and Nanodevices for Targeted Therapeutics
See more of this Group/Topical: Nanoscale Science and Engineering Forum

Hydrolytically Degradable Beta-Cyclodextrin-Based Nanoparticles Bearing Quaternary Amine Groups for Drug Delivery Across the Blood-Tissue Barriers

Linfeng Wu, Department of Pharmaceutical Sciences, Thomas Jefferson University, Philadelphia, PA

Tao L. Lowe, Department of Pharmaceutical Sciences, Thomas Jefferson University, Philadelphia, PA

The blood-tissue barriers such as the blood-brain barrier and the blood-retinal barrier are selective and protective barriers. Most of therapeutic agents developed can not pass these barriers. Therefore, these barriers become a formidable obstacle to overcome when trying to deliver therapeutic agents to treat the diseases occurring behind these barriers.

The purpose of this study is to develop novel hydrolytically degradable b-cyclodextrin-based nanoparticles to enhance the permeability of drugs across the blood-tissue barriers. Recently, poly(amino ester) polymers have been developed as carriers for gene delivery because of their biodegradability and cationic nature[1]. Previous work[2] in our group has demonstrated that quaternary ammonium β-cyclodextrin nanoparticles are nontoxic and can significantly enhance the permeability of doxorubicin (DOX) across the *in-vitro* blood-brain barrier. Based on these previous studies, we developed a new generation of β-cyclodextrin-based nanoparticles. In these nanoparticles, quaternary amine groups were conjugated to β-cyclodextrin and hydrolytically degradable esters were incorporated in the crosslinking structure in the form of poly(amino ester) or poly(ester).

The chemical, physical and degradation properties of the developed nanoparticles were characterized using attenuated total reflectance Fourier transform infrared spectroscopy (ATR-FTIR), nuclear magnetic resonance (NMR), dynamic light scattering (DLS), and atomic force microscopy (AFM). The MTT assay was used to evaluate the cytotoxicity of the nanoparticles to human brain microvascular endothelial cells (HBMVEC). HBMVEC monolayers were used as an *in vitro* blood-tissue barrier model to assess the effects of the nanoparticles on the blood-tissue barrier permeability of model drugs. Model drugs such as DOX were loaded into the nanoparticles in dimethyl sulfoxide by partition. Studies were also conducted on the release kinetics of model drugs from the nanoparticles in phosphate buffered saline solution (pH 7.4) at 37 °C. Our preliminary results suggested that these hydrolytically degradable nanoparticles have great potential for delivering drugs across the blood-tissue barriers to treat retinoblastoma, brain tumors and other diseases behind these barriers.

References:

[1] Lynn, D.M. and Langer, R. *J. Am. Chem. Soc.*, **2000,** *122 (44): 10761-10768.*

[2] Gil, E.S.; Li, J.; Xiao, H. and Lowe, T.L. Biomacromolecules, 2009, 10 (3): 505-516.

See more of this Session: Micro- and Nanodevices for Targeted Therapeutics
See more of this Group/Topical: Nanoscale Science and Engineering Forum

Development of Folate-PAMAM and Azithromycin-PAMAM Nanodevices for Biodistribution, Imaging, Cellular Uptake, and Targeted Delivery to Chlamydial Infections

Manoj K. Mishra, Department of Chemical Engineering and Material Sciences, Wayne State University, Detroit, MI

Kishore Kotta, Department of Chemical Engineering and Material Sciences, Wayne State University, Detroit, MI

Ilyes Benchaala, Department of Immunology & Microbiology, School of Medicine, Wayne State University, Detroit, MI

Susan Wykes, Department of Immunology & Microbiology, School of Medicine, Wayne State University, Detroit, MI

Mirabela Hali, Department of Immunology & Microbiology, School of Medicine, Wayne State University, Detroit, MI

Indrajit Sinha, Department of Immunology & Microbiology, School of Medicine, Wayne State University, Detroit, MI

Judith Whittum-Hudson, Department of Immunology & Microbiology, School of Medicine, Wayne State University, Detroit, MI

Rangaramanujam M. Kannan, Department of Chemical Engineering and Material Sciences, Wayne State University, Detroit, MI

Chlamydia trachomatis is an intracellular pathogen that causes major sexually transmitted diseases. Chlamydiae are very immunogenic bacteria and can cause inflammatory reactions in different parts of the body (1). Reactive arthritis (ReA) is one such example of an inflammatory reaction caused by chlamydial infection; ReA develops after dissemination of infected cells from the original site of infection (usually genital tract) (2). Since Chlamydiae can go into a persistent infection form, treatment with antibiotics such as azithromycin is a challenge. Dendrimers are emerging as a new class of nanoscale drug delivery vehicle, because of their well-defined structure, tailored surface properties, and ability to deliver drugs intracellularly (3). Since targeting ligands can be used to target dendrimers to the sites of inflammation it can be an effective delivery vehicle. This study seeks to develop, characterize and evaluate (in vitro and in vivo both) dendrimer-based targeted drug delivery nanodevices for therapy of chlamydial infections including targeting chlamydia-induced arthritic inflammation in a murine model (4).

It has been suggested that folic-acid conjugated nanoparticles can target inflammation in vivo. Therefore, we expect dendrimer-folic acid conjugates could also target inflammation. To explore this concept, we have synthesized two nanodevices for in vivo animal imaging experiments. We conjugated Cy5.5, a near IR imaging agent to the PAMAM dendrimer (PAMAM-Cy5.5) as a control, and a folic acid-PAMAM conjugate (PAMAM-Cy5.5-FA), by conjugating folic acid to the dendrimer and labeled it with Cy5.5. PAMAM-Cy5.5 was prepared by conjugating Cy5.5 NHS ester to the PAMAM dendrimer alone (G4-NH2). The purity of the nanodevices was determined by reverse phase HPLC. Folate mediated nanodevice was developed by reacting folic acid with PAMAM dendrimer (G4-NH2) under coupling reaction condition (5, 6). The resulting compound was purified by dialysis (MW cutoff 1000 Da) and lyophilized to get PAMAM-FA and characterized by proton NMR and MALDI TOF mass. The resulting compound has a comparatively stable amide bond and a multiplet at 6.6 ppm represents aromatic protons of folic acid and indicates the formation of compound. The loading of folic acid to the surface of dendrimer was calculated by proton NMR techniques by comparing amidic protons of dendrimer and aromatic protons of folic acid. Cy5.5 NHS ester was reacted to the above intermediate under basic conditions (7) to get folate mediated nanodevice, PAMAM-Cy5.5-FA.

The biodistribution of the nanodevices was imaged at the Cy5.5 wavelength (Ex 625 nm; Em 700 nm), using a Kodak whole animal imager 2h to 120h after injection at 7-14 dpi. The paws (major sites of inflammatory response), and the various organs were harvested for quantification of nanodevice uptake. PAMAM-Cy5.5-FA was taken up into inflamed paws at significantly higher levels than PAMAM-Cy5.5. All infected mice showed similar internal pattern of Cy5.5 distribution i.e., paws, genital tract, thymus, spleen, and lung in both cases. We did not see any significant localization in the liver, heart or the kidney. These results indicate that dendrimers are certainly playing a role in the delivery to the sites of inflammation, but folate-targeting (PAMAM-Cy5.5-FA) enabled higher accumulation and retention at the sites of inflammation (up to 120h). This lengthy time of detection further suggests that FA facilitates the retention of the dendrimer in the circulation and may be targeting to specific sites containing inflammatory cells which bear FA receptors. Analyses were performed with ImagePro Plus software (Media Cybernetics, Inc).

The therapeutic nanodevices based on PAMAM dendrimer conjugated to Azithromycin (PAMAM-AZ) were evaluated in vitro in C trachomatis infected cells. First we have reacted Azithromycin with glutaric acid under coupling reaction condition to get an acid functionality intermediate. A multiplet at 2.0 ppm represents methylene groups of glutaric acid and confirms the formation of the intermediate. This intermediate was further reacted with hydroxy group of PAMAM dendrimer under same reaction condition to get our therapeutic nanodevice (PAMAM-AZ). The reaction mixture was purified by dialysis and lyophilized to get PAMAM-AZ and characterized by proton NMR and MALDI TOF mass. The loading of Azithromycin to the surface of the dendrimer was calculated by comparing amidic proton of the dendrimer to aliphatic protons of the Azithromycin in between 4.0-5.0 ppm in NMR spectra and we conclude that 6.6 molecules were tagged per molecule of dendrimer. Chlamydia trachomatis-infected mammalian HEp2 cells were treated with either free AZ (control) or PAMAM-AZ (at equivalent AZ concentrations) at either the time of infection (t=0) or 24 hours post infection (t=24). PAMAM-AZ added at the time of infection (t=0) completely prevented infection of the cells, as measured by the total absence of Chlamydial inclusions. When PAMAM-AZ was added 24 hours after infection, it reduced the inclusion count by more than 30%. The therapeutic nanodevice delivers the drug in a more efficient way in a shorter span of time and maintains its effect for a longer time frame as seen by reduced inclusion size (area). This suggests that dendrimers transport the drug into the cells, and release it in a successful manner to reduce overall infectivity and infectious load within infected cells (size of inclusion).

In both imaging and biodistribution and in vitro cell entry experiments, we found a considerable level of targeting of the nanodevices to the sites of infection. The whole animal imaging data suggest a significantly higher uptake of folic acid mediated nanodevice compared to nanodevice having no folic acid. Based on the results of our experiments, we will design therapeutic nanodevices with folic acid as ligand which would target inflammation at sites of chlamydial infection and thereby significantly improve drug delivery to the infected cells. This is a noble step towards diagnosis and treatment of Chlamydia related infections and associated inflammatory diseases. It represents the first report of dendrimer-FA targeting of chlamydia-associated arthritic inflammation to the best of our knowledge.

References:

1. Bavoil, P.M. and Wyrick, P.B., eds., Chlamydia Genomics and Pathogenesis, eds Horizon Bioscience, Norfolk, UK, 2006.

2. Whittum-Hudson, J.A., Gérard, H.C., Schumacher, H.R. Jr., and Hudson A.P. In: Chlamydia Genomics and Pathogenesis, eds Bavoil, P.M., Wyrick, P.B. (Horizon Bioscience, Norfolk, UK) 2006, 475-504.

3. Lee, C.C., MacKay, J.A., Fréchet, J.M.J. and Szoka F.C.; Nature Biotechnology, 2005, 23, 1517-1526.

4. Whittum-Hudson, J.A., Gérard, H.C., Clayburne, G., Schumacher, H.R., and Hudson, A.P.; Revu de Rheum (English), 66 (1, Suppl), 1999, 50S-56S.

5. Shukla, S., Wu, G., Chatterjee, M.,Yang, W., Sekido, M., Diop, L. A., Muller, R., Sudimack, J. J., Lee, R. J., Barth, R. F. and Tjarks, W.; Biocojugate Chem., 2003, 14, 158-167.

6. Kano, K., Liu, M. and Frechet, M. J.; Biocojugate Chem., 1999, 10, 1115-1121.

7. Chen, Z., Levi, J., Xiong, Z., Gheysens, O., Keren, S., Chen, X., and Gambhir, S. S.; Bioconjugate Chem., 2006, 17, 662-669.

See more of this Session: Micro- and Nanodevices for Targeted Therapeutics
See more of this Group/Topical: Nanoscale Science and Engineering Forum

Biomimetic Spleen for Sepsis Therapy

Chong Wing Yung, Surgery - Vascular Biology Program, Harvard Medical School, Children's Hospital Boston & Wyss Institute, Boston, MA

Donald E. Ingber, Surgery & Pathology, Harvard Medical School, Children's Hospital Boston & Wyss Institute, Boston, MA

Sepsis is a major killer of man because a high microbial load in blood overcomes the immune system resulting in multi-systems failure. This is a particularly significant problem in premature infants and immunocompromised patients with fungal infections since antibiotic therapies are dose-limiting. Development of an extracorporeal blood cleansing device that can rapidly clear the blood of pathogens without loss or alteration of normal blood components could help remedy this problem. We have previously developed a prototype microfluidic-micromagnetic blood cleansing device that rapidly clears living Candida albicans pathogens from flowing human whole blood with high efficiency (80%) and throughput (20 mL/hr/device) in a single pass. Magnetic opsonins were created by coating 1-μm microbeads with pathogen-specific antibodies. One drawback of this device is that some blood is loss in the saline stream (~13%) during the separation process. More recently, we have developed an advanced design inspired by the native architecture of the spleen that virtually eliminates blood loss while further enhancing separation efficiency. This Biomimetic Spleen has three microfluidic layers that replicate the microvasculature, sinusoids, and lymphoid follicles of the organ. The result is a simple and yet robust device that can clear 85-95% of C.albicans at similar flow rates without any blood loss or dilution. (Funded by CIMIT, DoD, Wyss Institute)

See more of this Session: Micro- and Nanodevices for Targeted Therapeutics
See more of this Group/Topical: Nanoscale Science and Engineering Forum

Incorporation of Reactive Silver-Tricalcium Phosphate Nanoparticles Into Polyamide (Nylon) Allows Preparation of Self-Sterilizing Fibers

Lukas C. Gerber, Department of Chemistry and Applied Biosciences, ETH Zurich, 8093 Zurich, Switzerland

Dirk Mohn, Department of Chemistry and Applied Biosciences, ETH Zurich, 8093 Zurich, Switzerland

Wendelin J. Stark, Department of Chemistry and Applied Biosciences, ETH Zurich, 8093 Zurich, Switzerland

Polymer fibers are ever-present in our world. In 2002 the global production of manufactured fibers was 36 million metric tons: Fiberglass pads build modern cars, polyesters form textile goods and polyamide fibers are used as bristles for brooms and toothbrushes. Since bacterial contamination may cause diseases or even death, it is desirable to provide fibers with antibacterial properties.

The present contribution investigates the ability of silver-tricalcium phosphate (Ag/TCP) nanoparticles to build a reactive antibacterial system in polyamide fibers. These nanoparticles consist of 20-50 nm sized carriers, based on calcium phosphate, a major nutrient of bacteria. This bait is spiked with 1-2 nm silver particles. In the presence of growing bacteria, the tricalcium phosphate (TCP) carrier particles are biodegraded and trigger the release of the antibacterial silver [1]. Applying silver in this complex form allows significant increases in antibacterial efficiency due to a larger contact area [2] and triggered release of the active silver exclusively when TCP is consumed by bacteria [3]. This reduces the required amount of silver if compared to non-reactive systems.

Ag/TCP nanoparticles (1.3 wt% silver) were produced by flame spray synthesis [4] and subsequently extruded with polyamide 6 yielding a composite polymer with 2 wt% filler. Following melt-spinning of the extrudates resulted in fibers (~110 μm thick; containing 260 ppm silver) for antibacterial testing. Fibers were contaminated with the clinically relevant strains *Escherichia coli* or *Streptococcus sanguinis*, respectively. The reactive fibers demonstrated significantly reduced plate count within 24 hours if compared to pure polyamide reference fibers (several orders of magnitude). More specifically, compared to pure polymer bristles, the number of colony forming units was reduced by 99.999 % with *E. coli* and 99.6 % with *S. sanguinis*.

Figure: Antibacterial activity of silver nanoparticles incorporated into a reactive system with tricalcium phosphate nanoparticles extruded with polyamide 6 and comparison to pure polymer. Temporal evolution of the total bacteria number during the course of the day after contamination with *E. coli* or *S. sanguinis* is shown. (Dark: pure PA6 fibers; Light: Ag/TCP fibers).

References: 1. Q.L. Feng, J. Wu, G.Q. Chen, F.Z. Cui, T.N. Kim, and J.O. Kim, J. Biomed. Mater. Res., 52, 4 (2000). 2. I. Sondi and B. Salopek-Sondi, J. Colloid Interface Sci., 275, 1 (2004). 3. S. Loher, O.D. Schneider, T. Maienfisch, S. Bokorny, and W.J. Stark, Small, 4, 6 (2008). 4. W.J. Stark, S. Pratsinis, M. Maciejewski, S.F. Loher, and A. Baiker, WO2005087660-A1 (2005).

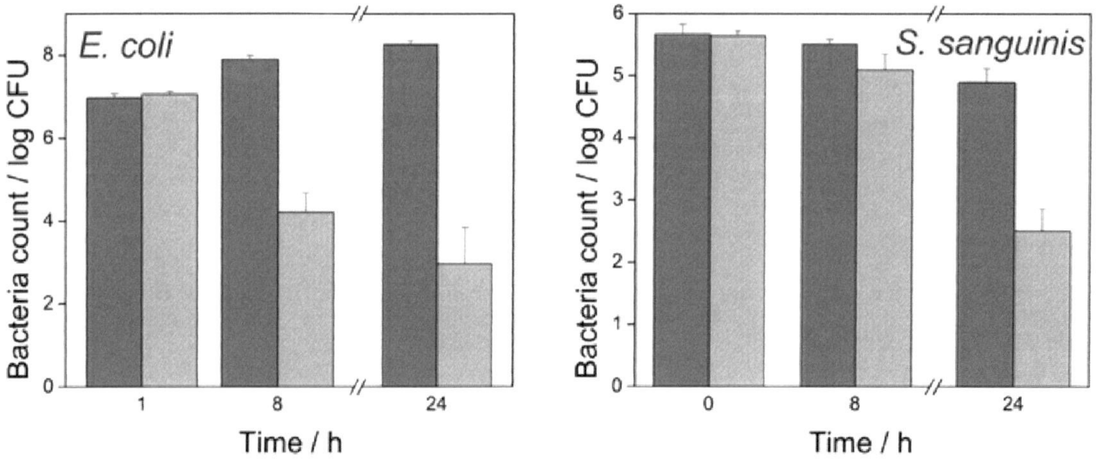

See more of this Session: Micro- and Nanodevices for Targeted Therapeutics
See more of this Group/Topical: Nanoscale Science and Engineering Forum

Size-Selective Fractionation of Nanoparticles at Significant Scales Using CO_2-Expanded Liquids

Steven R. Saunders, Department of Chemical Engineering, Auburn University, Auburn, AL

Christopher B. Roberts, Department of Chemical Engineering, Auburn University, Auburn University, AL

Size-based fractionation of nanoparticles remains a non-trivial task for the preparation of significant quantities of well-defined nanomaterials for certain applications and fundamental studies. Typical fractionation techniques prove to be inefficient for larger nanoparticle quantities due to the expense of equipment, low-throughput, and/or the amount of organic solvent waste produced. Through the use of the pressure-tunable physico-chemical properties of CO_2-expanded liquids, a rapid, precise, and environmentally sustainable size-selective fractionation of ligand-stabilized nanoparticles is possible through simple variations in applied CO_2 pressure. An apparatus capable of fractionating large quantities of nanoparticles into distinct fractions with targeted mean diameters with narrow distributions (± 1 nm) has been developed. This apparatus consists of three vertically mounted high pressure vessels connected in series with high pressure needle valves. This process at current design scales, operated at room temperature, and CO_2 pressures between 0 and 50 bar, results in a batch size-selective fractionation of a concentrated nanoparticle dispersion. This paper presents this technique and method of size-selectively fractionating large quantities of metallic nanoparticles from various solvents, as well as a computational model of this system.

See more of this Session: Nanofabrication and Nanoscale Processing
See more of this Group/Topical: Nanoscale Science and Engineering Forum

Self-Assembled Templates for Device Fabrication Using Si Wafer and Roll-Roll Process Platforms

Vikram Daga, Chemical Engineering, University of Massachussets-Amherst, Amherst, MA

Curran Chandler, Polymer Science and Engineering, University of Massachussets-Amherst, Amherst, MA

Ying Lin, Polymer Science and Engineering, University of Massachussets-Amherst, Amherst, MA

James J. Watkins, Polymer Science and Engineering, University of Massachussets-Amherst, Amherst, MA

The fabrication of nanotechnology enabled devices requires not only the creation and functionalization of well defined nanostructures, but also practical routes for the two and three dimensional integration of these structures with components and systems across multiple length scales. Approaches for the use of self-assembled block copolymer templates to achieve these goals for sub-10 nm elements using either Si wafer or roll to roll processing platforms will be discussed.

While the potential for self assembled templates for patterning precision electronics, including high density data storage is an area of active study and great promise, less work has been carried out to adapt this approach for lower cost per function devices, including solar energy applications and flexible electronics. For example, practical use of self assembly in roll to roll processes requires the development of versatile, low cost templates. Recently we reported that nanostructured templates with sub-10 nm domains can be easily obtained through the blending of disordered polymer surfactants containing poly(ethylene oxide) (PEO) as the hydrophilic block with homopolymers including poly (acrylic acid), poly (4-vinyl phenol) and poly (styrene sulfonate) that selectively associate with the PEO block through hydrogen bonding. These inexpensive blends are strongly segregated, yielding well ordered domains. Moreover, the functionalities imparted by the homopolymers provide convenient handles for binding active materials such as nanoparticles and for promoting in situ, phase selective reactions to produce hierarchical metal oxide polymer composites. We have now extended our approach to demonstrate that ordering can likewise be induced by small molecule additives that can undergo multi-point hydrogen bonding with the surfactants. The use of small molecule additives offers additional structural, chemical and functional diversity. The behavior of the template systems and their use for the fabrication of well ordered polymer/nanoparticle, metal/oxide polymer, and metal oxide/polymer/nanoparticles will be discussed.

See more of this Session: Nanofabrication and Nanoscale Processing
See more of this Group/Topical: Nanoscale Science and Engineering Forum

Synthesis of Vapor Grown Carbon Fibers (VGCF) On Sintered Metal Fibers (SMF) for Air-Filtration

Amogh N. Karwa, Department of Chemical Engineering, Auburn University, Auburn, AL

Virginia A. Davis, Department of Chemical Engineering, Auburn University, Auburn, AL

Bruce J. Tatarchuk, Department of Chemical Engineering, Auburn University, Auburn, AL

Air filters need to be optimized for three properties: (1) filtration efficiency, (2) pressure drop across filter and (3) dirt-holding capacity. Air filtration properties are dependent on various factors like void volume of filter fabric, fiber diameter, filter thickness, filter media pleating etc. Non-woven fabrics are typically used as air filter media. The preform metal fiber fabric was made by wet-lay process employing metal fibers which yielded a non-woven fabric. The non-woven preform was sintered at high temperature of ~1000°C in reducing environment. This SMF fabric was used as substrate for synthesis of VGCF without any additional deposition of catalyst. SMF fabrics were made of different metal fibers such as nickel and stainless steel. The synthesis of VGCF allows for changes in the morphology of metal fibers by changing the surface roughness of fibers. The increase in surface roughness of metal fibers has increased the air-filtration efficiency without substantial increase in pressure drop across the filter media. This translates into energy savings for the lifetime of the filter. The effect of increase in fiber roughness has been studied for SMF fabrics made of different fiber diameter. The results of filtration efficiency tests and growth of VGCF on SMF will be presented.

See more of this Session: Nanofabrication and Nanoscale Processing
See more of this Group/Topical: Nanoscale Science and Engineering Forum

Novel Metal Nanoporous Structures Fabricated From Reduction Reaction On Aluminum Nanoparticle Templates through Galvanic Replacement Mechanism

Qingzhou Cui, Chemical Engineering, University of Massachusetts Lowell, Lowell, MA

Julie Chen, Mechanical Engineering, University of Massachusetts Lowell, Lowell, MA

Zhiyong Gu, Chemical Engineering, University of Massachusetts Lowell, Lowell, MA

With the development of nanoscience in the last two decades, our ability to manipulate nanostructures and nanomaterials has improved significantly. Excellent examples include the synthesis and novel property characterization of carbon nanotubes, semiconducting or metallic nanowires, and quantum dots. Reduction reaction directly on nanoparticles through galvanic replacement reaction has been used to synthesize many novel nanostructures including nanoshell, nanobox, nanorattle, etc. However, only a few template materials including silver (Ag) and palladium (Pd) have been reported so far. This is because the redox potentials for the currently used template materials are too positive to reduce many other metals ions. In the present study, we report the use of aluminum (Al) nanoparticles for the reduction process through galvanic replacement reaction. The advantage for using Al nanoparticles is its large negative redox potential that could reduce many types of metal ions and thus make numerous nanostructures for new metal species. We have demonstrated that nanoporous structures of nickel, copper, cobalt can be successfully fabricated from the Al nanoparticles. These nanostructures have been characterized by SEM, TEM, EDAX and XRD. The applications of these novel materials with high surface area, such as catalysts, are under investigation.

See more of this Session: Nanofabrication and Nanoscale Processing
See more of this Group/Topical: Nanoscale Science and Engineering Forum

Engineered Microfluidic Mixing for Green Nanocrystal Fabrication

Noah Malmstadt, Mork Family Department of Chemical Engineering and Materials Science, University of Southern California, Los Angeles, CA

Shih-Ju Yang, Mork Family Department of Chemical Engineering and Materials Science, University of Southern California, Los Angeles, CA

Peichi Hu, Mork Family Department of Chemical Engineering and Materials Science, University of Southern California, Los Angeles, CA

Metal oxide nanocrystals are finding increasing application as photocatalysts, sensor elements, piezoelectric transducers, and pigments. Traditional methods for fabricating nanocrystals rely on costly, difficult-to-scale, high-temperature batch methods. Solution-phase techniques for fabricating these crystals, while promising in terms of scalability, require toxic, flammable, and unrecoverable organic solvents. We are developing a route to the high-yield production of zinc oxide (ZnO) and titanium dioxide (TiO2) nanocrystals based on microfluidic reactions in environmentally friendly ionic liquid solvents. These techniques take advantage of the unique mixing properties of microfluidic reactors, and are scalable by simple microfluidic parallelization.

Ionic liquids (ILs) are nonvolatile, polar, organic salts that are liquid at room temperature. Their lack of volatility, non-flammability, and capacity for recycling have made them attractive solvents for green chemistry applications. They are especially attractive for nanocrystal fabrication reactions since they can act simultaneously as solvents and surface passivation agents. They are ideal in this role because they have low interfacial tensions, leading to high nucleation rates and small, uniform particles. While ILs have been successfully utilized in the batch syntheses of ZnO and TiO2 nanocrystals, their use has not been translated to a microfluidic platform until now. By transferring IL-based fabrication technologies to a microfluidic system, we are able to leverage the advantages of microfluidic mixing in an environmentally friendly reaction that is easily optimized to form small, homogeneous nanocrystals.

Nanoparticle formation is a nucleation-and-growth process, with nucleation occurring at a higher concentration of nanoparticle precursor (solute) than growth. Since the growth rate is constant, this kind of system can only produce monodisperse particles when the nucleation occurs during a short-lived burst. That is, since particles nucleated at different times will grow to different final sizes, monodispersity can only be obtained if nucleation ends very quickly after beginning. In bulk reaction systems, this is the case only if nucleation can be treated as a critical process, with a sudden concentration-dependant transition between nucleating and non-nucleating states and the process of nucleation itself rapidly consuming free solute. In bulk, therefore, the propensity of a given system to form monodisperse particles is entirely a function of its thermo¬dynamics. Microfluidic systems, however, allow for precise manipulation of the conditions under which reagents come into contact and mix, thereby allowing for control over the nucleation and growth phases of particle formation.

We are developing two mixing schemes for IL-based nanofabrication reactions in microfluidic channels, each with its distinct advantages. In laminar flow mixing, reactants flow through the microchannel side-by-side, with mixing occurring by diffusion between the streams. In droplet mixing, reactants are co-encapsulated in moving droplets in the microchannel, with mixing occurring by convection within each droplet.

A laminar flow configuration is the typical flow profile formed at a junction of two converging continuous microfluidic flow streams. Since the Reynolds number in a microchannel is low, there is no turbulent mixing between the two streams. Rather, they flow side by side and mixing occurs via diffusion between the two streams. This configuration allows for slow stream mixing and confinement of reactions to the interface between the streams. Diffusional mixing in laminar flow is a potentially ideal configuration for nanoparticle fabrication reactions with a thermodynamically constrained nucleation burst. Nucleation occurs at the interface as the reactant streams begin to interdiffuse. If the nucleation burst is rapid, the concentration of reactants at the interface region drops rapidly to below the level necessary for further nucleation. Particle growth is fueled and limited by the further diffusion of reactants into the zone surrounding the interface. Particles have relatively low diffusivities and remain in the interface region. Final particle size can be controlled by total residence time in the device. Since the interface region is small, with a very high surface area-to-volume ratio, heat flow is very fast and temperature disturbances do not persist.

There are a number of disadvantages to laminar flows, however. Chief among these is axial dispersion; the parabolic flow profile leads to mixing in the direction of the flow. This means that rather than having a single residence time, the fluid in the device is described by a residence time distribution. This limits the uniformity of particles that can be obtained in laminar systems. Laminar flow also limits the ability to rapidly quench a reaction (either the nucleation or growth phase of particle fabrication), since mixing between reactants and quenching agents will necessarily be slow.

Microfluidic systems also allow for flow in a droplet, or plug, profile. We are developing a system in which droplets of IL-solvated nanocrystal reactants flow down a microchannel separated by an inert fluorocarbon carrier phase. No-slip boundary conditions at channel walls induce convective flows within individual droplets. This convection leads to fast mixing within the droplet. Since droplet contents are quickly homogenized, they are potentially excellent environments for nanoparticle fabrication: nucleation begins at a single time point throughout the droplet, and solute concentration is the same throughout the droplet as growth proceeds. Droplet flow configurations also allow for rapid quenching of reactions by droplet merger, since upon merger the contents of a droplet containing a quenching agent will mix convectively with the reactants. Controlled quenching of nucleation soon after initial droplet mixing has the potential to improve particle monodispersity.

In this work, laminar and droplet-based mixing are compared for the synthesis of ZnO and TiO2 nanocrystals. Several laminar schemes are explored. These include approaches in which nanocrystal precursor reagents are introduced as solvents in IL streams and approaches in which these reagents are introduced either neat or in water solution in laminar flow adjacent to an IL stream. The effect of IL viscosity on the flow configuration is explored. Nanocrystal products are analyzed by transmission electron microscopy and X-ray diffraction.

See more of this Session: Nanofabrication and Nanoscale Processing
See more of this Group/Topical: Nanoscale Science and Engineering Forum

Fabrication and Transfer of Aligned Carbon Nanotube-Polymer Nanostructures

Yu Mao, Biosystems Engineering, Oklahoma State University, Stillwater, OK

Yumin Ye, Biosystems Engineering, Oklahoma State University, Stillwater, OK

Composite films of carbon nanotubes (CNTs) have important applications in functional devices such as sensors and conductive membranes. Aligned CNTs, with the ordered structure, offer unique advantages in collective manipulation and easy integration of nanotubes into such devices. We report a novel method to fabricate aligned CNT-polymer nanostructures. This method allows transferring of the aligned CNTs from the original silicon substrates to target substrates via a flip-over technique. The nanotube alignment in the polymer-nanotube nanostructures was well retained. In addition, the transferred polymer-nanotube nanostructure is electrically conductive and mechanically strong, with a conductivity of 76.7 siemens/cm, an elastic modulus enhancement of 650% and a hardness increase of 750% over the original polymer. Characterizations using SEM, TEM, and FT-IR indicated uniform polymer nanocoatings covering each individual CNTs. The polymer-nanotube nanostructures also demonstrate outstanding stability during wetting and drying processes because of the robustness and rigidity provided by the polymer.

See more of this Session: Nanofabrication and Nanoscale Processing
See more of this Group/Topical: Nanoscale Science and Engineering Forum

Genetically Diverse Biosensors and Bionanoanalytical Methods

Sylvia Daunert, Chemistry, University of Kentucky, Lexington, KY

TBA

See more of this Session: Bionanotechnology: Plenary Session I
See more of this Group/Topical: Nanoscale Science and Engineering Forum

Dynamic Peptide Assemblies for Sensing Applications

Raymond Tu, Chemical Engineering, City College of New York (of CUNY), New York, NY

TBA

See more of this Session: Bionanotechnology: Plenary Session I
See more of this Group/Topical: Nanoscale Science and Engineering Forum

Micro-Rheology of Biologically Relevant Interfaces

Todd M. Squires, Chemical Engineering, University of California, Santa Barbara, Santa Barbara, CA

TBA

See more of this Session: Bionanotechnology: Plenary Session I
See more of this Group/Topical: Nanoscale Science and Engineering Forum

Award Submission: Tracking Biodistribution of NIR-Labeled Polymeric Micelles of Varying Shape and Surface Charge

David A. Christian, Chemical & Biomolecular Engineering, University of Pennsylvania, Philadelphia, PA

Dennis E. Discher, Chemical & Biomolecular Engineering, University of Pennsylvania, Philadelphia, PA

Polymeric micelles synthesized from the self-assembly of amphiphilic block copolymers in solution have received increasing attention for applications in drug delivery. While the effect of size of these micelles on biodistribution and circulation kinetics has been well characterized, questions of the effects of micelle shape and surface charge remain largely unanswered. In this study, the shape of micellar carriers can be tuned from spheres to short and long cylinders (filomicelles) to bilayer vesicles (polymersomes) with simple changes in block copolymer architecture. In addition, the surface charge of polymersomes can be tuned to mimic the charge of red blood cells without chemical modification of the block copolymers. To efficiently track the biodistribution of these polymeric micelles in vivo, a near infrared fluorophore (NIRF) was non-covalently incorporated without affecting micelle morphology or surface charge. Preliminary results for the effect of surface charge on polymersomes indicate a distinct shift in the carrier biodistribution where anionic polymersomes are directed almost exclusively to the liver rather than the spleen and liver for neutral polymersomes. The tracking of NIRF-labeled filomicelles in healthy mice shows that the majority of carriers - independent of length - remained in circulation after 24 hours. In tumor-bearing mice, NIRF imaging suggests that filomicelles convect into a solid tumor and some fragments can penetrate into the tumor stroma, suggesting a mechanism by which filomicelles can effectively deliver encapsulated anti-cancer therapeutics to tumors [1].

[1] Christian DA, Cai S, Garbuzenko OB, Harada T, Zajac AL, Minko T, Discher DE. Molecular Pharmaceutics, DOI: 10.1021/mp900022m, (2009).

See more of this Session: Bionanotechnology Graduate Student Award Session
See more of this Group/Topical: Nanoscale Science and Engineering Forum

Award Submission: Environmentally Responsive Gold Nanorod-Polypeptide Assemblies

Huang-Chiao Huang, Department of Chemical Engineering, Arizona State University, Tempe, AZ

Kaushal Rege, Chemical Engineering, Arizona State University, Tempe, AZ

Environmentally responsive nanoassemblies based on polypeptides and nanoparticles are promising in a number of biological/biomedical applications. We demonstrate the formation of gold nanorod (GNR)-elastin-like polypeptide (ELP) nanoassemblies whose optical response can be remotely controlled based on near-infrared (NIR) light exposure. In this study, cysteine-containing ELPs (C2ELP) were self -assembled on gold nanorods (GNRs) leading to GNR- C2ELP nanoassemblies. Exposure of GNR- C2ELP assemblies to NIR laser resulted in the heating of GNRs due to surface plasmon resonance. Heat transfer from the GNRs resulted in a temperature increase of the self-assembled C2ELP above its transition temperature (Tt), leading to a detectable phase transition and aggregation of the GNR- C2ELP assemblies. This phase transition was quantified using an optical readout, based on an increase in the optical density of the GNR- C2ELP dispersion. This optical response was seen only in the case of GNR- C2ELP nanoassemblies. While a photothermal response was observed in the case of a dispersion consisting of GNRs alone, no optical response could be seen due to the absence of the C2ELP. In addition C2ELP solutions demonstrated neither a photothermal nor an optical response. The reproducibility and reversibility of optical response of GNR-C2ELP nanoassemblies was demonstrated across multiple cycles following exposure and removal of the laser irradiation. In summary, our results indicate that polypeptides can be interfaced with GNRs resulting in optically responsive nanoassemblies. Current work involves an investigation into the optical and photothermal properties of nanoassemblies using higher cysteine-containing ELP homologues. We anticipate that these nanoassemblies can be useful for sensing, tissue engineering, and drug delivery applications.

See more of this Session: Bionanotechnology Graduate Student Award Session
See more of this Group/Topical: Nanoscale Science and Engineering Forum

Award Submission: Radiofrequency Actuation of Iron Oxide-Hydrogel Nanocomposites: Experimental Analysis and Modeling

Nitin S. Satarkar, Chemical & Materials Engineering, University of Kentucky, Lexington, KY

Samantha A. Meenach, Chemical & Materials Engineering, University of Kentucky, Lexington, KY

Christopher R. Barton, Chemical & Materials Engineering, University of Kentucky, Lexington, KY

Kimberly W. Anderson, Chemical & Materials Engineering, University of Kentucky, Lexington, KY

J. Zach Hilt, Chemical & Materials Engineering, University of Kentucky, Lexington, KY

Hydrogels and hydrogel nanocomposites are attractive materials for a variety of biomedical/pharmaceutical applications. In recent studies, magnetic hydrogel composites of N-isopropylacrylamide (NIPAAm) hydrogels have been demonstrated as pulsatile drug delivery systems with radiofrequency(RF) at 293 kHz as a trigger.1 In another study, magnetic nanocomposites were demonstrated as actuators in microfluidic device, with ability to control the flow by application of RF.2 Some of the other applications being pursued with these systems is use of RF induced heating for hyperthermia cancer treatment. All these applications rely heavily on design of the nanocomposite systems. In order to achieve desired response from hydrogel nanocomposite system, it is very important to understand the heating induced by application of RF and resultant change in the hydrogel properties.

Goal of this study was to analyze and predict nanocomposite response based on hydrogel composition, nanoparticle loading, RF strength, and nanocomposite dimensions. Magnetic nanocomposites of poly (ethylene glycol) hydrogels were synthesized with iron oxide (Fe3O4) nanoparticles. Different nanocomposite systems were obtained by variation of nanoparticle loadings and hydrogel composition. 293 kHz RF was applied to nanocomposite discs and surface temperatures were analyzed using infrared (IR) thermography. The strength of RF was varied to further analyze the heating effect.

A heat transfer model was developed to account heat generation due to RF as well as heat loss to surroundings. By fitting the temperature data, a correlation was obtained for dependence of heat generation on nanoparticle loadings, and RF strength. It was demonstrated that the model could successfully predict the resultant temperatures by using a hydrogel system with different swelling properties. The heating analysis was further extended for potential hyperthermia cancer treatment applications. The bioheat equation was used to account the heat transport in vivo, and resultant nanocomposite temperatures were predicted.

In the case of NIPAAm nanocomposites, RF heating is accompanied by collapse, which forces the fluids out of hydrogel network. Magnetic NIPAAm nanocomposites of different dimensions (thickness) were synthesized and subjected to RF heating. The resultant collapse and recovery process was quantified by weight measurements at different time points. The heat transfer model was modified to account change in the mass as hydrogel collapsed. Furthermore, a mass transfer model was developed for transport of liquid as the nanocomposite collapsed and recovered.

1. Satarkar, N. S.; Hilt, J. Z. J. Control. Rel. 2008, 130, 246-251.

2. Satarkar, N. S.; Zhang, W.; Eitel, R. E.; Hilt, J. Z. Lab Chip 2009, In Press.

See more of this Session: Bionanotechnology Graduate Student Award Session
See more of this Group/Topical: Nanoscale Science and Engineering Forum

Award Submission: Activation by and Processing of Polyanhydride Nanospheres by Antigen Presenting Cells

Bret D. Ulery, Chemical and Biological Engineering, Iowa State University, Ames, IA

S. Avanti Sarkar, Veterinary Microbiology and Preventative Medicine, Iowa State University, Ames, IA

Bryan H. Bellaire, Veterinary Microbiology and Preventative Medicine, Iowa State University, Ames, IA

Michael J. Wannemuehler, Veterinary Microbiology and Preventative Medicine, Iowa State University, Ames, IA

Balaji Narasimhan, Chemical and Biological Engineering, Iowa State University, Ames, IA

Improving vaccine efficacy is a major area of recent research. It has been shown that biodegradable polyanhydride particles possess the capacity to function both as superb adjuvants and as single dose delivery vehicles. Currently, it is unknown how polymer particles facilitate both antigen presenting cell (APC) activation and particle internalization and processing. The method by which encapsulated antigen goes from being internalized within a particle to being processed and effectively presented on the surface of APCs plays a crucial role in the initiation of an immune response. By observing how polyanhydride nanospheres are uptaken and processed by APCs using microscopy and evaluating activation of these cells by cell surface marker expression and cytokine secretion, we can better understand the potential of these materials as vaccine delivery vehicles. In addition, polyanhydride chemistry effects on APC uptake mechanisms will provide valuable information assisting in the rational design of vaccine delivery platforms. Copolymers of 1,6-bis(p-carboxyphenoxy)hexane and sebacic acid and copolymers of 1,8-bis(p-carboxyphenoxy)-3,6-dioxaoctane and 1,6-bis(p-carboxyphenoxy)hexane were synthesized and fabricated into nanospheres by anti-solvent precipitation. Murine bone-marrow derived dendritic cells (BMDCs) were differentiated and incubated with polyanhydride nanospheres in vitro to allow for particle internalization and cellular activation. Laser scanning confocal microscopy was utilized to visualize particle uptake and processing by BMDCs. Flow cytometry was employed to investigate the capacity for polyanhydride nanospheres to increase cell surface marker expression on these cells. A LUMINEX bead assay was used to measure the secretion of multiple cytokines from APCs. As polymer hydrophobicity decreased, cell surface marker expression was enhanced and more polyanhydride nanospheres were internalized and processed. Since cellular processing is dependent on polymer chemistry and particle size, knowledge of how these variables can be exploited to best design delivery systems that facilitate desired immune responses for a variety of vaccine candidates is highly desirable.

See more of this Session: Bionanotechnology Graduate Student Award Session
See more of this Group/Topical: Nanoscale Science and Engineering Forum

'Award Submission': Trafficking of Nanoparticles to Intracellular Aggresome-Like Strucutres: Implications for Non-Viral Gene Delivery

Sutapa Barua, Chemical Engineering, Arizona State University, Tempe, AZ

Kaushal Rege, Chemical Engineering, Arizona State University, Tempe, AZ

Nanoscale therapeutics, including nonviral gene delivery systems, have been investigated for a variety of cancer cell types. However, the efficacy of these therapeutics is compromised by cancer cell resistance mechanisms. The inefficient sub-cellular delivery of nanoparticles is of particular relevance to nanoscale therapeutics. Consequently, investigations into the role of cancer cell phenotype in the cellular uptake, intracellular sorting, transport, and sub-cellular localization can lead to improved efficacies of nanoscale therapeutics.

Here we show that phenotypic differences in human prostate cancer cells lead to dramatic differences in intracellular sorting and localization of exogenous nanomaterials including nanoparticles (quantum dots), polymer:DNA, and polymer:siRNA complexes. These nanoparticles demonstrate a punctated intracellular localization throughout the cytoplasm in PC3 cells. In contrast, the materials were taken up by clathrin-mediated endocytosis and trafficked along microtubules, resulting in their localization in aggresome-like compartments in PC3-PSMA cells (sub-clones of PC3 cells retrovirally transduced to stably express PSMA receptors).

We hypothesized that localization of polymer:DNA complexes in aggresome-like compartments was a major barrier to the transport of genes to the nucleus and therefore, is responsible for low transfection efficacies. We have recently generated polymers that show increased transfection efficacies and lower cytotoxicities than polyethyleneimine, a current standard for polymer-mediated gene delivery. Intracellular trafficking studies indicated co-localization of quantum dots and polymer:plasmid DNA polyplexes in juxtanuclear aggresome-like structures in PC3-PSMA cells. Confocal microscopy studies were carried out to characterize the nature of the aggresome-like compartment. In contrast, QDs and polyplexes colocalized in punctate vesciles throughout the cytoplasm of PC3 cells. Comparison between PC3 and PC3-PSMA cells indicated higher transfection efficacies in PC3 cells compared to PC3-PSMA cells, which correlated with the formation of aggresome-like structures. Strategies for overcoming this intracellular barrier to gene and siRNA delivery were investigated. Our results provide a platform to understand mechanisms behind the resistance of certain cancer cell types to gene and siRNA delivery and describe opportunities for overcoming these resistance mechanisms for nanoscale therapeutics.

See more of this Session: Bionanotechnology Graduate Student Award Session
See more of this Group/Topical: Nanoscale Science and Engineering Forum

Award Submission: PEG-Based Magnetic Hydrogel Nanocomposites for Combined Chemotherapy and Hyperthermia Treatment of Cancer

Samantha A. Meenach, Chemical & Materials Engineering, University of Kentucky, Lexington, KY

Chinedu G. Otu, Department of Biology, University of Kentucky, Lexington, KY

Christopher R. Barton, Chemical & Materials Engineering, University of Kentucky, Lexington, KY

J. Zach Hilt, Chemical & Materials Engineering, University of Kentucky, Lexington, KY

Kimberly W. Anderson, Chemical & Materials Engineering, University of Kentucky, Lexington, KY

Hyperthermia, the heating of tissue to 41 to 45°C, has been shown to improve the efficacy of cancer therapy when used in conjunction with irradiation and/or chemotherapy. Current limitations include but are not limited to non-homogeneous distribution of temperature over the cancer site, low specificity, patient discomfort, and lack of localized delivery of heat. These shortcomings can be overcome through the utilization of polymeric biomaterials which can be delivered at tumor sites and remotely heated from outside the body while simultaneously delivering therapeutic drugs. Hydrogel nanocomposites which are composed of biocompatible crosslinked polymeric material and nanoparticulates capable of heating upon exposure to an electromagnetic field are a potential platform of polymeric biomaterials for this application. In addition to the remote heating, the network structure of the hydrogel allows for the tailored delivery of therapeutic agents. In this work, hydrogel nanocomposites have been developed that can be implanted or injected within or near a tumor site and deliver both heat and a chemotherapeutic agent (e.g., paclitaxel). In previous studies, the efficacy of paclitaxel has been proven to increase when combined with elevated temperatures. Here, the nanocomposites studied involve a stealth, poly(ethylene glycol) and iron oxide nanoparticle-based system. The iron oxide nanoparticles were physically entrapped in the hydrogel matrix which was comprised of poly(ethylene glycol) methyl methacrylate macromer and poly(ethylene glycol) dimethacrylate as the crosslinker.

Thermal analysis showed the capability of the hydrogels to be heated in an alternating magnetic field to various target surface temperatures depending on the strength of the field and composition of the hydrogel matrix. Hydrogels with lower swelling ratios proved to heat better than those with higher ratios due to the greater polymeric content in the gels which corresponds to higher iron oxide loading in the gels. The heating of the hydrogel nanocomposites was controlled by the strength of the magnetic field and hyperthermia-specific temperature ranges were reached for all hydrogel systems at field strengths ranging from 12.7 to 17.4 kA/m. To examine the cytocompatibility of the hydrogels, NIH 3T3 murine fibroblasts were exposed to the hydrogel systems, and the viability of the cells was not affected by leachants from the hydrogels. The delivery of a model drug, rifampicin, from the hydrogel nanocomposites was investigated and was shown to be controlled via the hydrogel matrix. This drug was chosen due to its similarity in hydrophobicity and molecular weight to paclitaxel. The delivery of paclitaxel from the hydrogels is also being investigated. An initial proof-of-concept demonstration was completed where M059K glioblastoma multiforme cells were heated to thermoablative temperatures (above 55°C) with the hydrogel nanocomposites exposed to the AMF and cell death was demonstrated. Studies with both M059K glioblastoma and MDA MB 231 breast adenocarcinoma cells where the cells are exposed to both hyperthermia-specific temperatures and paclitaxel (i.e., delivered directly in the cell media) to study if the efficacy of paclitaxel in inducing cell death with and without moderate heat are underway. In summary, the ability of these systems to heat and their capability to deliver drugs to specific sites make them viable systems for treating of deep-seated tumors or reoccurring cancer.

1. Meenach SA, Hilt JZ, Anderson KW, Poly(ethylene glycol)-Based Magnetic Hydrogel Nanocomposites for Hyperthermia Cancer Therapy, Acta Biomaterialia, submitted.

2. Babincova M, Leszczynska D, Sourivong P, Cicmanec P, Babinec P. Superparamagnetic gel as a novel material for electromagnetically induced hyperthermia. Journal of Magnetism and Magnetic Materials 2001;225:109-112.

3. Hergt R, Dutz S. Magnetic particle hyperthermia - biophysical limitations of a visionary tumour therapy. Journal of Magnetism and Magnetic Materials 2007;311:187-192.

4. Moroz P, Jones SK, Gray BN. Magnetically mediated hyperthermia: current status and future directions. International Journal of Hyperthermia 2002;18(4):267-284.

See more of this Session: Bionanotechnology Graduate Student Award Session
See more of this Group/Topical: Nanoscale Science and Engineering Forum

Award Submission: Design and Tailoring the Self-Assembly of Amphiphilic Oligopeptide Nanostructures as Versatile Biomaterials for Delivery of Anticancer Drugs, Genes, or Both, for Improved Cancer Treatment

Nikken Wiradharma, Drug Gene and Protein Delivery; Chemical and Biomolecular Engineering, Institute of Bioengineering and Nanotechnology; National University of Singapore, Singapore, Singapore

Yen Wah Tong, Department of Chemical and Biomolecular Engineering, National University of Singapore, Singapore, Singapore

Yi Yan Yang, Drug and Gene Delivery, Insititute of Bioengineering and Nanotechnology, Singapore, Singapore

The number of clinical trials for gene therapy has increased world-wide, with 66.5% of its target were aimed at cancer therapy.As opposed to viral based delivery vector, cationic polymers offers benefits like easy manufacturing and scale up, low immunogenicity and high amount of genes it can carry. One of the major barriers in cancer treatments is drug resistance, which has lately been suggested to be the result of chromosomal alterations leading to genetic malfunctions in cancer cells. As such, delivery of siRNA or anticancer genes have been attempted to sensitize cancer cells towards anticancer drug treatments.

In this study, we had designed various amphiphilic oligopeptides with different blocks to achieve different self-assembly properties desirable for different types of therapeutic cargo molecules. Versatility of oligopeptide to form a novel class of biodegradable biomaterials as carriers of therapeutics were demonstrated by tailoring different types of amino acids in the design of the carrier molecules. The oligopeptides designed contained both hydrophobic and hydrophilic amino acids to provide driving force for micellar nanoparticle formations. Various types of hydrophobic amino acids with different degrees of hydrophobicity were used to tailor different properties and strength of the hydrophobic interactions in forming the nanoparticles. On the other hand, the compositions of the amino acids in the hydrophilic block were designed based on the different therapeutic cargos to be delivered. Lysine residue was chosen to provide gene binding and protection, histidine residue was chosen to provide endosomal escape mechanism upon uptake of nanoparticles, and Glycine or Serine residues were used as neutral hydrophilic shell to encapsulate and solubilize hydrophobic drugs inside the core. Tailoring the amino acids was also shown to affect the particle stability, drug encapsulation and release properties, but without affecting gene binding capacity. This allows the use of the oligopeptide nanostructures for simultaneous delivery of drugs and genes to achieve synergistic effect for cancer treatment.

See more of this Session: Bionanotechnology Graduate Student Award Session
See more of this Group/Topical: Nanoscale Science and Engineering Forum

Capture and Release of Cardiac Fibroblasts in Microfluidic Devices Using Peptide-Functionalized Alginate Gels

Brian D. Plouffe, Chemical Engineering, Northeastern University, Boston, MA

Melissa A. Brown, IBBME/Chemical Engineering and Applied Chemistry, University of Toronto, Toronto, ON, Canada

Milica Radisic, IBBME/Chemical Engineering and Applied Chemistry, University of Toronto, Toronto, ON, Canada

Shashi K. Murthy, Department of Chemical Engineering, Northeastern University, Boston, MA

Microfluidic devices with surface-immobilized ligands have recently emerged as an attractive tool for cell separation by positive selection with applications ranging from cancer diagnostics and clinical genomics to tissue engineering. A critical unmet need in these applications, however, is the ability to detach the captured cells non-destructively. We describe the application of peptide functionalized alginic acid via carbodiimide chemistry as a methodology for cell capture and release in microfluidic channels. An alginate-peptide complex was adsorbed on the surface of the channels and rinsing with Ca2+ ionic solutions results in a thin hydrogel layer on the channel surface. Primary rat cardiac fibroblasts were injected into the device and subsequently captured from the flow stream by the adsorbed ligand. The conjugation of Arg-Gly-Asp-Ser (RGDS), a tetrapeptide to the alginate backbone showed a two-fold and four-fold increase in fibroblast adhesion compared to unconjugated alginate and bare glass controls, respectively. Rinsing with EDTA solution resulted in over 95% cell release within the device. The detached fibroblasts had the same viability and spreading characteristics as freshly harvested cells. This method is simple and amendable to all micro-scale devices that operate at low fluid shear stresses. In addition, the conjugation chemistry utilized in this work can be applied to a wide range of molecules containing primary amines and cellular release requires no application of external forces such as heat, electrical potential, or photo-activation.

See more of this Session: Polymers as Functional Components of Micro- and Nanodevices
See more of this Group/Topical: Nanoscale Science and Engineering Forum

ATRP Synthesized Environmentally Responsive Micro- and Nanostructured Hydrogels Over Biomedical Devices

Hariharasudhan D. Chirra, Department of Chemical and Materials Engineering, University of Kentucky, Lexington, KY

J. Zach Hilt, Chemical and Materials Engineering, University of Kentucky, Lexington, KY

Surfaces that are spatially functionalized with intelligent hydrogels, especially at the micro- and nanoscale, are of high interest in the biomedical field. Though patterned brushes can be successfully employed for biodiagnostic applications, the advantages of a crosslinked polymer network such as mechanical robustness, material entrapment, and tunable transport properties make hydrogels useful for both sensing and therapeutic applications. Herein, microcontact printing (μCP) for XY control, followed by atom transfer radical polymerization (ATRP) for Z/thickness control was applied to synthesize a controlled array of environmentally responsive hydrogels (e.g., N-isopropylacrylamide) over material surfaces/devices. The nanoscale drug/particulate loading, response to external stimuli and release were characterized using atomic force microscopy (AFM), quartz crystal microbalance with dissipation (QCM-D) and scanning electron microscopy (SEM). Using such arrays of responsive hydrogels, it is possible to fabricate rapid point of care biomedical devices for diagnostic and therapeutic applications.

See more of this Session: Polymers as Functional Components of Micro- and Nanodevices
See more of this Group/Topical: Nanoscale Science and Engineering Forum

Novel Composite Polymer Electrode for Enzyme-Based Biofuel Cell

Sunmook Lee, Institute of Industrial Science, The University of Tokyo, Tokyo, Japan

Bokkyu Choi, Institute of Industrial Science, The University of Tokyo, Tokyo, Japan

Atsushi Tsutsumi, Institute of Industrial Science, The University of Tokyo, Tokyo, Japan

Novel composite polymer layers were used for enzyme-based biofuel cell electrode platform and their electrochemical properties were investigated. Polyaniline, one of the most widely used conductive polymers, was electropolymerized on a planar gold electrode in the presence of carboxydextran and/or carboxydextran-gold hybrid nanomaterials. These composite films showed the redox activity in neutral pH regions, which was confirmed with cyclic voltammetry. The composite films showed an excellent electrocatalytic activity toward the oxidation of ascorbic acid, indicating that these composite films are a good candidate for enzyme-based biofuel cell electrode.

See more of this Session: Polymers as Functional Components of Micro- and Nanodevices
See more of this Group/Topical: Nanoscale Science and Engineering Forum

Antimicrobial Biomaterials Based On Poly(lactic-co-glycolic acid) Dispersed with Carbon Nanotubes

Seyma Aslan, Yale University, New Haven, CT

Codruta Zoican, Yale University, New Haven, CT

Nan Li, Yale University, New Haven, CT

Salim Derrouiche, Yale University, New Haven, CT

Xiaoming Wang, Yale University, New Haven, CT

Seoktae Kang, Yale University, New Haven, CT

Menachem Elimelech, Yale University, New Haven, CT

Lisa Pfefferle, Yale University, New Haven, CT

Paul R. Van Tassel, Yale University, New Haven, CT

In this study, we assess the antimicrobial behavior of single-walled carbon nanotubes (CNT) within poly(lactic-co-glycolic acid) (PLGA) polymeric biomaterials. CNT of diameter 0.85 to 1.2 nm are synthesized via CO decomposition at 1073 K and 6 atm over a Co incorporated MCM-41 catalyst (with a 3% metal loading). CNT length is characterized via Raman spectroscopy and Transmission Electron Microscopy. CNT-PLGA thin films of thickness 150-200 nm are realized via a spin coating method. Contact angle measurements confirm the surface homogeneity of these thin film morphologies. Antimicrobial activity is assayed by exposing films to Escherichia coli (E. coli). We find E. coli viability and metabolic activity to be highly correlated with CNT length and concentration. Up to 86% of bacteria die following one hour of exposure on CNT containing samples, whereas 12% die on pure PLGA (control) films. Shorter CNT generally exhibit higher toxicity, most likely due to an increased density of open tube ends. This study demonstrates the potential usefulness of CNT-PLGA as an antimicrobial biomaterial.

See more of this Session: Polymers as Functional Components of Micro- and Nanodevices
See more of this Group/Topical: Nanoscale Science and Engineering Forum

Size-Selective Mass Transport Via a Block Polymer / Micromachined Silicon Composite Membrane

Eric Nuxoll, Chemical and Biochemical Engineering, University of Iowa, Iowa City, IA

Marc A. Hillmyer, Chemistry, University of Minnesota, Minneapolis, MN

Ronald A. Siegel, Pharmaceutics, University of Minnesota, Minneapolis, MN

One challenge in implantable chemical sensing is blocking immunoproteins and other fouling agents from entering the device while simultaneously facilitating fast diffusion of target solute into the device. As the immunoproteins are typically much larger than the target solute, one strategy is size exclusion, but this requires short, straight, densely-packed pores tens of nanometers across to simultaneously block large molecules while offering a short, wide cross-section for solute transport.

By integrating a triblock polymer film with a microfabricated silicon support, we have constructed an asymmetric membrane in which the polymer forms a size-selective layer less than 100 nm thick with densely-packed, uniform 40 nm pores. This polymer film is supported by 100 um thick silicon with uniform 20 um wide pores, providing mechanical support while also keeping a short, wide path for solute transport. Diaphragm cell studies were performed to establish the size selectivity of the membrane for two model solutes, methyl orange (MW = 367 Da) and blue dextran (MW = 2 MDa). The fifty-fold favoring of the smaller solute confirms the membrane's size selectivity, while the microfabricated support provides a convenient handle for incorporating the membrane into MEMS-style sensors and drug-delivery systems.

See more of this Session: Polymers as Functional Components of Micro- and Nanodevices
See more of this Group/Topical: Nanoscale Science and Engineering Forum

Photothermal Ablation and Non-Viral Gene Delivery of Cancer Cells Using Polyelectrolyte-Stabilized Gold Nanorods

Huang-Chiao Huang, Department of Chemical Engineering, Arizona State University, Tempe, AZ

Kaushal Rege, Chemical Engineering, Arizona State University, Tempe, AZ

Gold nanoparticles possess a number of attractive properties including enhanced surface electric field, high biocompatibility, and ease of chemical conjugation. Consequently, gold nanoparticles are being increasingly investigated as potential therapeutics, drug delivery vehicles, imaging agents and diagnostics as part of the expanding field of nanomedicine. In particular, gold nanorods demonstrate a tunable photothermal response to near infrared (NIR) light have been employed for the ablation of cancer cells, biomolecule detection and the generation of functional nanoscale assemblies. We demonstrate that cetyltrimethyl ammonium bromide (CTAB)-based gold nanorods, prepared by the commonly used seed-mediated method, demonstrate poor stability in phosphate buffered saline (PBS), serum-free media (SFM) and serum-containing media (SCM), depending on the number of centrifugation steps carried out to remove excess CTAB from the dispersion. We employed the layer-by-layer deposition of recently generated cationic polymers in our laboratory in order to increase the short-term and long-term stability of gold nanorods in biological media, Polyelectrolyte-coated gold nanorods exhibit excellent long term optical stability in all three biologically media even after four weeks of storage. In addition, stabilization of the optical response resulted in a reliable, Arrhenius-like photothermal response of the well-dispersed gold nanorods over the period of investigation. Cytotoxicity evaluation indicated that gold nanorods coated with our polymers were less cytotoxic to mammalian cells than those covered with 25-kDa polyethyleneimine (pEI25). Stable and biocompatible polyelectrolyte-coated nanorods were successfully employed for the photothermal ablation of PC3-PSMA cells using NIR laser irradiation. In addition, sub-toxic concentrations of these nanorods were also employed for delivering plasmid DNA to these cells. Our results indicate that molecular engineering of cationic polyelectrolytes can lead to stable, biocompatible gold nanorod assemblies that can be useful for a variety of applications including non-viral gene delivery and cancer cell photothermal ablation.

See more of this Session: Polymers as Functional Components of Micro- and Nanodevices
See more of this Group/Topical: Nanoscale Science and Engineering Forum

Macroporous Polymer-Based Color Reflective Displays

Hongta Yang, Chemical Engineering, University of Florida, Gainesville, FL

Chih-Hung Sun, Chemical Engineering, University of Florida, Gainesville, FL

Peng Jiang, Chemical Engineering, University of Florida, Gainesville, FL

Inexpensive fabrication of ultrathin flat-panel color displays is of great technological importance in a large variety of optoelectronic devices. Self-assembled diffractive colloidal photonic crystals provide a simple and inexpensive approach for developing all-color displays. Here we report a new concept that enables the realization of color reflective displays by using self-assembled macroporous polymer membranes. Macroporous polymers with three-dimensionally ordered voids are templated from self-assembled colloidal crystals. The resulting films exhibit brilliant iridescent colors due to the Bragg diffraction of visible light by the periodic voids. We show that the voids can be filled up by a refractive index-matched liquid using controlled capillary condensation, resulting in transparent films with no reflective colors. By using this simple approach, we have demonstrated that active-matrix reflective displays can be fabricated on both flat and curved surfaces.

See more of this Session: Polymers as Functional Components of Micro- and Nanodevices
See more of this Group/Topical: Nanoscale Science and Engineering Forum

Biofuel Cell Using Immobilization of Glucose Oxidase On Carbon Nanotubes Surface

Thang Ho, Chemical Engineering, University of Arkansas, Fayetteville, AR

Pratyush Rai, Departement of Electrical Engineering, University of Arkansas, Fayetteville, AR

Jining Xie, Departement of Electrical Engineering, University of Arkansas, Fayetteville, AR

Vijay Varadan, Departement of Electrical Engineering, University of Arkansas, Fayetteville, AR

Jamie A. Hestekin, Ralph E. Martin Department of Chemical Engineering, University of Arkansas, Fayetteville, AR

Biofuel cells are of great interest in alternative energy and miniaturization applications such as portable energy supplies, pacemakers, and glucose sensors. Biofuel cells can be constructed with enzymes and\or microorganisms. The most common biofuel cell uses glucose oxidase to produce electrons from the glucose to gluconic acid chemical reaction. The electrical potential of glucose oxidase for converting glucose to gluconic acid is 0.75V. Many researchers have developed methods to immobilize glucose oxidase enzymes onto different electrode surfaces; thus creating a fuel cell that could provide power in biological applications. There are some promising results; however, the disadvantages of the current methods are low power output and short electrode lifetime due to enzyme denaturization and product inhibition. Utilizing the high conductivity of carbon nanotubes (CNT) (>1000 X more conductive than copper), this paper modifies the surface of CNTs and covalently binds with glucose oxidase enzymes. With the high conductivity structure and stabilized enzyme immobilization, this paper explores electrical output and electrode lifetime in comparison with previous studies. Further, our studies look at using ion exchange mechanisms to remove products and extend lifetimes further. The optimal enzyme loading per mass of the CNT will be presented. Further, the paper provides a unique approach to screen printing miniaturization for use in mass production of biofuel cells with CNT and ion exchange functionality. Electric power is evaluated based on geometry and the area of active electron transfer surface. The results give a comprehensive view of the optimal design for a miniaturized bioelectrode, especially using a screen printing miniaturization process.

See more of this Session: Poster Session: Nanoscale Science and Engineering
See more of this Group/Topical: Nanoscale Science and Engineering Forum

Synthesis of SiO2 Inverse Opals in Supercritical Carbon Dioxide

Ah Ram Kim, Department of Chemical and Biomolecular Engineering, Sogang University, Seoul, South Korea

Jun Hyuk Moon, Department of Chemical and Biomolecular Engineering, Sogang University, Seoul, South Korea

Ki-Pung Yoo, Department of Chemical and Biomolecular Engineering, Sogang University, Seoul, South Korea

Jong Sung Lim, Department of Chemical and Biomolecular Engineering, Sogang University, Seoul, South Korea

SiO2 inverse opals were produced in supercritical carbon dioxide (scCO2) with latex arrays as templates. PS (Polystyrene) latex particles composed of different hydrophilic groups. Particles were sank into BSA (benzenesulfonicacid), which activates the condensation process of silica precursors. The polymeric templates were reacted with TEOS (tetraethyl ortho-silicate) and H2O in scCO2 at certain conditions. After calcination of this template, highly porous materials duplicating the structure of the original template were obtained. The porosity of the materials obtained for each template is different. We could get the scanning electron microscopy (SEM) images which show the particle reaction on the surface of template in scCO2. This new method of synthesizing inverse opals in scCO2 could make it possible to overcome some of the limitations of the liquid-phase techniques, being a faster method of synthesis and rendering materials of unique properties.

See more of this Session: Poster Session: Nanoscale Science and Engineering
See more of this Group/Topical: Nanoscale Science and Engineering Forum

Formation of Spherical Nanoparticle Arrays Via Electron-Beam Lithography Using Electroless Gold Plating

Phillip Blake, Department of Chemical Engineering, University of Arkansas, Fayetteville, AR

Wonmi Ahn, University of Arkansas, Fayetteville, AR

D. Keith Roper, University of Arkansas, Fayetteville, AR

Ordered arrays of metallic nanoparticles (NPs) are being developed for applications ranging from waveguides, nano-antennae, and optical switches to enhanced fluorescence sensors, surface enhanced Raman spectroscopy (SERS), and biosensors enhanced electromagnetic response. Recent theoretical analyses show that for arrays of spherical gold (Au) NP, the electromagnetic field enhancement can be up to 107 relative to an isolated NP. Current fabrication methods, such as nanosphere lithograhpy (NSL), focused ion beam (FIB) milling, and electron beam lithography (EBL) rely on thermal evaporation or sputtering to form the metallic NPs. The directional nature of this approach limits particle shapes that can be created. We have created gold nanoparticle arrays using electroless Au plating and successfully transformed the produced cylinders into spheroid particles.

See more of this Session: Poster Session: Nanoscale Science and Engineering
See more of this Group/Topical: Nanoscale Science and Engineering Forum

Study of Macroscopic Uniformity and Scale-Ability of Continuous Flow Electroless Plated Au Film

Gyoung-Gug Jang, Ralph E. Martin Department of Chemical Engineering, University of Arkansas, Fayetteville, AR

D. Keith Roper, University of Arkansas, Fayetteville, AR

Continuous flow (CF) electroless (EL) plating deposits thin gold (Au) films that exhibit increased continuity and uniformity relative to batch immersion plating as measured by scanning electron microscopy and atomic force microscopy. The enhanced mass transfer rate and steady, higher average local Au ion concentration achieved during CF plating improves continuity and uniformity of the films. Microscopic uniformity as well as macroscopic uniformity and scale-extension of CF-EF deposited Au film are evaluated by transmission UV-vis spectrtroscopy. The mass transfer models and material balances for CF and batch immersion plating are utilized to estimate the maximum scale extension for uniformly deposited Au films. Uniformly deposited CF-EL metal films with high scale-ability have exciting applications in many fields including aerospace and automotive technology, chemical processing, oil and gas production and food processing.

See more of this Session: Poster Session: Nanoscale Science and Engineering
See more of this Group/Topical: Nanoscale Science and Engineering Forum

Laser-Induced Plasmon Excitation in Gold Nanoparticles to Evaporate Water and Increase Hydrogen Fuel Cell Efficiency

Aaron G. Russell, Department of Chemical Engineering, University of Akansas, Fayetteville, AR

D. Keith Roper, Chemical Engineering, University of Akansas, Fayetteville, AR

Flooding of recombined water in the cathode of polymer electrolyte fuel cells results in the necessity of excessive platinum (Pt) and reduced catalytic efficiency. Thermal sensitivity of the fuel cell membrane does not allow bulk heating of the cell as a means of combating flooding. However, we anticipate that flooding can be mitigated and the cell efficiency increased by the addition of targeted heating at the catalyst surface phonon decay of photon-excited plasmons in gold (Au) nanoparticle (NP) arrays imbedded within the cathode. We are currently using an Argon ion laser system to characterize the ability of plasmon heating in solid-state Au NP arrays to evaporate fluid from the array surface. We have expanded an energy balance to predict both the fluid evaporation rate and the effect of evaporative cooling on the local temperature profile of surfaces adjacent to the array. We compare the predicted thermal profiles using the model in its current state with those obtained experimentally. We believe the difference in equilibrium temperatures will be resolved by accounting in the model for mass change of the system due to the evaporating water.

See more of this Session: Poster Session: Nanoscale Science and Engineering
See more of this Group/Topical: Nanoscale Science and Engineering Forum

Plasmonics-Enhanced Microalgal Growth in Mini-Photobioreactors

Sarah Torkamani, Department of Energy, Environmental and Chemical Engineering, Washington University, St. Louis, MO

Satvik Wani, Department of Energy, Environmental and Chemical Engineering, Washington University, St. Louis, MO

Yinjie Tang, Department of Energy, Environmental and Chemical Engineering, Washington University, St. Louis, MO

Radhakrishna Sureshkumar, Department of Energy, Environmental and Chemical Engineering, Washington University, St. Louis, MO

Metal nanoparticles can couple with different wavelengths of the electromagnetic spectrum to produce strong absorption. This interaction can be tuned to enhance the absorption in a specific region. As is shown in this paper, silver nanoparticles can enhance the irradiance provided for microalgae and hence, increase the microalgal biomass growth. Microalgal mini-photobioreactor configuration can be designed using a model that takes plasmonics into account. We designed a mini-photobioreactor which consists of an additional section containing a dispersion of silver nanoparticles in water and ethylene glycol. This section scatters the light back to the medium and therefore enhances the microalgal growth. The proposed system has the potential to be applied in biosensors. Following figure depicts the experimental results.

See more of this Session: Poster Session: Nanoscale Science and Engineering
See more of this Group/Topical: Nanoscale Science and Engineering Forum

Growth of Single-Walled Carbon Nanotube Carpets From Alumina-Supported Fe Catalysts: Role of Alumina Type

Placidus B. Amama, Materials & Manufacturing Directorate, RXBN, UDRI / Wright-Patterson Air Force Research Laboratory, Dayton, OH

Cary L. Pint, Richard E. Smalley Institute for Nanoscale Science and Technology, Rice University, Houston, TX

Seung Min Kim, School of Materials Engineering and Birck Nanotechnology Center, Purdue University, West Lafayette, IN

Robert H. Hauge, Richard E. Smalley Institute for Nanoscale Science and Technology, Rice University, Houston, TX

Eric A. Stach, School of Materials Engineering and Birck Nanotechnology Center, Purdue University, West Lafayette, IN

Benji Maruyama, Materials & Manufacturing Directorate, RXBN, Wright Patterson Air Force Research Laboratory, Dayton, OH

The important role alumina plays as a catalyst support in single-walled carbon nanotube (SWNT) carpet growth is unsurprising as it is a well known structural modifier or catalyst promoter in hydrocarbon reforming and ammonia synthesis reactions. Although a lot of studies have been carried out so far to explain the role of alumina and the Fe-alumina interaction in the these reactions, an important area of SWNT growth that has not yet been studied explicitly is how the type of alumina affects the activity and lifetime of the catalyst, and the growth termination process. Here, we explore the Ostwald ripening behavior, catalyst lifetime, and mass loss (or sub-surface diffusion) of Fe catalysts deposited on different alumina supports: (a) sapphire; 10 nm thick alumina film deposited by (b) atomic layer deposition (ALD), (c) e-beam, (d) annealed e-beam, and (e) magnetron sputtering. The study reveals a strong dependence of the Ostwald ripening behavior, SWNT carpet height or the catalyst lifetime on the type of alumina used as catalyst support. TEM, ellipsometry, and XPS have been used to account for the marked differences observed.

See more of this Session: Poster Session: Nanoscale Science and Engineering
See more of this Group/Topical: Nanoscale Science and Engineering Forum

Enzyme Immobilization On Magnetic Nanoparticles for Cellulose Hydrolysis

Patrick A. Johnson, Chemical & Petroleum Engineering, University of Wyoming, Laramie, WY

Hee Joon Park, Chemical & Petroleum Engineering, University of Wyoming, Laramie, WY

Joshua T. McConnell, Chemical & Petroleum Engineering, University of Wyoming, Laramie, WY

Cellulosic ethanol has received a lot of recent attention as an alternative transportation fuel to reduce global dependence on oil. However, due to the high cost of enzymes, the process is not currently economically feasible. Our approach is to formulate recyclable enzyme constructs by immobilizing cellulases on magnetic nanoparticles. Glucose oxidase was selected as a model enzyme to demonstrate the immobilization strategies. Three different size magnetic nanoparticles (5 nm, 25 nm and 50 nm) were fabricated to explore the effect of particle size on diffusion efficiency. Two different methods, co-precipitation and oxidation of Fe(OH)2 were used to fabricate different sizes of magnetic nanoparticles. Magnetic nanoparticles were functionalized with amine groups by 3-(amino propyl) triethoxysilane. Glutaldehyde was then used as a cross-linking agent between functionalized magnetic nanoparticles and glucose oxidase. We have found that activity of glucose oxidase immobilized on various size magnetic nanoparticles retained activity after more than 40 days. The analysis of transmission electron microscopy (TEM) images showed that the morphology of magnetic nanoparticles was spherical and that sizes agreed with results from those of Brunauer, Emmett, Teller (BET) method analysis. The magnetic strength of the nanoparticles was analyzed via a Physical Properties Measurement System (PPMS). X-ray photoelectron spectroscopy (XPS) confirmed each step of the magnetic nanoparticle surface modification and successful enzyme immobilization. Recycling stability studies showed that only 20% of activity loss occurred during 10 consecutive recycle for large (50 nm) and medium (25 nm) size enzyme magnetic nanoparticles. Cellulase multi-enzyme mixtures along with extra β-glucosidase are the key enzymes for biomass degradation. We are applying the immobilization methods to these biomass related enzymes to test the feasibility of recycling the biocatalysts during cellulose hydrolysis.

See more of this Session: Poster Session: Nanoscale Science and Engineering
See more of this Group/Topical: Nanoscale Science and Engineering Forum

TurboBeads Llc. Making Chemistry Magnetic

Robert N. Grass, Department of Chemistry and Applied Biosciences, ETH Zurich, 8093 Zurich, Switzerland

Wendelin J. Stark, Department of Chemistry and Applied Biosciences, ETH Zurich, 8093 Zurich, Switzerland

Making chemistry magnetic is the core vision of TurboBeads Llc. which is based on metal nanomagnets developed by ETH Zurich, Switzerland. We drive the commercialization of specific magnetic nanoparticles for chemical separation applications. This talk will give a brief overview of TurboBeads' technology and nanoparticle synthesis strategies. There will be a strong focus on the application driven introduction of magnetic particles to the markets of biochemistry, metal refining, magnetic catalysis and magnetic polymers, giving specific application examples and discussing pricing and production volumes.

TurboBeads Llc. was founded in 2007 and is privately owned and financed. The company has development contracts with two Global Fortune 500 companies as well as several ongoing smaller projects. The presenter is founder and CEO of the company and has been involved in the AIChE annual conference throughout his career.

See more of this Session: Poster Session: Nanoscale Science and Engineering
See more of this Group/Topical: Nanoscale Science and Engineering Forum

MWCNT-Hydrogel Nanocomposites: Synthesis, Characterization, and Radiofrequency Heating

Nitin S. Satarkar, Chemical & Materials Engineering, University of Kentucky, Lexington, KY

Don Johnson, Chemical & Materials Engineering, University of Kentucky, Lexington, KY

Brock Marrs, Center for Applied Energy Research, Lexington, KY

Rodney Andrews, Center for Applied Energy Research, Lexington, KY

Belal Gharaibeh, Mechanical Engineering, University of Kentucky, Lexington, KY

Churn Poh, Mechanical Engineering, University of Kentucky, Lexington, KY

Kozo Saito, Mechanical Engineering, University of Kentucky, Lexington, KY

Samantha A. Meenach, Chemical & Materials Engineering, University of Kentucky, Lexington, KY

Kimberly W. Anderson, Chemical & Materials Engineering, University of Kentucky, Lexington, KY

J. Zach Hilt, Chemical & Materials Engineering, University of Kentucky, Lexington, KY

Nanocomposites obtained by incorporation of nanoparticulates in hydrogel matrix are attractive biomaterials for a variety of biomedical applications including drug delivery, hyperthermia cancer treatment, sensors and actuators, and tissue engineering. In this study, we present a nanocomposite of hydrogels with multi-walled carbon nanotubes (MWCNT). MWCNT were dispersed in temperature responsive N-isopropylacrylamide hydrogels. Lower critical solution temperature (LCST) of the nanocomposites was tailored for physiological applications by addition of varying amounts of acrylamide. LCST shifted to higher temperatures with increasing amounts of acrylamide. Nanocomposites were characterized for dispersion, temperature responsive swelling, mechanical properties, and response to radiofrequency (RF) field at 13.56 MHz. Addition of nanotubes contributed to interesting properties including reduction in swelling due to MWCNT hydrophobicity. Addition of MWCNT increased mechanical strength and was dependent on loadings. The effect of MWCNT addition on mechanical properties of the nanocomposites was characterized over a range of temperatures. Application of RF resulted into nanocomposite heating and resultant temperatures were measured using infrared thermography. The heating ability of nanocomposites was proportional to amount of MWCNT.

In addition to N-isopropylacrylamide, poly(ethylene glycol) hydrogel nanocomposites were also synthesized. Swelling, mechanical, and RF heating properties of the nanocomposites were studied for different MWCNT loadings. Cytotoxicity of the nanocomposites was evaluated with direct contact method.

See more of this Session: Poster Session: Nanoscale Science and Engineering
See more of this Group/Topical: Nanoscale Science and Engineering Forum

Life Cycle Assessment of Multi-Megawatt Wind Turbines with Carbon Nanofiber-Modified Rotors

Laura A. Merugula, Lowrie Department of Chemical and Biomolecular Engineering, The Ohio State University, Columbus, OH

Vikas Khanna, Lowrie Department of Chemical and Biomolecular Engineering, The Ohio State University, Columbus, OH

Bhavik R. Bakshi, Lowrie Department of Chemical and Biomolecular Engineering, The Ohio State University, Columbus, OH

It is generally accepted that wind turbine farms are zero-carbon emitters in the use phase, and most emissions are attributed to manufacturing and construction of the parts, periodic maintenance, and decommissioning. Life cycle assessments in open literature have supported this systematically and have allowed conclusions to be drawn from effects of material and site selections. Such assessments are based on a choice of fundamental unit by which systems serving equal purposes can be compared. Wind farms are consistently assessed by impact per kilowatt-hour of electricity produced. To effectively normalize an evaluation with regard to the functional unit, a life span of the materials and a maintenance schedule must be assumed.

For wind turbines, fatigue failure remains as a great uncertainty in materials selection for the rotor, which are composite jointed materials undergoing millions of flexural cycles in the operational phase. While the trend is towards glass fibers in a resin matrix as the main load-bearing materials, carbon fiber-epoxy systems are being considered more for their strength-to-weight ratio as turbine size increases. However, this electrically conductive system makes blades more susceptible to lightning strikes, which can cause generator shorts or turbine fires if not effectively grounded. The incidence of lightning is, coincidentally, higher in the offshore environment for which such larger turbines are slated. Stated research community goals are to develop models of turbines having individual ratings of 8-10 MW and rotor diameters over 120 m. With increase in rotor size, there is concomitant weight increase making resilience to fatigue even more necessary. Maintenance and repair from premature failure can drive down cost-effectiveness, if not make a project cost-prohibitive, so expected life span and presumed downtime must be modeled with such issues in mind.

This current project aims to contribute to modeling by using systems analysis techniques to evaluate emerging methods and materials meant to modify the glass fiber-resin interface of turbine blades by addition of carbon nanofibers. The modifications are meant to improve characteristic fatigue response by incorporating an ultra-high strength nanoscale reinforcing material at the interface so that it better reinforces the resin at the molecular level without creating an additional phase in the composite, which is one way to design damage tolerance into the structure. Also, the modifications are meant to be amenable to current production techniques so that the maturing system of manufacturing is hardly affected, except for the modification of the glass fibers. Thus, with the consistency of functional unit and candidates in the 2-5 MW range, a life cycle assessment is performed to compare the material and energy uses and impacts assuming a replacement of a percentage of glass fibers by an equal volume of carbon nanofibers. The resulting impacts will be compared to those of a conventional wind turbine farm.

Increased energy payback periods are due to energy-intensiveness of carbon nanofiber production. There is also an introduction of solvents and related emissions during the incorporation of the nanofibers, the effects of which may improve as laboratory methods mature. Methods of electrical grounding will be considered in the event of a change in related materials resulting from the addition of the conductive nanofibers. The major method of analysis is process-scale life cycle assessment. Exergy and ecological life cycle analyses are also undertaken to explore the implications of wind energy and the use of fossil fuel-based materials in achieving zero-emission electricity production versus unmodified wind turbines, i.e., without nanomaterials, and traditional electricity generation from non-renewable resources.

See more of this Session: Poster Session: Nanoscale Science and Engineering
See more of this Group/Topical: Nanoscale Science and Engineering Forum

Quantifying Nano-Scale Interactions of Chemically-Modified Nanoparticles with Surfaces: A QCM-D Study

Mark A. Poggi, Q-Sense, Glen Burnie, MD

Archana Jaiswal, Q-Sense, Glen Burnie, MD

Bartosz Grzybowski, Department of Chemical Engineering, Northwestern University, Evanston, IL

Stoyan K. Smoukov, Department of Chemical and Biomolecular Engineering, North Carolina State University, Raleigh, NC

There has been an increasing demand for analytical tools that can quantify the interaction and/or reaction of nano-scale particles, polymers as well as bio-molecules, with various surfaces. In-situ monitoring of changes of nano-scale materials, is of critical importance. If scientists and engineers have a more complete understanding of the behavior of materials at the nano-scale, new materials can be built from the nano-scale up.

Quartz Crystal Microbalance with Dissipation Monitoring (QCM-D), a nanomechanical acoustic-based analytical technique, provides in situ, real-time characterization of the interactions of various polymeric and bio-materials at variety of interfaces. With QCM-D, simultaneous measurement of resonance frequency change (delta F) and energy dissipation change(delat D) is performed by periodically switching off the driving power of oscillation of the sensor crystal and recording the decay of damped oscillation as the adsorption and/or structural changes takes place at sensor crystal surface (Figure 1). While change in frequency provides information about mass changes, changes in dissipation (D) provides structural information about the viscoelastic properties of adsorbed films in real time.

This presentation will be focused on recent research that we have done where we have followed in real-time the grafting of chemically-modified nanoparticles onto surfaces.

See more of this Session: Poster Session: Nanoscale Science and Engineering
See more of this Group/Topical: Nanoscale Science and Engineering Forum

Catalytic Decomposition of Alcohols over Size-Selected Pt Nanoparticles Supported on ZrO_2: A study of activity, selectivity, and stability

*S. Mostafa[a,b], J. R. Croy[a], H. Heinrich[a,c], B. Roldan Cuenya[a,b,d]**

[a]Department of Physics, [b]Department of Civil, Environmental, and Construction Engineering, [c]Advanced Materials Processing and Analysis Center, [d]Nanoscience and Technology Center, University of Central Florida, Orlando, Fl 32816

** Corresponding Author*
Email: roldan@physics.ucf.edu

Abstract

This article discusses the performance of ZrO_2-supported size-selected Pt nanoparticles for the decomposition of methanol, ethanol, 2-propanol, and 2-butanol. The potential of each alcohol for H_2 production and other relevant products in the presence of a catalyst is studied in a packed-bed mass flow reactor operating at atmospheric pressure. All the alcohols studied show some decomposition activity below 200°C which increased with increasing temperature. In all cases, high selectivity towards H_2 formation is observed. With the exception of methanol, all alcohol conversion reactions lead to catalyst deactivation at high temperatures (T >250°C for 2-propanol and 2-butanol, T >325°C for ethanol) due to carbon poisoning. However, long-term catalyst deactivation can be avoided by optimizing reaction conditions such as operating temperature.

Introduction

The decomposition of alcohols over metal and metal oxide catalysts has been the subject of numerous studies due to its applicability to a variety of chemical synthesis processes [1-5]. In recent years, this subject has gained particular interest due to growing environmental, economic, and political concerns regarding energy production [2]. Safe and efficient in-situ hydrogen generation from alcohols (i.e. methanol, ethanol, propanol, butanol) can promote the use of fuel cells and other clean technologies as a source of energy for mobile applications. Alcohols can serve as H_2 carriers that are compatible with current infrastructures for liquid fuels and can be catalytically decomposed on-site in order to minimize energy input requirements and operating temperatures [6]. In order to assess the potential of catalyzed alcohol decomposition we have conducted a systematic study of these reactions over size-selected Pt nanoparticles (NPs) supported on ZrO_2. The objective of this study is to assess the catalytic performance (activity, selectivity, and lifetime) of novel, micelle-synthesized, Pt catalysts. The conversion and selectivity of each C1-C4 alcohol was evaluated over a range of temperatures (100°C-300°C) under atmospheric pressure in a packed-bed mass flow reactor interfaced to a mass spectrometer. Catalyst stability was monitored through a prolonged exposure to the reactants and insight into the origin of their deactivation at high temperature due to carbonaceous species poisoning was evaluated by X-ray photoelectron spectroscopy (XPS). The by-products of the decomposition reactions are also investigated in order to assess the potential emissions associated with these processes as well as opportunities for material recovery in industrial applications.

Experimental

Size-selected platinum nanoparticles were synthesized using inverse-micelle encapsulation [7-9]. Pt particles were deposited via impregnation on our nanocrystalline ZrO_2 support with a nominal Pt loading of 2% wt Pt. The Pt/ZrO_2 catalyst powder was then annealed in an oxygen environment (50% O_2, 50% He) in a packed-bed reactor at 500°C. This treatment results in the removal of the polymeric particle shell. Transmission electron microscopy (TEM, FEI Tecnai F30) of our powder samples was done after annealing in O_2 as well as after exposure to reactants under the experimental conditions described below. Additionally, chemical characterization of our Pt NP catalysts, including their oxidation state, was conducted by XPS (SPECS GmbH) before and after the exposure to different alcohols using a monochromatic X-ray source (Al Kα, 1486.6eV) operating at an energy of 350W. The catalytic performance of our zirconia-supported Pt NPs in the decomposition of C1-C4 alcohols was evaluated in a packed-bed mass flow reactor. The reactor consists of a 7.4 mm (inner diameter) stainless steel (SS) tube where 100 mg of the catalyst sample is placed and supported by quartz wool. The composition of the gas exiting the reactor is monitored using a quadrupole mass spectrometer (QMS, HIDEN, HPR-20) which is connected to the reactor outlet via a heated capillary tube. Changes in the partial pressures of the alcohols and their associated products were used to determine activity (alcohol conversion) and selectivity over temperatures ranging between 100°C and 300°C.

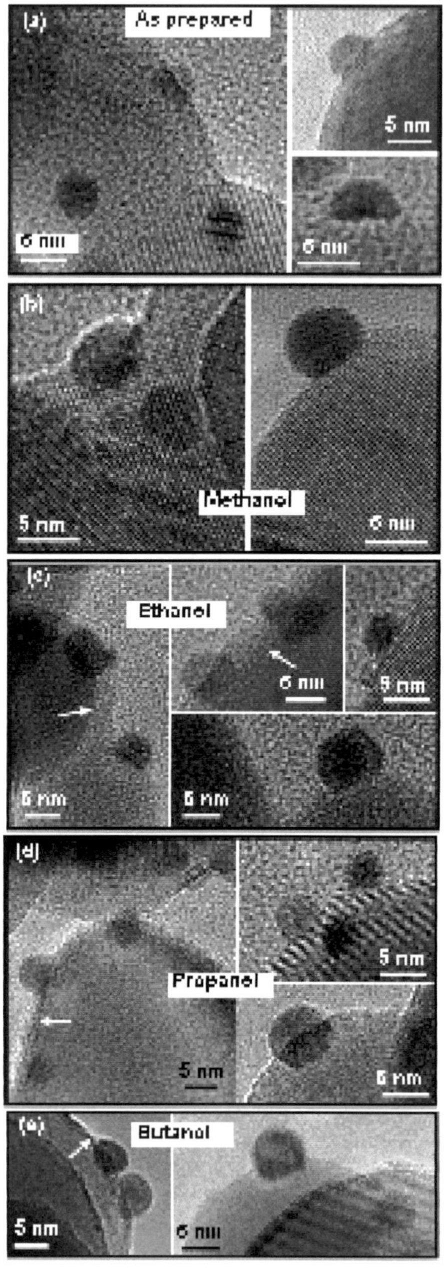

Figure 1. High resolution TEM images of the Pt/ZrO_2 catalysts as-prepared (only exposed to a pre-treatment in O2 at 500°C for four hours) (a) and after exposure to methanol (b), ethanol (c), 2-propanol (d) and 2-butanol (e)

Catalyst characterization

The TEM images shown in Fig. 1 show the as-prepared Pt NPs deposited on the ZrO_2 support after annealing in O_2 and before reaction with the different alcohols [1(a)], and after reaction with methanol [1(b)], ethanol [1(c)], 2-propanol [1(d)], and 2-butanol [1(e)]. Our TEM images demonstrate a narrow particle size distribution over the zirconia surface before and after the different reactions. Based on these high resolution TEM images, the average particle diameter in these samples is 4.8 ± 1.0 nm. No significant change in the average size is observed when comparing samples before and after exposure to reactants at temperatures up to 400°C, indicating that coarsening is not taking place during the different reactions. However, an amorphous layer not present in the fresh sample [after O_2 annealing,

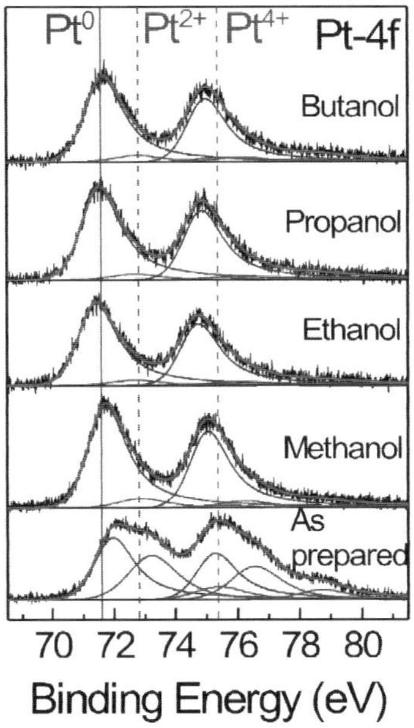

Figure 2. XPS spectra of the Pt-4f core level region of a fresh Pt/ZrO_2 sample after a pre-treatment in oxygen at 500°C (as prepared sample), and after exposure to reactants

Fig. 1(a)], is observed by TEM in the catalyst sample exposed to ethanol, Fig. 1(c), and to a certain extent also in those exposed to 2-propanol and 2-butanol [Figs. 1(d) and (e), respectively]. This layer is believed to be the result of carbonaceous species deposited during C2-C4 decomposition, as will be shown in connection with XPS data.

XPS measurements were used to determine the oxidation state of the as-prepared Pt nanocatalysts and to monitor possible changes in this state after reactions. Fig. 2 displays fitted XPS spectra of the Pt-4f core level region of the Pt/ZrO_2 catalyst before and after reactions with alcohols (at temperatures up to 400°C) as given by the right-hand side labels. Vertical lines in Fig. 2 denote the approximate position of the Pt-$4f_{7/2}$ components of Pt^0 in metallic Pt at 71.7 eV (solid line) [9], Pt^{2+} in PtO at 73.0 eV (dashed line) [10], and Pt^{4+} in PtO_2 at 75.3 eV (dashed line) [10]. A shift with respect to the position of bulk metallic platinum from 71.1 eV to ~ 71.7 eV is observed for the metallic Pt component, which is consistent with previous observations by our group [9] and is attributed to particle size effects [7, 11-13] and/or the formation of Pt-Zr, Pt-Zr-O alloys [14]. Catalyst exposure to all alcohols, and the associated sample heating up to 400°C, resulted in a strong reduction of Pt oxide species, although complete reduction was never observed.

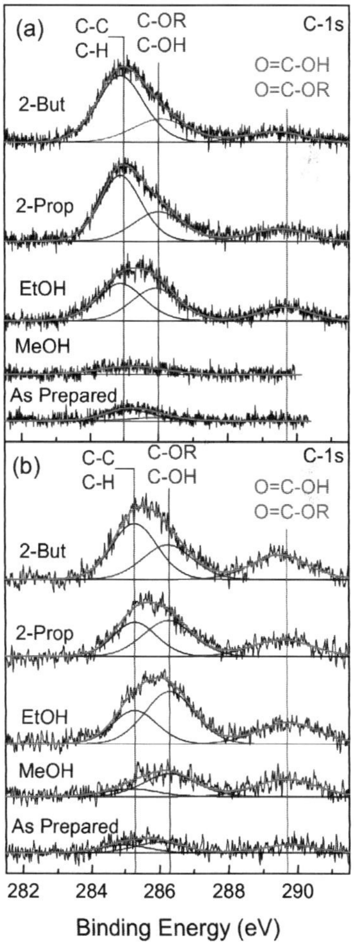

Figure 3. XPS spectra of the C-1s core level region of (a) our nanoscale Pt/ZrO_2 catalysts and (b) the Pt-free ZrO_2 support

XPS measurements from the C-1s region conducted on the Pt/ZrO_2 catalyst and the Pt-free ZrO_2 support are displayed in Fig. 3(a) and Fig. 3(b), respectively. The bottom spectra in these figures represent the fresh Pt/ZrO_2 and ZrO_2 samples and the remaining spectra correspond to measurements conducted after exposure to the different alcohols as indicated by the left-hand side labels. All spectra are plotted using the same scale but are displaced vertically for clarity. Only a very small C signal, associated with adventitious carbon, is observed in the fresh samples (labeled "as prepared"), indicating the removal of the encapsulating polymer from the Pt/ZrO_2 sample by the O_2 annealing pre-treatment [15]. Interestingly, after exposure to MeOH the ZrO_2 and Pt/ZrO_2 samples remain carbon-free, while a large increase in the C-1s intensity is observed after ethanol, 2-propanol, and 2-butanol exposure, possibly due to the deposition of carbonaceous species on the catalyst surface. The C-1s spectra of all the samples were fitted using three different photoelectron peaks, Fig. 3. The first peak at ~285 eV corresponds to the presence carbon species predominantly containing bonds of the type C-C and C-H (e.g. hydrocarbons, C_xH_y) [16, 17]. The second peak at ~286 eV is assigned to the presence of oxygenated carbonaceous species (C-OH/C-OR) [17], and the third peak at ~289.8 eV is assigned to carboxyl and/or carbonate species over the catalyst surface [16-19]. The C signals detected in the Pt/ZrO_2 samples exposed to C2-C4 alcohols are in agreement with the amorphous thin layers observed in the TEM images of some of these samples, Fig. 1, and they also correlate to the trends in activity and selectivity that will be discussed in the next sections.

Catalyst performance

In order to evaluate the activity of our catalyst for alcohol decomposition, the conversion taking place over a range of temperatures between 100°C and 300°C was calculated. The conversion for each of the alcohols as a function of temperature is shown in Fig. 4(a) for the Pt NPs supported on ZrO_2, and in Fig. 4(b) for the Pt-free nanocrystalline ZrO_2 support, both after exposure to an identical thermal pre-treatment in oxygen. As expected, based on previous studies by our group on similarly prepared catalysts [8, 20], methanol was efficiently decomposed over the Pt/ZrO_2 catalyst. The onset of methanol conversion occurs at a temperature of about 175°C, and the activity was found to increase with temperature, reaching over 90 % conversion at 300°C. Beyond that temperature, complete conversion and no deactivation for at least 10 hours (time frame of our reactivity experiment) were observed. Ethanol conversion over Pt/ZrO_2 was found to start at ~150°C, increasing with increasing temperature up to about 47 % conversion at 300°C. Catalytic activity decreases sharply above 325°C (not shown). 2-propanol and 2-butanol displayed the lowest onset temperature of ~125°C and were effectively decomposed throughout the entire temperature range showing similar trends in terms of conversion. After reaching maxima in activity at ~ 250°C, some catalyst deactivation takes place as demonstrated by a decrease in the conversion activity of both alcohols with increasing temperature up to 300°C. The conversion of the different alcohols over the Pt-free ZrO_2 support was relatively low (2-20%) as compared to the activity observed in the presence of Pt (47-92%).

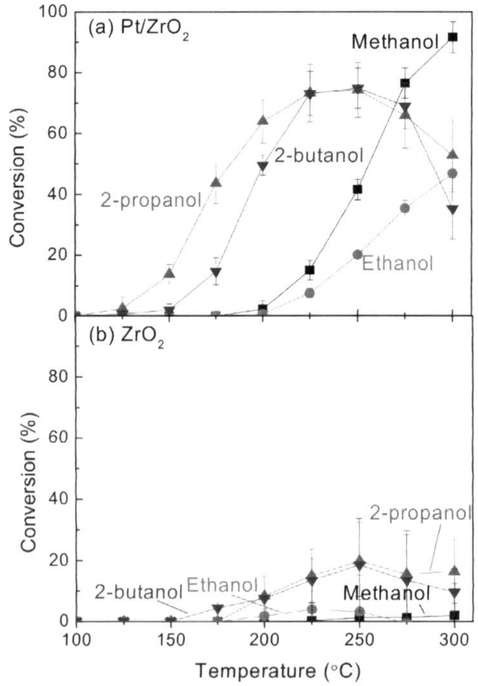

Figure 4. Temperature-dependent alcohol decomposition over (a) Pt nanoparticles deposited on nanocrystalline ZrO_2 and (b) Pt-free ZrO_2 support.

Catalyst selectivity was monitored with a particular interest in its ability to generate H_2 from C1-C4 alcohols. The dominant reaction pathways observed in the decomposition of each of the alcohols studied are described below. Equations (1-4) correspond to methanol decomposition, (5-9) to ethanol, (10-11) to 2-propanol, and (12-13) to 2-butanol decomposition.

Methanol decomposition and related reactions:

$CH_3OH \rightarrow CO + 2H_2$ (1)

$CH_3OH + H_2O \rightarrow CO_2 + 3H_2$ (2)

$CO + H_2O \rightarrow CO_2 + H_2$ (3)

$CO + 3H_2 \rightarrow CH_4 + H_2O$ (4)

Ethanol decomposition and related reactions:

$C_2H_5OH \rightarrow CO + CH_4 + H_2$ (5)

$C_2H_5OH + H_2O \rightarrow CO_2 + CH_4 + 2H_2$ (6)

$C_2H_5OH \rightarrow C_2H_4O + H_2$ (7)

$C_2H_5OH \rightarrow C_2H_4 + H_2O$ (8)

$C_2H_4O \rightarrow CH_4 + CO$ (9)

2-propanol dehydrogenation and dehydration:

$C_3H_7OH \rightarrow C_3H_6O + H_2$ (10)

$C_3H_7OH \rightarrow C_3H_6 + H_2O$ (11)

2-butanol dehydrogenation and dehydration:

$C_4H_9OH \rightarrow C_4H_8O + H_2$ (12)

$C_4H_9OH \rightarrow C_4H_8 + H_2O$ (13)

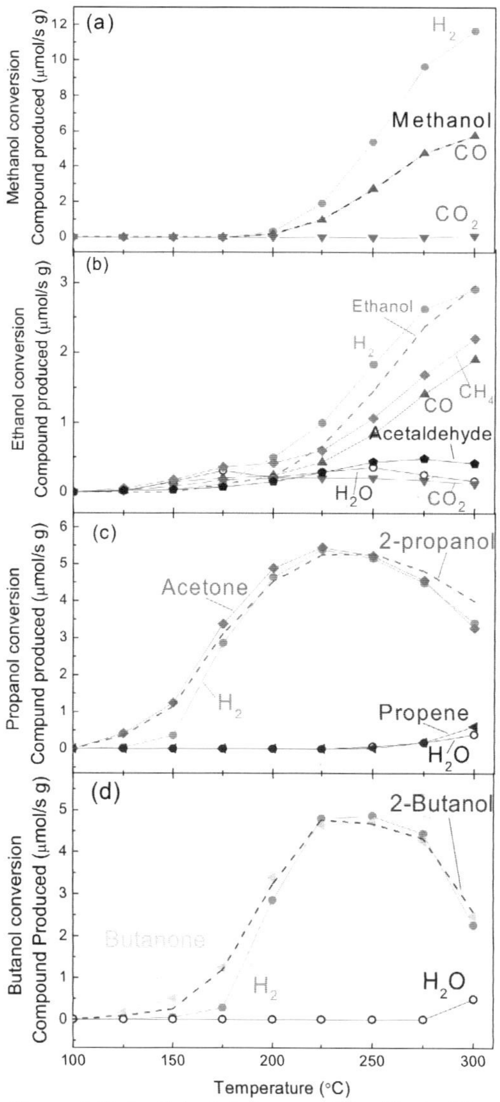

Figure 5. Rate of alcohol decomposition (dashed lines) and production of related by-products over Pt nanoparticles supported on nanocrystalline ZrO2 (solid lines and closed symbols).

Selectivity towards the different products was determined based on our mass spectrometry data. Figure 5 shows the rate of (a) methanol, (b) ethanol, (c) 2-propanol, and (d) 2-butanol conversion in units of μmol/s g catalyst (0.1 g Pt/ZrO_2) (dashed lines) together with the rate of production of H_2 and a variety of reaction by-products (methane, acetone, butanone, etc., solid symbols connected by solid lines). In the case of methanol, Fig. 5(a), the direct decomposition reaction described by eq. (1) was greatly favored, with H_2 and CO accounting for nearly 100 % of the products generated up to 300°C.The conversion data for ethanol and the associated reaction by-products are shown in Fig. 5(b) and may be explained in terms of the pathways suggested above. Ethanol decomposition shows significant H_2 production accompanied by the generation of CO, CH_4, as well as smaller amounts of CO_2, water, and acetaldehyde (C_2H_4O) for temperatures up to 300°C. This suggests that reaction (5) is dominant for this temperature range, while pathways (4), (6) and (7) may also occur to a lesser degree.

The selectivity of 2-propanol and 2-butanol decomposition are shown in figures 5(c) and 5(d), respectively. For temperatures below 250°C high selectivity for H_2 production is observed. The reactions observed in this temperature range result in the generation of acetone (C_3H_6O) in the case of 2-propanol [reaction (10)], and butanone (C_4H_8O) for the decomposition of 2-butanol [reaction (12)]. However, the formation of water is observed above 250°C, indicating the onset of the dehydration reactions [reactions (11) and (13)] for both reactants.

The peak in activity in the decomposition of 2-propanol and 2-butanol (T ~ 250°C) and the drop in activity that follows is an indication of catalyst deactivation. In the case of ethanol, a similar behavior was observed at T > 325°C (not shown). These trends can be explained in conjunction with XPS data, as resulting from the blocking of catalytically active sites by carbonaceous deposits. On the other hand, the catalytic activity for methanol decomposition remained steady after reaching a conversion > 95 % for the remainder of the temperatures studied (from T ~ 300°C up to 400°C). Therefore, the temperature at which our Pt/ZrO_2 catalyst deactivated was found to be a function of the alcohol being decomposed, possibly related to the temperature at which C-C bond scission takes place, as well as the formation of specific reaction intermediates. In order to distinguish between temperature and time-on-stream effects, stability experiments were conducted where the catalysts were exposed to the reactants for ~ 17 hours at temperatures below which short-term deactivation was observed for each alcohol. The temperatures selected for these experiments were 225°C for methanol, 250°C for ethanol, and 200°C for both 2-propanol and 2-butanol. The catalyst's activity in the decomposition of all the alcohols remained constant throughout the duration of each experiment. In addition, selectivities were similar to those reported in the low temperature region (<250°C) of Fig. 5.

The potential use of these reactions may be studied according to the specific application being considered. The dehydrogenation of 2-propanol will occur at the lowest temperature over the Pt/ZrO_2 catalyst as compared to the other reactions, generating H_2 and acetone. This makes 2-propanol a viable candidate for energy applications, particularly in stationary sources as well as industrial settings where both acetone and hydrogen could be recovered and reused. 2-butanol displays similar properties, generating butanone. However, at approximately 300°C, methanol decomposition will generate more than twice the amount of hydrogen from the same amount of reactant as any of the other alcohols studied here.

Conclusions

The main results derived from the analysis of our data include: (i) the onset temperature for alcohol decomposition (and H_2 generation) increases from propanol ≤ butanol < methanol ≤ ethanol, ranging from 125°C to 200°C; (ii) the maximum alcohol conversion was found to increase from ethanol < 2-butanol ≤ 2-propanol < methanol, ranging from ~ 47 % to 100 % in a temperature range of 100-300°C; (iii) the maximum H_2 production rates followed the same trend as the conversion rates, with a maximum of ~ 12 μmol(H_2)/s g catalyst during methanol decomposition, down to ~ 3 μmol(H_2)/s g catalyst for the ethanol reaction. High selectivity toward hydrogen production was observed for all the alcohols being decomposed over the Pt/ZrO_2 catalyst. With the exception of methanol, all the studied alcohols caused the catalyst to become deactivated at high temperatures (T >325°C for ethanol and T >250°C for 2-propanol and 2-butanol) due to the deposition of carbonaceous species over the Pt particles and their ZrO_2 support. The onset of dehydration reactions, possibly over the ZrO_2 support, was observed following indications of catalyst deactivation during 2-propanol and 2-butanol reactions. Optimum temperatures with the highest alcohol-to-H_2 conversions are 300°C for methanol and ethanol, and 250°C for 2-propanol and 2-butanol. Under these conditions, the Pt/ZrO_2 catalyst is expected to perform well for an extended period of time. It should also be noted that the oxidized Pt species originally present in the as-prepared samples were found to become significantly reduced in the course of the reactions with the different alcohols.

Acknowledgement

This work was possible thanks to the funding of the Office of Basic Sciences of the US Department of Energy (DE-FG02-08ER15995).

References

[1] J.D. Aiken and R.G. Finke, J. Mol. Catal. A **145** (1999) 1.
[2] G.C. Bond (Ed.), Metal-Catalyzed Reactions of Hydrocarbons (Springer, New York, NY, 2005).
[3] F.R. Keene, Coordin. Chem. Rev. **187** (1999) 121.
[4] B.Z. Zhan and A. Thompson, Tetrahedron **60** (2004) 2917.
[5] H. Hattori, Appl. Catal. A **222** (2001) 247.
[6] J.R. Rostrup-Nielsen, Phys. Chem. Chem. Phys. **3** (2001) 283.
[7] L.K. Ono, D. Sudfeld and B. Roldan Cuenya, Surf. Sci. **600** (2006) 5041.
[8] J.R. Croy, S. Mostafa, J. Liu, Y.H. Sohn and B. Roldan Cuenya, Catal. Lett. **118** (2007) 1.
[9] J.R. Croy, S. Mostafa, L. Hickman, H. Heinrich and B. Roldan Cuenya, Appl. Catal. A **350** (2008) 207.
[10] NIST X-ray Photoelectron Spectroscopy Database. Version 3.4 (Web Version), http://srdata.nist.gov/xps/index.htm
[11] A.K. Shukla, A.S. Arico, K.M. El-Khatib, H. Kim, P.L. Antonucci and V. Antonucci, Appl. Surf. Sci. **137** (1999) 20.
[12] F. Sen and G. Gokagac, J. Phys. Chem. C **111** (2007) 5715.
[13] W. Eberhardt, P. Fayet, D.M. Cox, Z. Fu, A. Kaldor, R. Sherwood and D. Sondericker, Phys. Rev. Lett. **64** (1990) 780.
[14] S.T. Srinivas and P.K. Rao, J. Catal. **179** (1998) 1.
[15] J.R. Croy, S. Mostafa, H. Heinrich, B. Roldan Cuenya, Catal. Lett. **131** (2009) 21
[16] Z. Paal, U. Wild, M. Muhler, J.M. Manoli, C. Potvin, T. Buchholz, S. Sprenger and G. Resofszki, Appl. Catal. A **188** (1999) 257.
[17] B.V. Crist (Ed.), On-Screen PDF Handbook of Monochromatic XPS Spectra - Commercially Pure Binary Oxides, XPS International, Inc., 2005, Vol.2, p 828 2005).
[18] F.B. Noronha, E.C. Fendley, R.R. Soares, W.E. Alvarez and D.E. Resasco, Chem. Eng. J. **82** (2001) 21.
[19] M.D. Appay, J.M. Manoli, C. Potvin, M. Muhler, U. Wild, O. Pozdnyakova and Z. Paal, J. Catal. **222** (2004) 419.
[20] J.R. Croy, S. Mostafa, J. Liu, Y.H. Sohn, H. Heinrich and B. Roldan Cuenya, Catal. Lett. **119** (2007) 209.

pH-Induced Self-Assembly of Partially Hydrolyzed Polyacrylamide with Low Molecular Weight

Yun FANG, Ting WANG*

School of Chemical and Material Engineering, Jiangnan University, Wuxi, Jiangsu, China

Abstract

Partially hydrolyzed polyacrylamide (HPAM) with low molecular weight (Mw=28000) is synthesized in this paper and the influence of environmental factors, such as temperature, concentration of NaOH and hydrolysis period on the hydrolysis degree of HPAM is studied. The self-assembly behavior of HPAM in aqueous environment was investigated. It is found out that micellization of HPAM could occurred merely through changing the pH in aqueous environment. TEM shows that HPAM self-assembled into spherical micelles with an average particle size of 130 nm under alkaline conditions, while in acidic conditions it turned to be "nuclear - shell" spherical micelles (average diameter 100 nm) and large complex vesicles (150-350 nm) with pore size of 2-10 nm.

Introduction

During the past decade, Self-assembly of highly regular macromolecules such as amphiphilic block or graft copolymers has been extensively studied because of the potential applications of the resultant aggregates[1]. The morphologies of aggregates usually includes sphere, vesicle, rod-like, cylinder, lamella, large compound micelle and many others[2-7]. Nowadays, random copolymers trends to be used in self-assembly because of the lower price and easier synthesis, and it can also self-assemble into various nanostructures. Faysal et al[8] gained giant vesicle from self-assembly of complementary random copolymers through hydrogen bonding. Liu et al[9] studied on self-assembly of random copolymers of poly(styrene-*co*-methacrylic acid) and obtained special bowl-shaped aggregates. Guo et al[10] gained spherical nanoparticles from Self-assembly of pH-sensitive random copolymers Poly (styrene-*co*-4-vinylpyridine) in a pH range of about 5–3 and a multicore structure at lower pH. In this paper, we report a pH induced self-assembly of HPAM (Mw=28000) in aqueous environment. The experimental result shows that HPAM can self-assemble into different morphologies under alkaline or acid conditions.

Experimental

Materials Acrylamide(AM, Shanghai Chemical Reagent Co., Ltd.) was recrystalized from chloroform. Ammonium persulfate (Shanghai Chemical Reagent Co., Ltd.) was recrystalized from isopropanol. Methanol, NaOH and HCl (Shanghai Chemical Reagent Co., Ltd.) were analytical grade and were used

* Corresponding author: yunfang@126.com. Supported by the National Natural Science Foundation of China (20871059).

without further purification. Deionized water with a resistance of 18 MΩ was used.

Preparation of HPAM copolymers Low-molecular-weight PAM was synthesized by solution polymerization using isopropanol as chain transfer agent according to the reference[11]. HPAM was obtained for example: 2 g of low-molecular-weight polyacrylamide (PAM) was mixed with 40 ml solution of 2.5 mol/L NaOH. The mixture was stirred with a mechanical stirrer and at 60 °C for 2 h. Then cooled mixture was precipitated with 300 ml purified methanol, filtered, and washed thoroughly with methanol until pH 7 was attained. The resulting product was dried in a 40 °C vacuum oven then kept in a desiccator.

HPAM with same molecular weight (Mw=28000) but different hydrolyzing degree (HD) was obtained by monitoring hydrolyzing conditions such as hydrolyzing time, hydrolyzing temperature and the concentration of sodium hydroxide.

Characterizations of the copolymers The infrared spectra of the copolymers were recorded on a FTLA2000-104 FT-IR spectrometer.

Preparation of HPAM micelles A 5 mg/mL stock solution of HPAM (HD=36%) in water was slowly (1 drop per min) diluted to 1 mg/mL and the required pH value with 0.1 mol/L NaOH or HCl.

Characterizations of the HPAM micelles TEM observations were performed on a Philips Tecnai CM12 electron microscope at an acceleration voltage of 120 KV. For all TEM observations, carbon-coated copper grids were dipped in the solutions of micelles at room temperature.

Results and Discussion

Characterizations of PAM and HPAM copolymers In a film of PAM, the amide I peak (C=O) and amide II peak (N-H) appear at 1665 and 1616 cm^{-1}, respectively. While in the film of HPAM, besides the two peaks of amide there were two peaks (COO^-) at 1560 and 1408 cm^{-1} (see Figure 1).

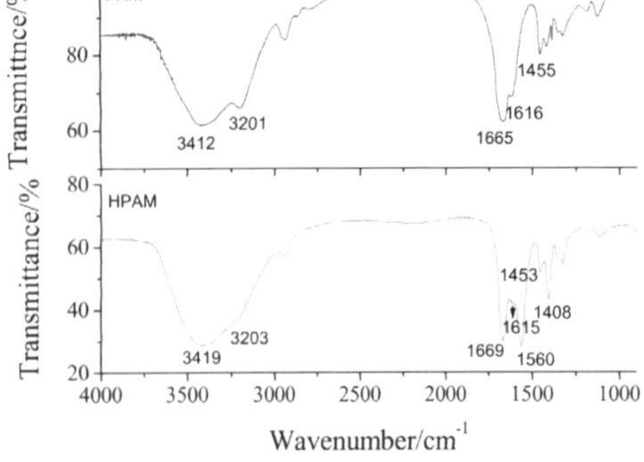

Figure 1 FTIR spectra of PAM and HPAM

Influence of environmental factors on HD of HPAM Figure 2 (A) shows that the hydrolysis equilibrium is basically approached after 4 h at lower temperature, and faster at higher temperature. Figure 2 (B) shows that HD of HPAM basically linearly increases with the temperature increase. Figure 2 (C) indicates that HD of HPAM is lower when c_{NaOH}<2.50 mol/L and higher when c_{NaOH}>2.50 mol/L. The HD of HPAM could be controlled between 35%~53% as one need just by running the hydrolyzing reaction at 40~80 °C for 0.5~4h maintaining NaOH concentration between 1.25~5.00 mol/L.

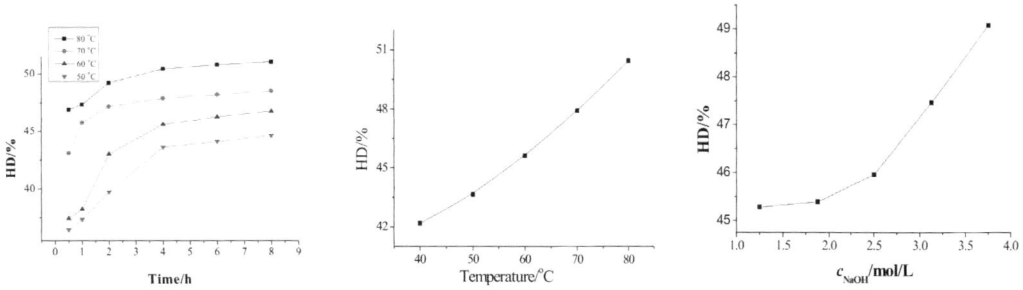

(A) (B) (C)

Figure 2 Effects of different conditions on hydrolysis degree of HPAM

(A) c_{NaOH}=1.25mol/L; (B) c_{NaOH}=1.25mol/L, 4h; (C) 60 °C, 4h

Aggregation of HPAM in aqueous environment Figure 3 shows the TEM image of the HPAM micelles. As shown in Figure 3 (A) at pH 12.0, HPAM forming the spherical micelles with an average particle size of 130 nm, and with a core of PAM/PAM complex due to hydrogen-bonding interaction between themselves, and with a compound shell of sufficiently dissociated PAA segments. While in pH 1.3, PAA are sufficiently protonated and forming the micelles with a core of PAA/PAM and PAA/PAA complex due to hydrogen-bonding interaction between them, and with a compound shell of sufficiently dissociated PAM. Interestingly, it turned to be "nuclear - shell" spherical micelles with an average diameter of 100 nm as shown in Figure 3 (D) combined with lots of large nonaxisymmetrical complex vesicles with average outer diameter of 150×350 nm and tiny vesicle pore size of 2-10 nm as shown in Figure 3 (B) and (C), which is reasonable considered as the result of secondary assembly of small spherical micelles with the average size of 10 nm.

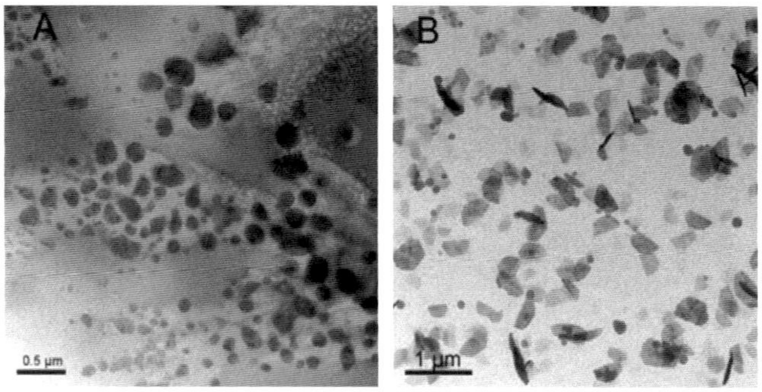

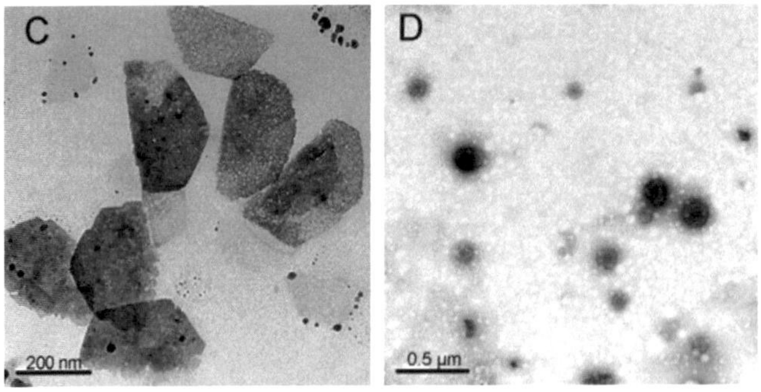

Figure 3 TEM image of the HPAM micelles

(A) pH=12.00, (B) and (C) pH=1.3, and (D) pH=1.3 and negative staining by phosphotungstic acid for 1 min

In conclusion, a series of HPAM with same molecular weight (Mw=28000) but different hydrolyzing degree (HD) was synthesized through monitoring hydrolyzing conditions. The experimental results show that the environmental factors such as hydrolyzing time, hydrolyzing temperature and concentration of NaOH have influence on HD of HPAM. The HPAM micelles self-assembled into spherical micelles with an average particle size of 130 nm at Ph 12.0 due to the hydrogen-bonding interaction between PAM/PAM segments themselves; while in acidic conditions it turned to be "nuclear - shell" spherical micelles with an average diameter 100 nm combined with lots of large nonaxisymmetrical complex vesicles with average outer diameter of 150×350 nm and tiny vesicle pore size of 2-10 nm at pH1.3 due to the hydrogen-bonding interaction between PAA/PAA and PAA/PAM.

References

1. N. Hadjichristidis; S. Pispas; G.A. Floudas. Block Copolymers: Synthetic Strategies, Physical Properties, and Applications, John Wiley & Sons: New Jersey, 2003.
2. L. F. Zhang, A. Eisenberg. Multiple Morphologies and Characteristics of "Crew-Cut" Micelle-like Aggregates of Polystyrene-*b*-poly (acrylic acid) Diblock Copolymers in Aqueous Solutions. *J. Am. Chem. Soc.*1996, 118 : 3168-3181.
3. L.B. Liu, X. Gao, Y. Cong, B.Y. Li, Y.C. Han. Multiple Morphologies and Their Transformation of a Polystyrene-block-poly (4-vinylpyridine) Block Copolymer. *Macromol. Rapid Commun.* 2006, 27: 260-265.
4. P. Bhargava, Y.F. Tu, J.X. Zheng, H.M. Xiong, R.P. Quirk, S.Z.D. Cheng. Temperature-Induced Reversible Morphological Changes of Polystyrene-block-Poly (ethylene Oxide) Micelles in Solution. *J. Am. Chem. Soc.* 2007, 129: 1113-1121.
5. T.J. Martin, K. Prochazka, P. Munk, S.E. Webber. pH-Dependent Micellization of Poly (2-vinylpyridine)-*block*-poly(ethylene oxide). *Macromolecules*, 1996, 29: 6071-6073.
6. N. Xu, F.S. Du, Z.C. Li. Synthesis of Poly (L-lysine)-graft-Polyesters through Michael Addition and Their Self-Assemblies in Aqueous Solutions. *J. Polym. Sci. Part A Polym. Chem.* 2007, 45: 1889-1898.

7. G. Mountrichas, S. Pispas. Synthesis and pH Responsive Self-Assembly of New Double Hydrophilic Block Copolymers. *Macromolecules*, 2006, 39: 4767-4774.
8. Faysal I., Trent H. G., et al. Giant Vesicle Formation through Self-Assembly of Complementary Random Copolymers. *J. Am. Chem. Soc.* 2000, 122(24): 5895-5896.
9. X.Y. Liu, J.S. Kim, J. Wu, A. Eisenberg. Bowl-Shaped Aggregates from the Self-Assembly of an Amphiphilic Random Copolymer of Poly (styrene-*co*-methacrylic acid). *Macromolecules*, 2005, 38: 6749-6751.
10. P.D. Guo, W.Y. Guan, L. Liang, P. Yao. Self-assembly of pH-sensitive random copolymers: Poly (styrene-*co*-4-vinylpyridine). *J. Colloid and Interface Sci.* 2008 323: 229-234.
11. X.Z. Zhao, D.M. Wang et al. Polymerization of Acrylamide Aaqueous Solution with Super Low Molecular Weight under High Monomer Concentration. *HuaXue Yu NianHe*, 2005, 27(4): 198-200.

Cryo-TEM Imaging of Carbon Nanotube / Superacid Systems

Nicholas G. Parra-Vasquez, Chemical and Biomolecular Engineering, Rice University, Houston, TX

Micah J. Green, Chemical and Biomolecular Engineering, Rice University, Houston, TX

Ellina Kesselman, Chemical Engineering, Technion-Israel Institute of Technology, Haifa 32000, Israel

Judith Schmidt, Chemical Engineering, Technion-Israel Institute of Technology, Haifa 32000, Israel

Yachin Cohen, Chemical Engineering, Technion-Israel Institute of Technology, Haifa 32000, Israel

Matteo Pasquali, Chemical and Biomolecular Engineering, Rice University, Houston, TX

Yeshayahu Talmon, Chemical Engineering, Technion-Israel Institute of Technology, Haifa 32000, Israel

The production of high-performance fibers, films, and composites from carbon nanotubes (CNTs) depends on the liquid-phase dispersion and processing of CNTs. However, CNTs are notorious for their low solubility in virtually all common solvents, because they naturally tend to form bundles and ropes due to van der Waals forces. In recent years we have explored the dissolution of carbon nanotubes in superacids, and have shown that such solutions can be processed into fibers and films. Here we utilize cryogenic-temperature transmission electron microscopy (cryo-TEM) to image nanotubes dispersed in chlorosulfonic acid and fuming sulfuric acid. Results indicate that nanotubes naturally dissolve as individuals in chlorosulfonic acid. To our knowledge, this is the first-ever imaging of any superacid solution, and these stunning images confirm that chlorosulfonic acid is a truly athermal solvent for CNTs, even for multi-walled CNTs and long (500 micron) CNTs.

See more of this Session: Poster Session: Nanoscale Science and Engineering
See more of this Group/Topical: Nanoscale Science and Engineering Forum

Dechlorination of Chlorinated Organic Compounds by Chemically Modified Zero Valent Iron (ZVI) Nanoparticles

Yunchul Cho, Department of Environmental Engineering, Kwangwoon University, Seoul, South Korea
Sang-Il Choi, Department of Environmental Engineering, Kwangwoon University, Seoul, South Korea
Donghyeok Park, Department of Environmental Engineering, Kwangwoon University, Seoul, South Korea
Doohyun Ryoo, Department of Environmental Health, Jeonju University, Jeonju, Jeollabuk-do, South Korea
Taesung Kim, Haechun ETS. Inc. Daejeon, South Korea
Sunkook Jung, ENPLUS Co. Ltd., Yeosu, Jeollanam-do, South Korea

Abstract

Batch kinetic tests were conducted to determine reductive dechlorination rates of TCE and 1,1,1-TCA by chemically modified zero valent iron (ZVI) nanoparticles. Chemical modification of ZVI nanoparticles was carried out using Pd as catalyst and CMC (carboxymethyl cellulose) as stabilizer. The average of TCE removal efficiency by CMC-modified FePd nanoparticles (FePd+CMC) was ~85%. This result suggests that CMC prevents FePd nanoparticles from aggregation, resulting in the increase of stability of the nanoparticles. For 1,1,1-TCA, the average of 1,1,1-TCA removal efficiency by CMC-modified FePd nanoparticles was ~56%. This result suggests that the structure of chemicals influences their degradation rate. Dechlorination of TCE and 1,1,1-TCA by FePd+CMC was well described by pseudo first order model. The observed rate constants (k_{obs}) of TCE and 1,1,1-TCA are 0.0228 min^{-1} and 0.0076 min^{-1}, respectively.

Introduction

Zero valent iron (ZVI) has been considered an effective reducing agent to degrade chlorinated organic compounds such as TCE and PCE [1,2]. Recently in order to enhance the reactivity of ZVI, reduction of particle size and usage of catalyst in the ZVI technology have been widely studied. The reactivity of nanoscale ZVI for chlorinated organic compounds is much higher compared to conventional microscale ZVI. This increase of reactivity may be attributed to the increased surface area of the nanoparticles. Also, for effective delivery of ZVI nanoparticles to hot spot zone, some stabilizers such as CMC (carboxymethyl cellulose) and starch have been used to enhance stability of ZVI for in-situ remediation. The stabilizers are found to prevent ZVI nanoparticles from forming aggregates, resulting in the increase in mobility of the nanoparticles in the subsurface environment [1].

In this study dechlorination rates of two chlorinated organic compounds (TCE and 1,1,1-TCA) by the chemically modified Fe-Pd nanoparticles were examined in laboratory batch systems. In order to describe the dechlorination reaction, pseudo first order model was applied.

Experimental

Preparation of Fe-Pd nanoparticles modified with CMC

Fe-Pd nanoparticles modified with CMC were prepared according to an oxidation-reduction method [1]. 1.244 g of $FeSO_4 \cdot H_2O$ was added in 0.1 L of deionized (DI) water to make a solution containing 0.1 g/L of Fe^{2+} ions, and then 10 mL of the solution was added to 236 mL of CMC solution (0.2% weight of Fe^{2+}). A certain amount of $NaBO_4$ was added to the mixture to yield a BH_4^-/Fe^{2+} mole ratio of 2. After the reducing reaction was allowed for 15 minutes, 1mL of K_2PdCl_6 (9.3×10^{-2} g/L) was added to the mixture to form a bimetallic Fe-Pd nanoparticles modified with CMC. All chemicals in this study were of analytical grade, and were used without further purification. Also, DI water used in this study was previously purged with N_2 to remove dissolved oxygen.

Dechlorination of TCE and 1,1,1-TCA

Batch tests were conducted to investigate dechlorination kinetics of TCE and 1,1,1-TCA by CMC- modified Fe-Pd nanoparticles. This bath experiment was carried out for 15 min, 30 min, 45 min, 1 h, and 2h. All experiments were performed in triplicate. Also, blank tests were conducted to evaluate loss of TCE and 1,1,1-TCA during the experiments. TCE and 1,1,1-TCA concentrations were analyzed by HPLC (LC-10AD, Shimazdu, Japan) with UV-VIS detector (SPD-10A, Shimazdu, Japan). Mobile phase was a mixture of acetonitrile and water (85:15 v/v) and Symmetric C18 column (Waters, USA) was used as HPLC column. In order to investigate chloride mass balance during dechlorination of TCE and 1,1,1-TCA by Fe-Pd nanoparticles, Cl^- ions released to solution were determined by UV spectrophotometry (UV-1650PC, Shimazdu, Japan). The sample preparation method for Cl^- ions measurement was as follows: 0.25 M ferric ammonium sulfate ($FeNH_4(SO_4) \cdot 12H_2O$) solution was prepared to add a certain amount of $FeNH_4(SO_4) \cdot 12H_2O$ to 9M HNO_3, and then stored in refrigerator. 2 g of $Hg(SCN)_2$ (mercury(II)thiocyanate) was added to 50 mL of ethanol (95%). Then the solution was shaken for 10 min and solution was allowed to stand for 1h at room temperature under atmosphere conditions. 160 μL of sample was reacted with 32 μL of $FeNH_4(SO_4) \cdot 12H_2O$ and 32 μL of $Hg(SCN)_2$ solutions. Chloride ions in the mixture were determined by UV spectrophotometry at 460nm.

Results and discussion

The dechlorination of TCE by CMC-modified Fe-Pd nanoparticles

TCE dechlorination rate by CMC-modified FePd nanoparitcles is shown in Fig 1a. All experiments were performed in triplicate runs and each value is the average from them. In the initial stage of TCE dechlorination CMC-modified Fe-Pd nanoparticles rapidly degraded TCE. After a 30min dechlorination reaction, TCE removal efficiency was around 76%. After 120 min dechlorination reaction, the removal efficiency reached upto ~85%. This rapid dechlorinatin is likely due to the increase in the amount of atomic hydrogen produced by formation of Fe-Pd nanoparticles. Also, CMC may contribute to enhance the reactivity and stability of the nanoparticles because it was found out that CMC prevents individual nanoparticles from forming aggregates.

Kinetic data were well fitted by pseudo first order model. The observed rate constant (k_{obs}) obtained from the fit of the dechlorination kinetics was 0.0228 min^{-1}. Fig 2b shows mass balance of chloride ions during TCE dechlorination, corresponding to Fig 1a. Theoretically 1mol of TCE can produce 3 mol of Cl^- after complete dechlorination of TCE to ethane. The theoretical amount of Cl was used as total amount of Cl (100 %) in the system (Fig 1b). "Cl in aqueous phase" represents a measured amount of Cl^- ions in the solution that was produced from TCE dechlorination at the certain time. "Cl in TCE" represents a calculated amount of Cl in the remaining TCE at the certain time. "unknown" represents experimental and analytical errors and Cl in byproducts such as DCE and VC. The amount of Cl^- ions in aqueous phase increased with the increase in reaction time (Fig 1b). The amount of Cl^- ions measured after 120 min reaction time was ~83% of theoretical amount of Cl^- ions produced from complete transformation of TCE to ethane. There is a good agreement between the measured value of the mass evolution of Cl^- ions and the corresponding one of TCE removal efficiency (~85%) after 120 min reaction.

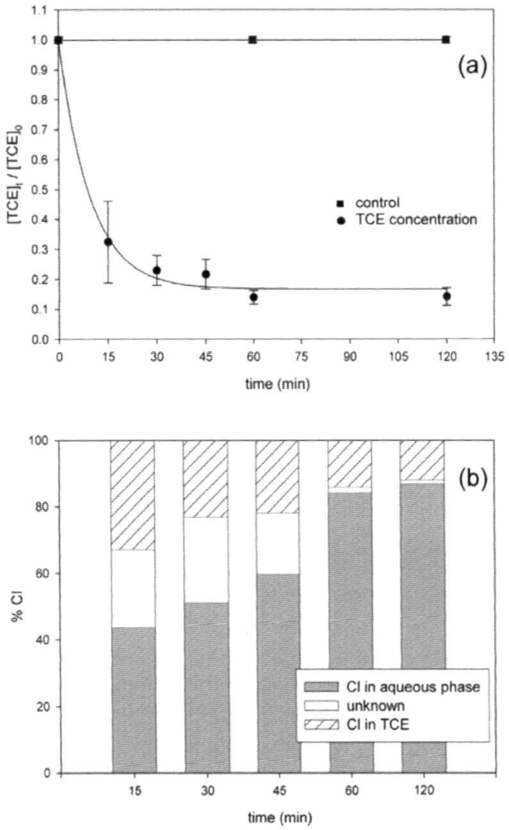

Figure 1. TCE dechlorination by CMC-modified Fe-Pd nanoparticles (a) and Cl mass balance (b)

The dechlorination of 1,1,1-TCA by CMC-modified Fe-Pd nanoparticles

1,1,1-TCA dechlorination rate by CMC-modified Fe-Pd nanoparitcles is shown in Fig 2a. After 30 and 120 dechlorination reactions, TCE removal efficiency was around ~37% and ~56%, respectively. In the initial stage of the dechlorination 1,1,1-TCA was rapidly degraded, however, the dechlorination rate constant (0.0076 min^{-1}) was smaller compared to TCE. It seems to be more difficult to dechlorinate 1,1,1-TCA due to the location of the chlorine. This result suggests that chlorine position of chlorinated organic compounds has a significant effect on the dechlorination kinetics.

On the other hand, in case of Cl mass balance (Fig 2b) "unknown" portion was small compared to TCE dechlorination. This result is likely due to no by-products such as DCE and VC unlike TCE.

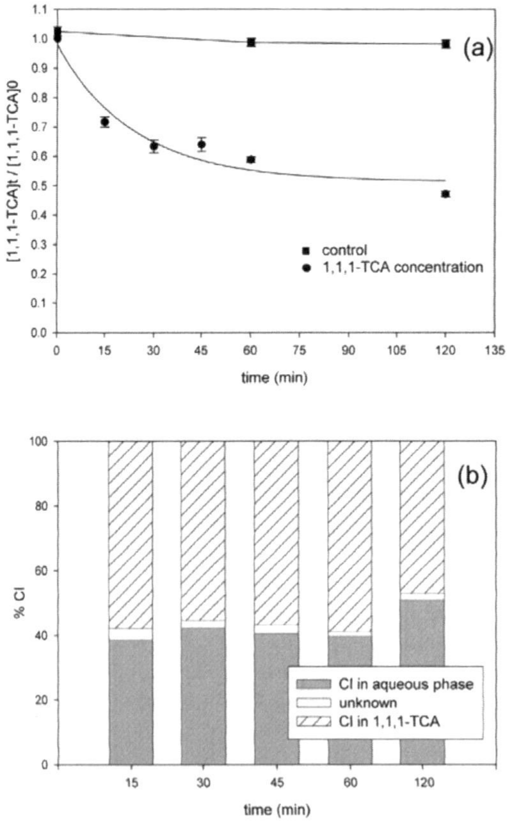

Figure 2. Dechlorination of 1,1,1-TCA by CMC-modified Fe-Pd nanoparticles (a) and Cl mass balance (b)

Conclusion

CMC-modified FePd nanoparticles were efficient reducing agents for treatment of TCE and 1,1,1-TCA. The amount of Cl ions in aqueous phase increased with the increase in the reaction time. This result suggested that main removal mechanism of two chlorinated compounds is reductive dechlorination. TCE dechlorination rate was faster compared to 1,1,1-TCA. It is likely due to difference of chemical structure between two chlorinated organic compounds.

Acknowledgement

This work was supported by Korea Ministry of Environment under GAIA (Geo-Advanced Innovative Action) research project

References

1. He, Feng, Dongye Zhao, Juncheng Liu, and Christopher B. Roberts (2007), "Stabilization of Fe–Pd Nanoparticles with Sodium Carboxymethyl Cellulose for Enhanced Transport and Dechlorination of Trichloroethylene in Soil and Groundwater" *Ind. Eng. Chem. Res.*, 46(1), pp. 29-34.
2. Lien, Hsing-Lung and Wei-Xian Zhang (2007) Nanoscale Pd/Fe bimetallic particles: Catalytic effects of palladium on hydrodechlorination. *Applied Catalysis B: Environmental* 77, pp. 110-116.

Fabrication of Long Sub-Micron/Nanochannels Via Coffee-Ring Effect

Yi-Je Juang, Chemical Engineering, National Cheng Kung University, Tainan, Taiwan

Kuo-Feng Lo, Chemical Engineering, National Cheng Kung University, Tainan, Taiwan

In recent years, micro/nanotechnology has been ubiquitous in our daily life ranging from electronic products, telecommunication to biomedical devices. One of the useful techniques in micro/nanotechnology is micro/nanopatterning, which is used to generate repetitive patterns of micro/nanofeatures on a relatively large area. Despite myriads of methods developed for micro/nanopatterning, the capillary force provides an easier, cheaper, and relatively rapid alternative. In this study, we proposed utilizing the coffee-ring effect and capillary force to perform large-area micro/nanopatterning using biomolecules. By means of replicating the pattern with polydimethyl siloxane (PDMS), we can obtain an array of long sub-micron/nanochannel at a relatively area. Several processing parameters such as solution concentrations, geometrical shapes and dimensions of the mold, and surface properties will be discussed.

Keywords: sub-micron/nanochannel, micro/nanopatterning, capillarity, coffee-ring effect, polydimethyl siloxane (PDMS)

See more of this Session: Poster Session: Nanoscale Science and Engineering
See more of this Group/Topical: Nanoscale Science and Engineering Forum

Self-Organized Nanoparticle-DNA Superstructures

Elizabeth J. Stewart, Chemical Engineering, University of Michigan, Ann Arbor, MI

Sudhanshu Srivastava, Chemical Engineering, University of Michigan, Ann Arbor, MI

Nicholas Kotov, Chemical Engineering, University of Michigan, Ann Arbor, MI

Understanding the self-organization of nanoscale materials is one of the key challenges to nanotechnology[1]. Of particular interest is the self-organization of biological materials with nanoparticles. Developing a fundamental understanding of the interactions between biological materials and nanoparticles will prove useful in the development of a novel class of materials as well as in providing insight onto the impact of nanomaterials on the body. This work considers the self-organization of semiconductor nanoparticles with deoxyribonucleic acid (DNA).

DNA based gene testing is used for diagnostic and presymptomatic testing of diseases such as Cystic Fibrosis, Lou Gehrig's disease, and Alzheimer's disease as well as for forensic and identity testing[2]. Nanostructures that contain DNA for sensing genes and semiconductor materials for monitoring signal changes after gene detection could prove useful for creating the next generation of DNA based biosensors for gene testing. In electronic devices, decreasing the size of components and increasing the uniformity of the pattern density on circuits will lead to more efficient and faster electronics[3].

In this study the electrostatic interactions of semiconductor nanoparticles and plasmid DNA or linear DNA are used to drive the self-organization of nanoparticle-DNA superstructures that could potentially be used in DNA based gene sensing or nanoelectronics. Specifically this research seeks to: (a) create stable superstructures with Cadmium Telluride (CdTe) and Zinc Oxide (ZnO) nanoparticles and plasmid and linear DNA, (b) charaterize the interactions between the nanoparticle-DNA systems, (c) determine the pair potential of various possible interactions within the system, and (d) investigate the possibility of manipulating stable superstructures to create dynamic architectures.

1. Mirkin, C.A., Programming the Assembly of Two- and Three-Dimensional Architectures with DNA and Nanoscale Inorganic Building Blocks. Inorganic Chemistry, 2000. 39(11): p. 2258-2272.

2. Human Genome Project Information: Gene Testing. US Department of Energy and Human Genome Project 2008; Available from: www.ornl.gov/hgmis.

3. Melosh, N.A., et al., Ultrahigh-Density Nanowire Lattices and Circuits. Science, 2003. 300(5616): p. 112-115.

See more of this Session: Poster Session: Nanoscale Science and Engineering
See more of this Group/Topical: Nanoscale Science and Engineering Forum

Mathematical modeling of transport and retention of nano-scale particles in homogeneous porous media.

Ashraf Aly Hassan, Zhen Li, Endalkachew Sahle-Demessie, George Sorial***

*** Civil and Environmental Engineering, University of Cincinnati, P.O. Box 21007, Cincinnati, OH 45221-0071, Office: 765G Baldwin Hall Tel: (513)556-2987, Fax: (513) 556-2599*

e-mail: George.Sorial@uc.edu

** U.S. EPA, NRMRL, 26 W. Martin Luther King Drive (MS 443), Cincinnati, Ohio 45268. (513) 569-7739, Fax: (513) 569-7677, e-mail: sahle-demessie.endalkachew@epa.govCincinnati, Ohio)*

The production, use, and disposal of nanomaterials could lead to their appearance in the environment, creating high exposure to humans and the ecosystem and affecting mobility and bioavailability of other toxics. Mathematical models were performed to validate experimental data investigating the transport and retention of nano-scale particles in water-saturated porous media.

The aqueous solution was introduced into several columns packed with sand at a flow rate of 10 ml/min. The sand columns were packed with sand obtained from AGSCO Co. 20-30 mesh (d_{50}= 0.71 mm). It is composed mainly of quartz and trace quantities of feldspar. Glass columns have an internal diameter of 2.5 cm and 45 cm length. The sand was thoroughly cleaned using sequential acid wash, water rinse and oven drying. The column had a void volume of 85 ml and a porosity of 0.37. Before introducing the nano-particle suspensions, the column was saturated and cleaned by Milli Q water. A schematic of the column experiments could be seen in figure 1.

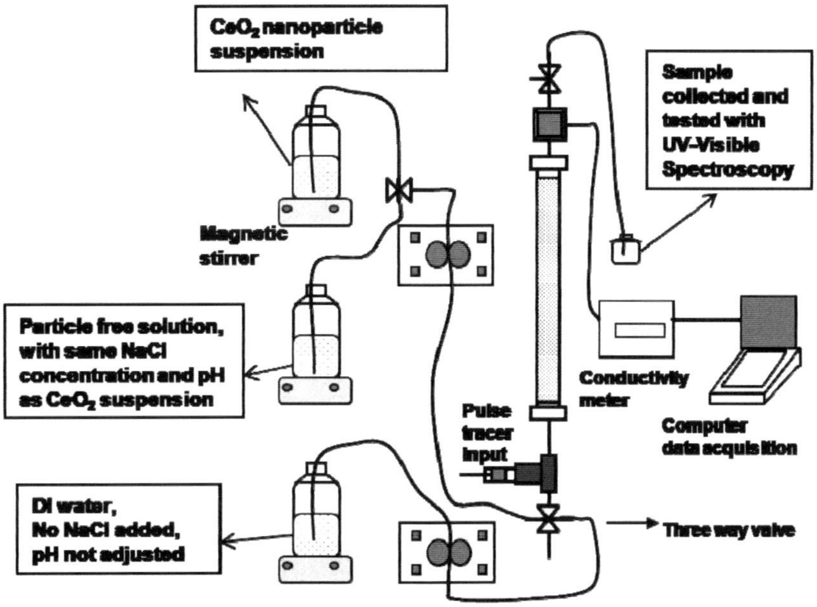

Figure 1: Transport experimental setup, Multi-column experimental system

The nano-particle suspension consisted of cerium oxide CeO_2. It is purchased from Alfa Aesar Co. in a suspension form, 18 wt % in water with dispersants and polishing additives. The different runs of Cerium oxide were introduced at different pH and background ionic strength. The pH was adjusted by adding appropriate amounts of HCl or NaOH solutions.

Effluent concentration and retention profile data were accurately simulated using a numerical model that solves conservation of mass and momentum, and the interaction and deposition of nano-particle with bed material. The mathematical model is based on porous media approach with a specific area-to-volume ratio pre-specified. The model predications were performed using computational fluid dynamic software, Fluent®. The model accounted for the attachment kinetics of nano-particle to the bed and a limiting retention capacity. The retention capacity was estimated using batch experiments, while the reaction kinetics is an approximate lump of all the mechanisms that occur to the nano-particle through its journey within the column.

Block Copolymer Assisted Carbon Nanotube Dispersion in Saturated Hydrocarbons

Xiaoming Wang, Department of Chemical Engineering, Yale University, New Haven, CT

Preetam K. Dutta, Department of Chemical Engineering, Yale University, New Haven, CT

Nan Li, Department of Chemical Engineering, Yale University, New Haven, CT

Salim Derrouiche, Department of Chemical Engineering, Yale University, New Haven, CT

Lisa D. Pfefferle, Department of Chemical Engineering, Yale University, New Haven, CT

Gary L. Haller, Department of Chemical Engineering, Yale University, New Haven, CT

Many of CNT's great applications require well-dispersed CNT in a certain media, but the tubes tend to bundle with each other due to the strong intertube van der Waals interaction. Research has been carried out to disperse CNT in polar solvents and aromatic non-polar solvents, with the assistance of surfactants and/or polymers. However, good dispersion of CNT in saturated hydrocarbon media has not yet been reported.

Here we report a block copolymer (BCP) assisted non-covalent method to disperse CNT in saturated hydrocarbons. A styrene-butadiene block copolymer (PS-b-PBD) was chosen to help the CNT dispersion because of the good compatibility between the styrene block and the CNT, and between the butadiene block and the solvent. Detailed research showed that this PS-b-PBD could help CNT to disperse in saturated hydrocarbons only when it forms micelles in the solvent. The star micelles with PS as the core and PBD as the corona could accommodate as much as 100 mg/L CNT in a hexane solvent. The CNT-BCP complex is very stable and would not dissociate upon sonication or dilution. After removing the solvent, the CNT-BCP complex forms a rubber-like film, and this material is readily re-dispersed in saturated hydrocarbon solvents.

See more of this Session: Poster Session: Nanoscale Science and Engineering
See more of this Group/Topical: Nanoscale Science and Engineering Forum

Fluorescent Polymerization-Based Signal Amplification of Antigenic Binding Events for Immunofluorescent Imaging of Cells

Heather J. Avens, Department of Chemical and Biological Engineering, University of Colorado, Boulder, CO

Brad J. Berron, Department of Chemical and Biological Engineering, University of Colorado, Boulder, CO

Chistopher N. Bowman, Department of Chemical and Biological Engineering, University of Colorado, Boulder, CO

Immunofluorescent imaging of cells is a widely used and critical tool for researchers and clinical pathologists to assess both protein expression and localization within cells. Applications have included detection of cancer biomarkers, study of protein localization during organogenesis, and imaging of cell surface proteins associated with drug resistance and cell signaling. Though several antigens are readily detected using fluorescently labeled antibodies, numerous lower abundance antigens are only detectable with improved microscopic techniques or signal amplification methods that increase the number of fluorophores coupled to each antigen/antibody binding event. Fluorescent polymerization-based amplification (FPBA) is presented as a signal amplification platform suitable for immunocytochemistry that enables highly sensitive fluorescent detection of cellular antigens, with advantages over many currently employed fluorescent signal amplification methods.

Previous reports have highlighted the use of FPBA for sensitive detection of DNA and proteins in a microarray format.[1,2] Here, we demonstrate that FPBA is a valuable method for immunofluorescent detection in cells. To achieve highly sensitive detection, FPBA takes advantage of the amplification inherent in radical chain polymerization. Specifically, photoinitiator molecules are coupled to protein probes, serving as a sensitive reporter for antigenic biodetection by initiating radical chain polymerization in the presence of monomer, light, and a fluorescent moiety, thereby forming a highly fluorescent gel specifically where an antigenic probe has bound its protein target. The work presented here employs a streptavidin-eosin conjugate (SA-eosin)[3] that couples the photoinitiation capacity of eosin to the biorecognition function of streptavidin, a protein with an extremely high affinity for biotin, a common label for biological probes. SA-eosin specifically binds biotin-labeled antibody probes, thereby localizing eosin photoinitiators to regions of the surface where the biotin-labeled antibody probes have coupled to their antigenic target. A monomer formulation comprising 420 mM poly(ethylene glycol) diacrylate, 35 mM vinyl pyrrolidone, 210 mM N-methyldiethanolamine as a coinitiator, and 0.05 wt% nile red fluorescent NPs in water is applied to the surface which is subsequently exposed for 20 minutes to 30 mW/cm^2, > 480 nm light, generating a highly fluorescent film only where antibody/antigen recognition has occurred.

To assess the sensitivity of FPBA for immunofluorescent imaging of cells, FPBA staining was performed with decreasing concentrations of primary antibody against von Willebrand Factor (vWF), a protein that is abundant in the cytoplasm of lung endothelial cells, but is absent from fibroblast cells. Although for standard indirect immunofluorescence detection this antibody is recommended for use at 1:200 dilution, with FPBA, we are able to achieve clear visual distinction between positive staining in endothelial cells and background staining in fibroblast cells down to at least a 1:50,000 dilution of primary antibody, constituting an estimated 250-fold improvement in sensitivity. Also, in the absence of primary antibody, the endothelial cells do not show significant staining under these conditions, further verifying that the observed staining is specific for vWF. Based on the extremely low antibody concentrations that yield suitable vWF staining with FPBA, it is expected that FPBA will be capable of detecting lower abundance antigens that are difficult to detect by other approaches.

Since FPBA achieves signal amplification by generating a fluorescent polymer film with some inherent thickness, the suitability of FPBA for staining smaller, more localized cellular structures was investigated. Using primary antibody against the nuclear pore complex (NPC) proteins, FPBA was able to achieve distinct and specific staining of the nuclear membrane of human lung endothelial cells. The NPC is estimated to have a length of approximately 80 nm spanning through the nuclear envelope, while the estimated thickness of the observed fluorescent stain is approximately a few microns. Although the observed polymer film is much thicker than the actual size of the protein, the nuclear membrane is indeed visualized as a clearly distinguishable cellular structure. Further, in many applications, a stain that increases the feature thickness may be desirable as it will render the feature more readily observable. However, in situations in which a thinner film is required, it is likely that shorter polymerization times may be employed to reduce film thickness, while maintaining a bright fluorescent stain.

In conclusion, FPBA is found to be a highly sensitive signal amplification detection platform for immunofluorescent imaging of cells. Unlike other signal amplification strategies for cellular applications, FPBA does not employ peroxidase enzymes, and therefore is not hindered by the endogenous peroxidase activity in cells. In addition to microarrays and immunocytochemistry, it is expected that FPBA will also be advantageous for immunohistochemical imaging in tissues and flow cytometry-based cellular imaging techniques.

References

1. Avens, H. J., Bowman C. N. Acta Biomater, submitted.

2. Hansen, R. R.; Johnson, L. M.; Bowman, C. N. Anal Biochem 2009, 386, 285-287.

3. Hansen, R. R.; Sikes, H. D.; Bowman, C. N. Biomacromolecules 2008, 9, 355-362.

See more of this Session: Poster Session: Nanoscale Science and Engineering
See more of this Group/Topical: Nanoscale Science and Engineering Forum

Magnetorheological Elastomers

Joko Sutrisno, Chemical Engineering, University of Nevada, Reno, Reno, NV

Alan Fuchs, Chemical Engineering, University of Nevada, Reno, Reno, NV

Yanming Liu, AMAD Inc., Reno, NV

Faramarz Gordaninejad, Mechanical Engineering, University of Nevada, Reno, Reno, NV

Barkan Kavlicoglu, AMAD Inc., Reno, NV

Bryce Wallis, AMAD Inc., Reno, NV

Xaioje Wang, Mechanical Engineering, University of Nevada, Reno, Reno, NV

Hussein Sahin, AMAD Inc., Reno, NV

Praveen Mysore, Mechanical Engineering, University of Nevada, Reno, Reno, NV

Nima Ghafoorianfar, Mechanical Engineering, University of Nevada, Reno, Reno, NV

A new class of materials termed MR elastomers (MRE) has been developed which responds to externally imposed magnetic fields. Magnetic particles are embedded in viscoelastic solids or liquid elastomeric precursors. This kind of composite demonstrates a unique combination of good magnetic controllability and elastic properties. A significant change in the structure of several MRE matrices was observed using AFM images. Modifications of conventional elastomers includes plasticizing the polymer matrices and particle-modified systems such as silane organic coatings and nanostructured polymer grafting techniques. In addition, a solvent vaporization technique was used for surface polymerization of iron particles with isobutyl trimethoxysilane as a coating agent. The thermal and mechanical behavior was evaluated with differential scanning calorimetry (DSC) and a mechanical testing device (Instron). The effects of plasticizer, particle coating modifications, and iron particle alignment were imaged and analyzed using scanning electron microscope (SEM) equipped with an electron probe microanalyzer energy dispersive X-ray spectrometer (EDS). Images of the particles/shells were taken to confirm the existence of the coating, and analysis was used to determine oxidative stability of the particle systems after a one month submersion in a simulated sea water environment.

See more of this Session: Poster Session: Nanoscale Science and Engineering
See more of this Group/Topical: Nanoscale Science and Engineering Forum

Optimization of Protease-Sensitivity of Polymer Vesicles

Chris W. K. Yeung, Bioengineering Graduate Program, The Hong Kong University of Science and Technology, Kowloon, Hong Kong

Ying Chau, Department of Chemical and Biomolecular Engineering, The Hong Kong University of Science and Technology, Kowloon, Hong Kong

Polymer vesicles can be functionalized protease-sensitive. The self-assembling unit (block copolymer) contains a peptide substrate of chymotrypsin which is linked between PEG and PCL blocks. The hydrophilic-hydrophobic balance required to form stable vesicles is destroyed upon proteolysis, and hence, the encapsulant inside the vesicles is released. The protease sensitivity is expected to be dependent on the surface area for protease actions and mole fraction of block copolymer containing the peptide substrate. To test our hypotheses, polymer vesicles are extruded to yield different sizes in order to vary the total surface area and co-assemble with block copolymer without the peptide substrate respectively. The protease-sensitivity is monitored by the release kinetics of encapsulated dye in the presence of the protease.

See more of this Session: Poster Session: Nanoscale Science and Engineering
See more of this Group/Topical: Nanoscale Science and Engineering Forum

Enhanced Emission of Gold Nanoparticles with Atypical Electron Transfer From Surface Bound Molecules to Gold Surface

Chang-won Lee,Bioengineering, University of Utah, Salt Lake City, UT

Curtis Takagi, Materials Science and Engineering, University of Utah, Salt Lake City, UT

Agnes Ostafin, Materials Science and Engineering, University of Utah, Salt Lake City, UT

Fluorescent system has long been used for the bio-imaging agent. Water-soluble gold nanoparticles with atypical fluorescence have been developed with mercaptooctanoic acid as a capping agent under reduction condition with sodium borohydride. The product was purified with ethanol trituration for further application or characterization. TEM analysis of the nanoparticle showed its diameter to be 2.2 ± 0.6 nm. The gold nanoparticle solution shows bright red emission with the excitation on UV plate. The fluorescence of the gold nanoparticle is dependent on the physiological pH range and has a possibility to be used as a nano-pH sensor due to its carboxylic acid containing capping molecules. In comparison to the widely used quantum dots or fluorescent dye molecules which usually have broad excitation spectra and overlapped region with emission spectra, it showed distinct excitation and emission spectra (excitation spectra peak @ 280 nm and emission spectra peak @ 610 nm) with 200 nm gap between them with large Stokes shift (Figure 1). It shows the possibility of two band gap modalities with responsible for excitation and emission and electron transfer between them. The loss during the electron transfer is considered to be the reason of distinct excitation and emission peak location with large Stokes shift. In this hypothetical diagram, excitation occurs in the left hand side band gap from surface covering mercaptooctanoic acid (MOA) for 280 nm excitation and transferred to the surface of the gold nanoparticles and the emission process occurs in the right hand side with relatively smaller band gap responsible for 610 nm emission. A computation simulation study was also performed for the change of the band gap of mercaptooctanoic acid with and without Au-S bond formation. Gaussian03 program was used with b3pw91/lanl2dz functional/basis set combination. The result showed similar trend of band gap change as it is experimentally confirmed by absorption band edge measurement. Detailed result and its hypothesis will be discussed in the presentation. The fluorescence of the gold nanoparticle is more efficient than other reported fluorescent gold nanoparticles with quantum yield (QY) of 0.015 measured with fluorescein dye molecule as a standard (QY=0.95) and the life time of the fluorescence is a lot longer than other reported fluorescent systems with 1.7 μs of half life. Along with these findings, the distinct excitation and emission without any overlaps between them with 200 nm gap suggested us a complex excited electron transfer from surface bound molecules to the core gold for the emission. This gold nanoparticle system also showed strong relationship with pH on its emission. The fluorescence intensity weakened at lower pH while it remained high at higher pH. The dependency of the fluorescence on the pH change also supports the proposed mechanism of excitation on surface capping mercaptooctanoic acid molecules. The charge state of carboxylic acid groups of mercaptooctanoic acid is believed to be responsible for this phenomenon.

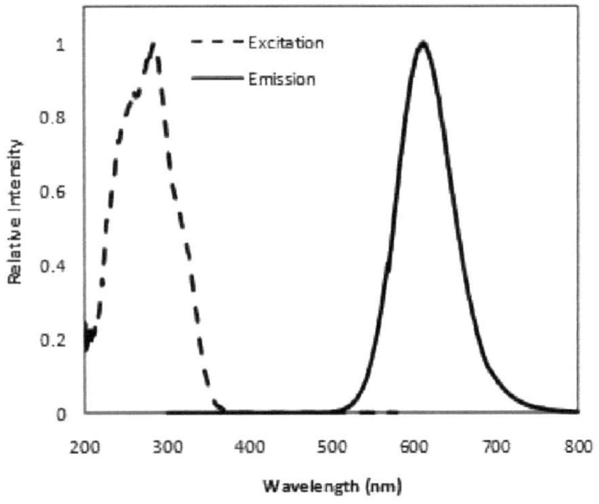

See more of this Session: Poster Session: Nanoscale Science and Engineering
See more of this Group/Topical: Nanoscale Science and Engineering Forum

Fabrication and Characterization of Electrospun Magnetic Polyacrylonitrile (PAN) Nanofibers for GMR Sensor Applications

Di Zhang, Dan F. Smith Department of Chemical Engineering, Lamar University, Beaumont, TX

Zhanhu Guo, Dan F. Smith Department of Chemical Engineering, Lamar University, Beaumont, TX

One-dimensional (1-D) nanomaterials have attracted increasing attention due to their characteristics combining the advantages of both polymer matrix and inorganic fillers. Transition metals or rare earth oxides are usually used as fillers because of their unique electrical, optical and chemical properties. A large number of synthesis and fabrication methods have already been used to produce 1-D nanomaterials in the form of nanofibers, nanorods, naowires, nanocables, nanobelts and nanotubes. But electrospining has been demonstrated that it is the simplest and low cost technique to produce nanocomposite fibers at a high speed. Electrospun fibers exhibit remarkable properties due to its large specific surface area, small diameter (50-500 nm) and extremely long length.

In this study, electrospun iron nanoparticles reinforced polyacrylonitrile (PAN) nanocomposite fibers are produced. Afterwards, electrspun nanocomposite fibers were annealed at nitrogen atmosphere under 800 oC for 2 hours to obtain the graphitic structure iron/carbon nanofibers. The crystallinity and magnetic properties of nanofibers will be reported.

See more of this Session: Poster Session: Nanoscale Science and Engineering
See more of this Group/Topical: Nanoscale Science and Engineering Forum

Multifunctional Polymer Nanocomposites: Case Study of Conductive Polypyrrole-SiC Particulate Nanocomposites

Pallavi Mavinakuli, Integrated Composites Laboratory, Dan F. Smith Department of Chemical Engineering, Lamar University, Beaumont, TX

Amar B. Karki, Department of Physics and Astronomy, Louisiana State University, Baton Rouge, LA

David P. Young, Department of Physics and Astronomy, Louisiana State University, Baton Rouge, LA

Jewel Andrew Gomes, Dan F. Smith Department of Chemical Engineering, Lamar University, Beaumont, TX

Zhanhu Guo, Dan F. Smith Department of Chemical Engineering, Lamar University, Beaumont, TX

Conducting polymers like polypyrrole have inspired great deal of interest due to their high and tunable electrical conductivity, unique physicochemical properties, easy preparation, high yield and environmental stability [1]. Electrical Conductivity of a conductive polymer mainly depends on the nature of the dopant. The material properties such as processability, mechanical strength and thermal stability can be improved by forming polymer nanocomposites [2]. Among the manipulated properties in the polypyrrole nanocomposites, the electrical conductivity can be tuned by using nanoparticles with different loadings.

Ceramic SiC is a wide and tunable band gap semiconductor, favorable material for high-temperature, high-frequency, excellent chemical resistance and high hardness applications. Due to its high mechanical strength, it has been used as filler for the preparation of strengthened polyurethane nanocomposite material [3].

By incorporating SiC nanoparticles into conductive polypyrrole matrix, we could create unique multifunctional polymer nanocomposites. These conductive polymer nanocomposites could have great potential to be used as sensors, biosensors, coating material, corrosion inhibitor, light emitting diodes (LED), and photovoltaic cells.

In this study, a conductive polymer nanocomposite containing polypyrrole and SiC nanoparticles was fabricated via oxidative polymerization. The effects of the particle loading and morphology (sphere and rod) of the nanoparticles on the physicochemical properties were investigated. Various characterization tools were carried out to explore the material properties. Thermal gravimetric analysis (TGA) was used to investigate the thermal stability of the polymer nanocompostes. The nanostructure of pure polymer and nanocomposites are investigated by scanning electron microscopy. Powder x-ray diffraction (XRD) was carried out to explore the effect of the nanoparticles on the crystal structure of PPy. Fourier transform infrared (FT-IR) spectrometry was used to investigate the interaction between the nanoparticles and the polymer matrix. Four-point probe measurement was carried out and the electron transport mechanism was explored.

[1]Z. Guo, K. Shin, A. B. Karki, D. P. Young, Fabrication and characterization of iron oxide nanoparticles filled polypyrrole nanocomposites, Journal of Nanoparticle Research, (2009) in press.

[2] M. Omastova, K. Boukerma, M. M. Chehimi, M. Trchova, Material research Bulletin 40 (2005)749-765.

[3] Z. Guo, T.Y. Kim, K. Lei, T. Pereira, J.G. Sugar, H.T. Hahn. Composite science and Technology 68(2008) 164-170.

See more of this Session: Poster Session: Nanoscale Science and Engineering

See more of this Group/Topical: Nanoscale Science and Engineering Forum

Nanostructured ZnO Thin Films for Sensor Applications

Hsin-Jung Hsieh, Chemical Engineering, Carnegie Mellon University, Pittsburgh, PA

James B. Miller, Chemical Engineering, Carnegie Mellon University, Pittsburgh, PA

Fabrication of sensitive and selective ZnO thin film gas sensors depends on the ability to control the composition and microstructure of the film. In this work, ZnO thin films were prepared by the sol-gel process using a Zn(CH3COO)2 2H2O precursor in a solution of isopropanol and ethanolamine. Transparent and stable films, ~25-250 nm in thickness, were deposited by spin-coating precursor solutions onto glass substrates, followed by heat treatment at temperatures through 500oC. For comparison, bulk ZnO xerogel powders were made by evaporation of the solvent from the sol-gel. Microstructure and composition were studied by X-ray Diffraction (XRD) and X-ray photoelectron spectroscopy (XPS). Scanning electron microcopy (SEM) was used to examine the surface morphology of the films. Both precursor concentration and heat-treatment conditions were found to strongly affect the crystallographic orientation and morphology of the ZnO films.

See more of this Session: Poster Session: Nanoscale Science and Engineering
See more of this Group/Topical: Nanoscale Science and Engineering Forum

Environmentally Responsive Hydrogel Nanocomposites for Biomedical Applications Synthesized Via Pcμcp

Hariharasudhan D. Chirra, Department of Chemical and Materials Engineering, University of Kentucky, Lexington, KY

J. Zach Hilt, Chemical and Materials Engineering, University of Kentucky, Lexington, KY

The incorporation of intelligent responsive hydrogel systems at the micro- and nanoscale has resulted in a wide variety of applications in the diagnostic and therapeutic field. Hydrogel nanocomposites can be tailored to exhibit combined effects of the component inside the crosslinked matrix and that of the response property of the hydrogel, thereby expanding their potential applications in the biomedical field. Herein, polymerization controlled by micro contact printing (PCmCP) was applied to synthesize micro- and nanostructured environmentally responsive hydrogel composites (e.g. crosslinked N-isopropyl acrylamide matrix with gold/iron oxide nanoparticles). XY control of the hydrogels was achieved using microcontact printing, and the Z/thickness control was achieved using a variety of polymerization techniques such as UV photopolymerization and atom transfer radical polymerization (ATRP). The responsive behavior of the hydrogels to external stimuli was characterized using optical microscopy, UV-visible spectroscopy, atomic force microscopy (AFM), and scanning electron microscopy (SEM). Using PCmCP based techniques; it is possible to synthesize controlled responsive hydrogel nanocomposites for potential applications in the biomedical field.

See more of this Session: Poster Session: Nanoscale Science and Engineering
See more of this Group/Topical: Nanoscale Science and Engineering Forum

Immobilized Enzymes for the Room Temperature Synthesis of TiO2 Nanostructures

Gregory J. Gibson, Department of Chemical and Environmental Engineering, University of California, Riverside, Riverside, CA

Yobed Woldeabzghi, University of California, Riverside, Riverside, CA

John Johnson, University of California, Riverside, Riverside, CA

David Kisailus, University of California, Riverside, Riverside, CA

Through billions of years of evolution, organisms have developed elegant mechanistic pathways to synthesize the full range of molecules necessary for life. Many of these molecules can be used outside of the cellular environment, their specific functionality being exploited to perform tasks beyond the reach or conventional chemistry. Because they are operational at atmospheric pressure and room temperature, adaptation of biomacromolecules in the form of enzymes can potentially replace current energy intensive synthesis methods for metal oxide semiconductors. In this work, the enzyme urease is used to induce localized pH changes in solution, thereby initiating the nucleation and growth of TiO_2 nanostructures with controlled size and morphology. These reactions use benign conditions, where temperatures of 150ºC or higher would normally be necessary. Controlled-size TiO_2 structures can be applied to numerous current engineering problems such as improving dye-sensitized solar cells and UV-initiated catalysis.

See more of this Session: Poster Session: Nanoscale Science and Engineering
See more of this Group/Topical: Nanoscale Science and Engineering Forum

Bio-Inspired SYNTHESIS and PROPERTIES of Photocatalytic METAL OXIDE Nanostructures

Nichola Kinsinger, Department of Chemical and Environmental Engineering, University of California, Riverside, Riverside, CA

Ashley Wong, University of California, Riverside, Riverside, CA

Eru Kyeyune-Nyombi Jr., University of California, Riverside, Riverside, CA

Fabian Villalobos, University of California, Riverside, Riverside, CA

Anthony Tantuccio, The Cooper Union, New York, NY

James C. Weaver, University of California, Riverside, Riverside, CA

David Kisailus, University of California, Riverside, Riverside, CA

Mineralizing biological systems demonstrate how nature can produce elegant structures at room temperature through controlled organic-mineral interactions. These organics (proteins and polysaccharides) exist as either soluble forms or as insoluble scaffolds that are often used to control size, shape and orientation of mineral.

Based on inspiration from Nature, we are using biologically-inspired scaffolds to template the nucleation and growth of inorganic materials such as TiO2. These scaffolds offer the potential to control surface energies that can dramatically influence crystal nucleation and growth behavior. In addition, these soluble ligands (modeled after specific mineral functionalities identified through the study of biological systems) that interact with minerals during homogeneous nucleation can be used to control the size, shape and phase of these particles and ultimately, their properties. In this work, we demonstrate size and phase control of metal oxide nanostructures and the resulting enhanced photocatalytic activity.

See more of this Session: Poster Session: Nanoscale Science and Engineering
See more of this Group/Topical: Nanoscale Science and Engineering Forum

Bisphosphonate Modified Gold Nanoparticles to Study Bone Resorption

Fedena Fanord, Chemical & Biomedical Engineering, University of South Florida, Tampa, FL

Korie Fairbairn, Chemical & Biomedical Engineering, University of South Florida, Tampa, FL

Harry K. W. Kim, Scottish Rite Hospital for Children, Dallas, TX

Venkat Bhethanabotla, Chemical & Biomedical Engineering, University of South Florida, Tampa, FL

Vinay K. Gupta, Chemical & Biomedical Engineering, University of South Florida, Tampa, FL

The bioavailability of a therapeutic agent is integral to its efficacy in treatment regimes. Low bioavailability necessitates increased dosage to achieve desired effects, in turn leading to increased risk and potential for adverse effects. One such example is bisphosphonates, which are a class of drugs used to treat diseases which induce increased bone resorption and bone loss. Low lipophilicity and high negative charge are the properties which most likely contribute to its low bioavailability. Localized delivery is one method of circumventing this effect. In vivo studies have shown localized delivery of bisphosphonates to be effective in inhibiting osteoclastic bone resorption, and thus prevention of femoral head deformity in Legg-Calvé-Perthes disease (LCPD). LCPD is a juvenile form of osteonecrosis of the femoral head, with no known preventive measures. Non-surgical treatment methods are limited. In our research, we have exploited conjugation of bisphosphonates with gold nanoparticles (GNP) and then explored delivery of these nanoparticles into osteoclasts. The optical properties of the GNP allow for in vivo detection of its distribution and retention as it inhibits bone resorption. Results from this work will be presented along with progress in development of an in vitro model to elucidate cellular interactions and response to the functionalized GNP. 09-->

See more of this Session: Bionanotechnology for Gene and Drug Delivery I
See more of this Group/Topical: Nanoscale Science and Engineering Forum

Transforming Growth Factor beta1 Releasing Degradable Nanogels for Chondrogenic Differentiation of Human Mesenchymal Stem Cells

Gauri P. Misra, Pharmaceutical Sciences, Thomas Jefferson University, Philadelphia, PA

Mugisha Niyibizi, Orthopaedics, Penn State College of Medicine, Hershey, PA

Jacquelyn Maddox, Orthopaedics, Penn State College of Medicine, Hershey, PA

Christopher Niyibizi, Orthopaedics, Penn State College of Medicine, Hershey, PA

Tao L. Lowe, Pharmaceutical Sciences, Thomas Jefferson University, Philadelphia, PA

Stem cell therapy has a promise for future medical treatment of a number of diseases due to the ability of the stem cell to differentiate with specialized functions and replace the ailing cells [1]. Growth factors are needed for stem cell differentiation. This can be achieved by the use of nano carriers [2]. Our group has been working on a novel polymeric nanogel system having both thermoresponsive and degradable properties [3]. In the current project, transforming growth factor beta1, TGF beta1, was encapsulated into the nanogels with high efficiency by UV photopolymerization at 45 °C using N-isopropyl acrylamide (NIPAAm) and a cross-linkable macromer, 2-hydroxyethylmethacrylate-lactate-dextran. Dynamic light scattering indicated hydrodynamic radii of the nanogels under 55 nm. Preliminary data indicated a slow release of TGF β1 in PBS (pH 7.4) from the nanogels. Fluorescence microscopy confirmed the up take of DTAF-labeled nanogels by the mesenchymal stem cells. The MTT cell viability data showed non-toxic nature of the nanogels for one week. These studies suggested that the current nanogels have potential for releasing growth factors for stem cell differentiation. Studies on optimization of TGF beta1 release kinetics from the nanogels and in vitro chondrogenic differentiation of the mesenchymal stem cells are underway.

See more of this Session: Bionanotechnology for Gene and Drug Delivery I
See more of this Group/Topical: Nanoscale Science and Engineering Forum

Development of a Lectin-Modified Core-Shell Paclitaxel Nanoparticle Based Drug Delivery System

Michael V. Pishko, Department of Chemical Engineering, The Pennsylvania State University, University Park, PA

Xiao Yu, Chemical Engineering, Texas A&M University, College station, TX

A novel lectin-mediated core-shell nanoparticle drug delivery system was fabricated with Paclitaxel as the model drug. This core-shell drug nanoparticle system has potential application in controlled and targeting colonic drug delivery. Surface modified paclitaxel nanospheres(approximately 200 nm in diameter) were used as the solid core on which synthetic and natural polyelectrolyte layers poly(allylamine hydrochloride) (PAH), poly(4-styrenesulfonic acid-co-maleic acid) (PSSCMA), chitosan and dextran-sulfate were electrostatically deposited by using layer by layer self-assembly method.[1] This nanoshell can provide specific sites for surface modification to improve biocompatibility and target cancer cells. Poly (ethylene glycol) (PEG) with a molecular weight of 3400 was covalently attached to the nanoshell surface forming a hydrophilic nanoparticle and avoid opsonization.[2] Fluorescent labeled wheat germ agglutinin (WGA) lectin moieties were then conjugated to the nanoshell to allow specific targeting to cancer cells.[3] Zeta potential measurement was used to confirm the successful addition of each polyelectrolyte layer. X-ray photoelectron spectroscopy (XPS) indicated the sulfur and nitrogen element composition change due to the increasing self-assembled layers. Analysis of high resolution spectra of C 1s and N 1s by XPS confirmed successful chemisorption of PEG onto the particle surface.[2] Successful conjugation of lectin WGA was confirmed by spectrofluorometer. Morphology of these core-shell nanoparticles was studied by transmission electron microscopy (TEM).

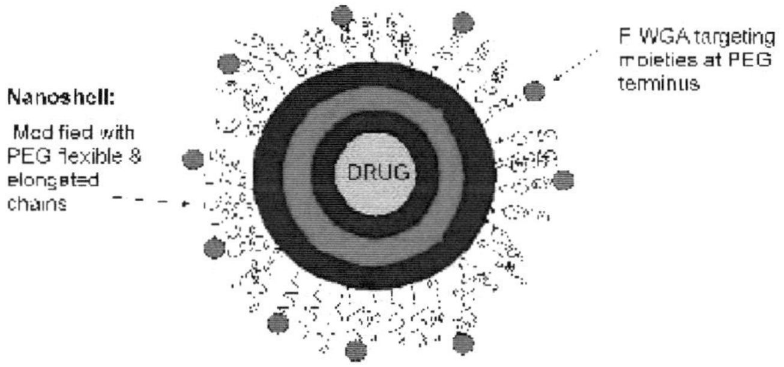

Scheme.1. Schematic representation of the lectin mediated core-shell paclitaxel nanoparticle based drug delivery system

Reference:

[1] Gero Decher, Fuzzy Nanoassemblies: Toward Layered Polymeric Multicomposites. Science, 1997, 277: p. 1232-1237.

[2] Alisar S. Zahr, Melgardt de Villiers, and Michael V. Pishko, Encapsulation of Drug Nanoparticles in Self-Assembled Macromolecular Nanoshells. Langmuir, 2005, 21:p.403-410.

[3] Padma V Devarajan, Ganeshchandra S Sonavane, Lectin mediated drug delivery systems. Express Pharma Pulse.

[Cited; Available from http://www.expresspharmaonline.com/20030605/ndds1.shtml]

See more of this Session: Bionanotechnology for Gene and Drug Delivery I
See more of this Group/Topical: Nanoscale Science and Engineering Forum

A Physiologically Based Pharmacokinetic (PBPK) Model for Evaluating the Efficacy of Liposomes at Treating Tricyclic Antidepressant or Local Anesthetic Overdoses

Brett A. Howell, Chemical Engineering, University of Florida, Gainesville, FL

Anuj Chauhan, Chemical Engineering, University of Florida, Gainesville, FL

Many drugs such as tricyclic antidepressants (TCA) and local anesthetics cause severe toxicity and/or death when taken at excessive and sometimes normal dosage levels. This study focuses on the use of physiologically based pharmacokinetic (PBPK) models for the development and optimization of liposomes for drug overdose treatment.

We have performed a series of in vitro binding studies using pegylated, anionic liposomes as drug sequestering agents. Drug affinities for various types and loadings of liposomes, as well as serum proteins, were quantified. PBPK models were then developed for predicting the efficacy of liposomes at overdose treatment in humans. Organ to blood partition coefficients for the drugs were calculated using published mechanistic equations. Hepatic extraction and renal elimination were obtained from previously published pharmacokinetic studies. Intravenous human data from various sources was compared to model predictions to serve as model validation. First order absorption was assumed and absorption rate constants for the oral absorption of antidepressants and non-oral absorption from the epidural, intercostal, or intraperitoneal spaces for anesthetics were obtained by best fits to published human studies. Thus, PBPK models capable of accurate drug concentration predictions in all tissues were constructed without the need for additional animal or human studies.

A series of overdose cases were then simulated with variations in drug and liposome dose, elapsed time between drug intake and liposome treatment, and patient specific input parameters. A single dose of the pegylated, anionic liposomes was predicted to be highly effective at treating antidepressant overdoses. A bolus dose of 1.44 g liposomes/L blood 2 hours after overdose reduced the area under the amitriptyline concentration curves (AUC) by 64% for the heart and brain. Heart and brain drug concentration peaks were predicted to drop by 20.1% and 20%, respectively. Although liposomes could potentially treat an anesthetic overdose, the drug redistribution was less effective, with bupivacaine AUC decreases of 15.4% and 15.3% for the heart and brain, respectively. Bupivacaine peak concentration reductions for both the heart and brain were 17.3%. Liposome doses were also increased twofold (2.88 g lipid/L blood as a single dose) to achieve such reductions for bupivacaine and the liposomes were assumed to be injected 15 minutes after the overdose.

The increased lipophilicity and proportion of molecules in the charged state at the physiological pH of 7.4 for antidepressants when compared to anesthetics, as well as their molecular structures, make them more susceptible to liposome sequestration. Nonetheless, the PBPK model showed that both antidepressant and local anesthetic overdoses may be successfully treated with liposome sequestration, with required doses drastically reduced compared to lipid emulsions currently used for local anesthetic overdose treatment. The predicted pharmacokinetic behavior after liposome administration was in reasonable agreement with available pharmacokinetic data from clinical studies where particles such as protein fragments were administered to patients for overdose treatment. Published data on ion channel function was used to connect the predicted concentrations in the body to local pharmacodynamic effects in the heart.

See more of this Session: Bionanotechnology for Gene and Drug Delivery I
See more of this Group/Topical: Nanoscale Science and Engineering Forum

PAMAM Dendrimer Conjugates for Intracellular Delivery of N-Acetyl-Cysteine

Yunus Emre Kurtoglu, Chemical Engineering, Wayne State University, Detroit, MI

Raghavendra Navath, Chemical Engineering/NICHD PRB, Wayne State University, Detroit, MI

Bing Wang, Pediatrics, NICHD PRB, Wayne State University, Detroit, MI

Hui Dai, Pediatrics, NICHD PRB, Wayne State University, Detroit, MI

Roberto Romero, National Institute of Child Health and Human Development, National Institutes of Health, Bethesda, MD

Sujatha Kannan, Pediatrics, NICHD PRB, Wayne State University, Detroit, MI

Rangaramanujam M. Kannan, Chemical Engineering and Materials Science, Wayne State University, Deroit, MI

N-Acetyl-L-Cysteine (NAC) is a drug frequently used as a mucolytic agent as well as acetaminophen overdose cases. The determination of its antioxidant, radical-scavenging properties, gave rise to investigation of its other therapeutic applications. Even though NAC proved to be a very therapeutic drug, delivery problems remain. The stability of NAC is low which increases the effective dose required. The high protein binding is another problem associated with NAC therapies since it affects activity and elimination of the drug. We proposed that the efficacy of NAC could be increased by protecting the drug from degradation, preventing protein binding and enhancing the cellular uptake, in addition to targeting it to tissue of interest in the body. For this purpose, we designed Polyamidoamine (PAMAM) dendrimer NAC conjugates as delivery devices that can address these issues. Evaluation of the nanodevices indicates that we successfully engineered Glutathione sensitive PAMAM dendrimer conjugates with high payload that release NAC specifically inside the cells and have higher in vitro efficacy compared to free NAC.

PAMAM dendrimers are highly hyper-branched synthetic polymers with well defined spherical structure and size. They have shown prospect as drug delivery vehicles with active and passive targeting capabilities, drug solubility and stability enhancement and vast opportunities for innovative designs at molecular level. In this study, we developed PAMAM dendrimer conjugates containing disulfide linkages. The choice of disulfide linkage was aimed at enhancing the intracellular availability of NAC by protecting it from degradation and protein binding and finally selectively releasing it intracellularly employing differences in Glutathione (GSH) levels. GSH is the most abundant thiol species in the cytoplasm and the major reducing agent in biochemical processes. The intracellular GSH concentration (1-10 mM) is substantially higher than extracellular levels (2uM in plasma), thus provides opportunities for intracellular cleavage of disulfide linkages.

In the present investigation, we have successfully synthesized two conjugates of NAC with G4-NH2 and G3.5-COOH PAMAM dendrimers. Additionally Fluorescein Isothiocyanate (FITC) was also conjugated for imaging and cellular uptake studies. The cell studies were carried out on Mouse microglial cells (BV-2). G4-NH2 and G3.5-COOH conjugates contained 16 and 18 NAC molecules per dendrimer respectively as determined by 1H-NMR and MALDI-TOF analysis. The drug release kinetics of the conjugates was evaluated at intracellular and extracellular GSH concentrations. The release mechanism was determined to be via disulfide exchange reactions with the thiol group of GSH. The conjugates released ~50% of their NAC payload within 1 hour at intracellular GSH concentrations whereas they were quite stable at extracellular GSH levels and did not release significant levels of NAC. The release rates indicate that the disulfide bond was easily and rapidly cleaved by GSH to a high extent within a short amount of time. These results have significant implications in achieving controlled release in dendrimer-based delivery systems as well as demonstrating that GSH can be used as a reliable in vivo releasing agent.

Cellular entry of FITC labeled G4-NH2-NAC conjugate was evaluated using flow cytometry. It was determined that most of the G4-NH2-NAC conjugate entered BV-2 cells within 15 minutes after incubation and uptake continued slowly for another 45 minutes. The cellular entry was also visualized by using Confocal Laser Scanning Microscopy. It is evident that the FITC labeled G4-NH2-NAC conjugate localized mostly in the cytoplasm and lysosomes, while the nucleus appears to be relatively free of the presence of any fluorescence two hours after incubation. The results indicate that PAMAM dendrimer conjugates are transported inside the cells efficiently by endocytosis mechanism.

Efficacies of conjugates were evaluated by measuring the Nitrite in BV-2 cell culture medium which indicates the level of anti-oxidation. Free NAC inhibited nitrite release in dose-dependent manner after 72 hours incubation, while only high concentration of free NAC (8mM) inhibited nitrite release significantly after 24 hour incubation. On the other hand both G4-NH2 and G3.5-COOH conjugates had more significant inhibition both at 24 and 72 hours compared to the same concentration of free NAC. The results indicated both dendrimer-NAC conjugates are more effective anti-oxidants compared to equivalent free NAC, especially at lower NAC doses. The results indicate that PAMAM dendrimer conjugates produce a higher local NAC concentration inside the cells by enhancing cellular uptake and blocking degradation of the drug. Our preliminary in vivo studies suggest that the efficacy of NAC can be increased 10 fold by using the described PAMAM dendrimer delivery nanodevice.

See more of this Session: Bionanotechnology for Gene and Drug Delivery I
See more of this Group/Topical: Nanoscale Science and Engineering Forum

Formation of Multifunctional Nanoparticles by Integrated Electrospray and Microfluidics for Cancer Imaging, Diagnostics and Therapy

Yun Wu, NanoScience and Engineering Center, The Ohio State University, Columbus, OH

Bo Yu, Department of Chemical and Biomolecular Engineering, The Ohio State University, Columbus, OH

Wei-Ching Liao, Mechanical Engineering, The Ohio State University, Columbus, OH

Andrew Jackson, NIST Center for Neutron Research, National Institute of Standards and Technology, Gaithersburg, MD

Weibin Zha, Nomacorc LLC, Zebulon, NC

Barbara E. Wyslouzil, Department of Chemical and Biomolecular Engineering, The Ohio State University, Columbus, OH

L. James Lee, Department of Chemical and Biomolecular Engineering, The Ohio State University, Columbus, OH

Multifunctional nanoparticles with combined imaging, diagnostic and therapeutic functions have attracted lots of research interest in cancer early detection and therapy. Here, we present a novel method to produce multifunctional lipoplex nanoparticles by the combination of electrohydrodynamics spraying (EHDS) and microfluidics (MHF). In this method, MHF is used to bring all components in a well controlled manner, and the pre-formed liquid mixture is dispersed into tiny droplets by EHDS. With quick evaporation of solvents due to the huge surface areas of the tiny droplets, multifunctional lipoplex nanoparticles are formed rapidly. In our work, oligonucleotide (ODN) was chosen as the model therapeutic agent and gold nanoparticles were chosen as model imaging or diagnostic agents. Transferrin was conjugated on surfaces of multifunctional nanoparticles to provide targeting ability. We investigated the effects of production parameters (flow rates, concentrations, etc.) on the size and structure of multifunctional lipoplex nanoparticles using dynamic light scattering, CryoTEM, small angle neutron scattering (SANS) and confocal microscopy. The biological activities of multifunctional lipoplex nanoparticles were further examined in K562 cells. The combination of MHF and EHDS provides a new, rapid and real -time method for production of multifunctional nanoparticles.

See more of this Session: Bionanotechnology for Gene and Drug Delivery I
See more of this Group/Topical: Nanoscale Science and Engineering Forum

Characterization of Remote Drug Release for Chemotherapy Using Magnetic Liposomes

Kyle Fugit, Department of Pharmaceutical Sciences, The University of Kentucky, Lexington, KY

J. Zach Hilt, Department of Chemical and Materials Engineering, The University of Kentucky, Lexington, KY

Brad Anderson, Department of Pharmaceutical Sciences, The University of Kentucky, Lexington, KY

Liposomal delivery systems are established formulations in the pharmaceutical industry. They take advantage of the enhanced permeation and retention effect associated with tumors. Evidence suggests that liposomes minimize severe side effects associated with systemic administration of chemotherapeutic agents, making them suitable vehicles for tumor-specific drug delivery. However, better control over the liposomal release characteristics is needed. An ideal release mechanism would retain the drug within the liposome until localization at the tumor; a subsequent triggering mechanism would allow drug release at the tumor site. This controlled drug delivery system would reduce adverse reactions while allowing a high dose of chemotherapy to reach the tumor site. The remote heating properties of magnetite nanoparticles in an alternating magnetic field cause a phase change in the liposome microenvironment. This phase change from the amorphous gel to the liquid crystalline phase is expected to increase the permeation and release of drug.

This project focused on characterizing drug release when magnetic magnetite nanoparticles were incorporated in a liposomal drug carrier. For this delivery system, magnetic magnetite nanoparticles were synthesized using a co-precipitation method, and liposomes were formulated through an extrusion process. Loading of the magnetic nanoparticles was carried out during liposome extrusion, and separation of unloaded magnetite was accomplished by precipitating the particles through acidification. Dynamic Light Scattering was used to confirm particle size of the liposomes and magnetic nanoparticles to be 200 nm and 15 nm in diameter respectively. Powder X-ray diffraction identified the composition of magnetic nanoparticles as magnetite. Finally, spectrophotometric analysis was used to determine the iron content inside the liposomes and validate no unloaded magnetite was present in solution. To ascertain changes in drug permeability, the development of a novel system to measure drug release at sink conditions in an alternating magnetic field was developed.

See more of this Session: Bionanotechnology for Gene and Drug Delivery I
See more of this Group/Topical: Nanoscale Science and Engineering Forum

Lipoplex and Polyplex Nanoparticles Based On Cholesterol Modified Oligonucleotides for Enhanced Gene Delivery

Bo Yu, Department of Chemical and Biomolecular Engineering, The Ohio State University, Columbus, OH

Weibin Zha, Department of Chemical and Biomolecular Engineering, The Ohio State University, Columbus, OH

Yicheng Mao, Division of Pharmaceutics, College of Pharmacy, the Ohio State University, Columbus

Robert J. Lee, Division of Pharmaceutics, College of Pharmacy, the Ohio State University, Columbus

L. James Lee, The Ohio State University, Columbus, OH

Non-viral nanoparticles such as lipoplex and polyplex are promising vesicles to efficiently deliver oligonucleotides (ONs) such as antisense ODN (AS-ODN) and small interfering RNA (siRNA) in vivo. However, it still remains a major challenge to produce ON nanoparticles with well-defined nanostructures and high drug loading for enhanced gene delivery. In this presentation, we demonstrate that cholesterol modified ON (Chol-ON) can be readily condensed and encapsulated inside liposomes through ionic complexation between calcium ions and phosphates on the backbone of ON. In contrast to typical "onion"-like structure, the core-shell structure of chol-ON/calcium/lipids nanoparticles was observed by cryogenic-temperature transmission electron microscopy (cryo-TEM). The resulting nanostructure was further investigated by atomic force microscopy (AFM). The changes of surface charge and particle size with/without calcium in the formulation were determined to explain the formation of this core-shell nanostructure. As an example, chol-G3139 (a 18-mer AS-ODN against Bcl-2 protein) were incorporated into liposomes containing lipids (DC-chol/EggPC/PEG-DSPE=30/68/2, molar ratio) in the presence of calcium. Comparing to regular G3139 or chol-G3139/lipids nanoparticles, the nanoparticles constructed from chol-G3139/calcium/lipids showed an increase in cell-uptake analyzed by confocal microscopy and Bcl-2 down-regulation determined by Western blotting in K562 and KB cells. This technology has also been extended to prepare chol-siRNA liposomal and chol-ON polyplex nanoparticles. The combination of calcium ions and chol-ON (eg. chol-ODN or chol-siRNA) provides a new strategy to control nanostructure of ON nanoparticles for cancer therapy.

See more of this Session: Bionanotechnology for Gene and Drug Delivery II
See more of this Group/Topical: Nanoscale Science and Engineering Forum

Trafficking of Nanoparticles to Intracellular Aggresome-Like Strucutres: Implications for Non-Viral Gene Delivery

Sutapa Barua, Chemical Engineering, Arizona State University, Tempe, AZ

Kaushal Rege, Chemical Engineering, Arizona State University, Tempe, AZ

Nanoscale therapeutics, including nonviral gene delivery systems, have been investigated for a variety of cancer cell types. However, the efficacy of these therapeutics is compromised by cancer cell resistance mechanisms. The inefficient sub-cellular delivery of nanoparticles is of particular relevance to nanoscale therapeutics. Consequently, investigations into the role of cancer cell phenotype in the cellular uptake, intracellular sorting, transport, and sub-cellular localization can lead to improved efficacies of nanoscale therapeutics.

Here we show that phenotypic differences in human prostate cancer cells lead to dramatic differences in intracellular sorting and localization of exogenous nanomaterials including nanoparticles (quantum dots), polymer:DNA, and polymer:siRNA complexes. These nanoparticles demonstrate a punctated intracellular localization throughout the cytoplasm in PC3 cells. In contrast, the materials were taken up by clathrin-mediated endocytosis and trafficked along microtubules, resulting in their localization in aggresome-like compartments in PC3-PSMA cells (sub-clones of PC3 cells retrovirally transduced to stably express PSMA receptors).

We hypothesized that localization of polymer:DNA complexes in aggresome-like compartments was a major barrier to the transport of genes to the nucleus and therefore, is responsible for low transfection efficacies. We have recently generated polymers that show increased transfection efficacies and lower cytotoxicities than polyethyleneimine, a current standard for polymer-mediated gene delivery. Intracellular trafficking studies indicated co-localization of quantum dots and polymer:plasmid DNA polylexes in juxtanuclear aggresome-like structures in PC3-PSMA cells. Confocal microscopy studies were carried out to characterize the nature of the aggresome-like compartment. In contrast, QDs and polyplexes colocalized in punctate vesciles throughout the cytoplasm of PC3 cells. Comparison between PC3 and PC3-PSMA cells indicated higher transfection efficacies in PC3 cells compared to PC3-PSMA cells, which correlated with the formation of aggresome-like structures. Strategies for overcoming this intracellular barrier to gene and siRNA delivery were investigated. Our results provide a platform to understand mechanisms behind the resistance of certain cancer cell types to gene and siRNA delivery and describe opportunities for overcoming these resistance mechanisms for nanoscale therapeutics.

See more of this Session: Bionanotechnology for Gene and Drug Delivery II
See more of this Group/Topical: Nanoscale Science and Engineering Forum

Nanoscale Directed Assembly to Prepare siRNA Containing Liposomes

Megan L. Cavanaugh, Chemical and Biomolecular Engineering, The Ohio State University, Columbus, OH

L. James Lee, Department of Chemical and Biomolecular Engineering, The Ohio State University, Columbus, OH

While liposomal nanoparticles represent a promising class of vectors for siRNA delivery, the attempts with conventional methods have not been successful in controlling the structure of siRNA containing liposomes. Most techniques for liposomal nanoparticle preparation rely on a self assembly process conducted on the macro-scale or in "bulk," which is much larger than the nanoparticles themselves and leads to limited control over particle size, structure, composition. We have proof of concept experiments using 1 um features which suggest that carefully controlling the liposome components' interactions create particles with better controlled structure and morphology because it is conducted on the same scale as the particles themselves. This work is being extended to the nanoscale using 200 nm wells in order to further control the particles at the ideal size range. The transfection efficiency of these particles is measured using a breast cancer model using siRNA targeted against the aromatase enzyme.

See more of this Session: Bionanotechnology for Gene and Drug Delivery II
See more of this Group/Topical: Nanoscale Science and Engineering Forum

pH Responsive Polypeptides for siRNA Complexation

Amanda C. Engler, Chemical Engineering, Massachusetts Institute of Technology, Cambridge, MA

Paula Hammond, Chemical Engineering, Massachusetts Institute of Technology, Cambridge, MA

One of the key challenges of the systemic delivery of siRNA is the ability to effectively complex the small charged oligonucleotides. Linear and branched cationic polymers are often selected for complexation because of their ability to bind and condense nucleic acids and stimulate endosomal escape. However, some of these polymers, such as poly(ethyleneimine), have limited success due to their high toxicity. A new polypeptide, poly(propargyl-L-glutamate) (PPLG) has been developed that can be easily functionalized with a wide range of amino groups by "click chemistry". PPLG is synthesized by well-studied N-carboxyanhydride (NCA) ring-opening polymerization of propargyl-L-glutamate NCA. Similar to natural polypeptides, these polymers possess the ability to assemble into well-defined, ordered structures, have low toxicity, and have well-controlled dimensions. Both PPLG and PEG-b-PPLG have been synthesized and various amino groups have been coupled to the PPLG to introduce charge and pH responsive behavior. Circular dichroism studies have been performed to observe alpha-helix to random coil transitions of the polymers as a function of pH. Turbidity studies have been performed to determine polymer solubility as a function of pH. Light scattering and AFM have been used to observe the self-assembly behavior of the PEG-b-PPLG at various pH levels. Initial studies have also been performed to confirm that these polymers are able to complex with siRNA.

See more of this Session: Bionanotechnology for Gene and Drug Delivery II
See more of this Group/Topical: Nanoscale Science and Engineering Forum

Development of Quantum Dot Labeling of Gene Delivery Vectors for Visualization of Viral Transduction Pathways

Kye Il Joo, Chemical Engineering, University of Southern California, Los Angeles, CA

Yun Fang, Chemical engineering, University of Southern California, Los Angeles, CA

Liang Xiao, Chemical engineering, University of Southern California, Los Angeles, CA

Yuning Lei, Chemical engineering, University of Southern California, Los Angeles, CA

Chi-Lin Lee, Chemical engineering, University of Southern California, Los Angeles, CA

Pin Wang, Chemical engineering, University of Southern California, Los Angeles, CA

γ-Retroviral and lentiviral vectors are currently the most commonly used gene delivery vehicles due to their ability to permanently integrate a therapeutic transgene into a target cell chromosome. Improved understanding of virus-host cell interactions can provide crucial insights for enhancing the efficacy of virus-mediated gene delivery as well as preventing virus-triggered diseases. The ability to track individual viruses is a powerful tool for investigating viral infection routes and characterizing the dynamic interactions between viruses and target cells. We report here a general method of labeling enveloped viruses with semiconductor quantum dots (QDs) for use in single virus trafficking studies. Retroviruses, including human immunodeficiency virus (HIV), could be successfully tagged with QDs through the membrane incorporation of a short acceptor peptide (AP) that is susceptible to site-specific biotinylation and attachment of streptavidin-conjugated QDs. We showed that this AP tag-based QD labeling had little effect on the viral infectivity and allowed for the study of the kinetics of the internalization of the recombinant lentivirus enveloped with vesicular stomatitis virus glycoprotein (VSVG) into the early endosomes. We further demonstrated by direct visualization of QD-labeled virus that VSVG-pseudotyped lentivirus enters cells independent of clatherin- and caveolin-pathways, while the entry of VSVG-pseudotyped retrovirus occurs via the clathrin pathway.

Adeno-associated virus (AAV), a member of the non-envelope parvoviridae family, has attracted considerable interest because it shows great promise for use in human gene therapy. However, the membrane penetration and uncoating mechanisms are poorly understood for these viruses, even though they are relatively small and simple in structure. Since the AAV virion is only about 20 nm in diameter, the number of dye molecules that can be attached to a single virus without causing self-quenching or affecting viral infectivity is very limited. Quantum dots are much brighter than the conventional fluorophores, which can allow the detection of viruses with much lower amounts of labeling molecules. This study also reports the general strategy for linking AAV virion with quantum dots through a coupling reaction for a single-virus tracking in target cells. Intracellular trafficking of AAV, including endosomal tracking and cytoskelton-dependent transport will be addressed in more detail.

See more of this Session: Bionanotechnology for Gene and Drug Delivery II
See more of this Group/Topical: Nanoscale Science and Engineering Forum

Gene Delivery Vehicles for Oral Inhalation Formulations

Denise S. Conti, Chemical Engineering and Materials Science, Wayne State University, Detroit, MI

Balaji Bharatwaj, Chemical Engineering and Materials Science, Wayne State University, Detroit, MI

Sandro R. P. da Rocha, Chemical Engineering and Materials Science, Wayne State University, Detroit, MI

Pressurized metered-dose inhalers (pMDIs) are the least expensive oral inhalation devices for the regional delivery of small therapeutics to the lungs. More recently, pMDIs have received renewed attention as it has been demonstrated that inhalation therapy also holds promise for the delivery of therapeutic biomolecules including proteins, oligonucleotides, and peptides to and through (systemically) the lungs. However, most therapeutics (hydrophobic, hydrophilic and biomacromolecules) have very low solubility in the propellants used in pMDIs, typically requiring the formulation of drugs as suspensions. Another challenge arises when attempting to deliver nanotherapeutics to the lungs, as in the case of polymeric nanoparticles for gene delivery applications, as they can be efficiently exhaled due to their small size. In this work we describe a novel approach to formulate polymer-DNA nanoparticulates in propellant-based inhalers for gene delivery applications. The formulation of stable dispersions of the polymer-DNA complexes, their aerosol characteristics and transfection efficiency in lung epithelial cells are discussed. The relevance of this work stems from the fact that gene therapy is a promising approach to treat a variety of medically relevant diseases, including lung cancer and cystic fibrosis, and also due to the fact that pMDIs are the most widely used inhalation formulations in the market.

See more of this Session: Bionanotechnology for Gene and Drug Delivery II
See more of this Group/Topical: Nanoscale Science and Engineering Forum

Design and Tailoring the Self-Assembly of Amphiphilic Oligopeptide Nanostructures as Versatile Biomaterials for Delivery of Anticancer Drugs, Genes, or Both, for Improved Cancer Treatment

Nikken Wiradharma, Drug Gene and Protein Delivery; Chemical and Biomolecular Engineering, Institute of Bioengineering and Nanotechnology; National University of Singapore, Singapore, Singapore

Yen Wah Tong, Department of Chemical and Biomolecular Engineering, National University of Singapore, Singapore, Singapore

Yi Yan Yang, Drugs, Proteins and Genes Delivery, Institute of Bioengineering and Nanotechnology, Singapore, Singapore

The number of clinical trials for gene therapy has increased world-wide, with 66.5% of its target were aimed at cancer therapy.As opposed to viral based delivery vector, cationic polymers offers benefits like easy manufacturing and scale up, low immunogenicity and high amount of genes it can carry. One of the major barriers in cancer treatments is drug resistance, which has lately been suggested to be the result of chromosomal alterations leading to genetic malfunctions in cancer cells. As such, delivery of siRNA or anticancer genes have been attempted to sensitize cancer cells towards anticancer drug treatments.

In this study, we had designed various amphiphilic oligopeptides with different blocks to achieve different self-assembly properties desirable for different types of therapeutic cargo molecules. Versatility of oligopeptide to form a novel class of biodegradable biomaterials as carriers of therapeutics were demonstrated by tailoring different types of amino acids in the design of the carrier molecules. The oligopeptides designed contained both hydrophobic and hydrophilic amino acids to provide driving force for micellar nanoparticle formations. Various types of hydrophobic amino acids with different degrees of hydrophobicity were used to tailor different properties and strength of the hydrophobic interactions in forming the nanoparticles. On the other hand, the compositions of the amino acids in the hydrophilic block were designed based on the different therapeutic cargos to be delivered. Lysine residue was chosen to provide gene binding and protection, histidine residue was chosen to provide endosomal escape mechanism upon uptake of nanoparticles, and Glycine or Serine residues were used as neutral hydrophilic shell to encapsulate and solubilize hydrophobic drugs inside the core. Tailoring the amino acids was also shown to affect the particle stability, drug encapsulation and release properties, but without affecting gene binding capacity. This allows the use of the oligopeptide nanostructures for simultaneous delivery of drugs and genes to achieve synergistic effect for cancer treatment.

See more of this Session: Bionanotechnology for Gene and Drug Delivery II
See more of this Group/Topical: Nanoscale Science and Engineering Forum

Principles for Nanoscale Science and Engineering using Liquid Crystals

Nicholas L. Abbott, Department of Chemical and Biological Engineering, University of Wisconsin-Madison, Madison, WI

This talk will present a broad perspective on recent advances in nanoscale science and engineering that involve the liquid crystalline state of matter, with a particular focus directed to formation of tunable nanostructured materials and amplification of nanoscopic phenomena. Key unresolved fundamental challenges as well as technological opportunities will be described. Examples of topics to be addressed include (i) the design of tailored liquid crystalline materials using anisometric organic and inorganic nanoparticles, (ii) the use of liquid crystalline solvents, and interparticle forces mediated by liquid crystals, to direct the assembly of nanoparticles into reconfigurable organized arrays, (iii) localized surface plasmonics of metallic nanostructures in liquid crystalline media, and (iv) reporting of nanoscale events at interfaces via ordering transitions induced in liquid crystalline phases.

See more of this Session: Chemical Engineering Principles for Nanotechnology (Plenary)
See more of this Group/Topical: Nanoscale Science and Engineering Forum

Semiconductor Nanowires: From Conception to Practice

Brian A. Korgel, University of Texas at Austin, Austin, TX

Semiconductor nanowires are threadlike crystals with nanometer-scale widths and lengths that can extend for millimeters. Their remarkably high surface area-to-volume ratios and nanoscale dimensions give rise to range of unique material properties that might be useful in a variety of new technologies, including printable electronics, solar cells, sensors and fabrics with unprecedented functionality. However, a significant roadblock to the development of semiconductor nanowire technologies has been the lack of production methods capable of producing industrial scale quantities of nanowires. After all, semiconductor nanowires are creations of chemistry-they are not found in nature, and must be synthesized in the laboratory.

During the past decade, we have been working to develop chemical methods capable of producing large quantities of silicon (Si) and germanium (Ge) nanowires with controlled dimensions. Our general approach relies on metal nanocrystals to seed nanowire growth from molecular reactants in a solvent. For example, Si nanowires are made in toluene using colloidal gold nanocrystals and monophenylsilane as a Si source. After nanowire formation, surfaces are modified with organic ligands to provide better chemical stability and improved electronic properties. We have also developed methods to chemically dope these nanowires with electrically-active impurities. The basic synthetic strategies and chemical and engineering lessons learned over the years will be presented, as well as a few examples of how these nanowires are being applied, as in printed nanowire field effect transistors, nanowire electromechanical resonators, and non-woven nanowire fabrics.

See more of this Session: Chemical Engineering Principles for Nanotechnology (Plenary)
See more of this Group/Topical: Nanoscale Science and Engineering Forum

Using Coarse-grained Computer Simulations to Probe Interactions between Nanoparticles and Lipid Bilayers

Anna C. Balazs, Chemical and Petroleum Engineering, University of Pittsburgh, Pittsburgh, PA

We use a coarse-grained numerical simulation-dissipative particle dynamics-to consider two different scenarios involving nanoparticles and lipid bilayers. In the first case, the particles are nanoscopic Janus beads that comprise both hydrophobic and hydrophilic portions.

We demonstrate that when the membrane rips and forms a hole due to an external stress, these nanoparticles diffuse to the edge of the hole and form a stable pore, which persists after the stress is released.

Once the particle-lined pore is formed, a small increase in membrane tension readily reopens the pore, allowing transport through the membrane. Besides the application of an external force, the membrane tension can be altered by varying, for example, temperature or pH.

Thus, the findings provide guidelines for designing nanoparticle-bilayer assemblies for targeted delivery, where the pores open and the cargo is released only when the local environmental conditions reach a critical value. In the second case, we examine the penetration of amphiphilic, nanoscale cylinders into lipid bilayers. We vary the physical features of these nanoparticles to determine the conditions where they will insert into and span the bilayer. The results yield design rules for creating artificial proteins that permit transport through the membrane.

See more of this Session: Chemical Engineering Principles for Nanotechnology (Plenary)
See more of this Group/Topical: Nanoscale Science and Engineering Forum

Particle ALD - Academic Invention to Commercial Development

Alan Weimer, Chemical and Biological Engineering, University of Colorado, Boulder, CO

Michael Masterson, ALD NanoSolutions, Inc., Broomfield, CO

This paper will describe the path from the discovery that fine primary fine particles could be functionalized with conformal and pinhole-free nano-coatings to the commercial development of the Particle Atomic Layer Deposition (ALD) process. Significant events include establishing the intellectual property, the company management, the funding, the scientific core, and the targeted markets. This is the continuing story of "boot-strapping", conservative management, no layoffs, no defections, and a seemingly bright future ahead for an 8 yr old university spinoff in the black as it navigates through a second economic downturn.

Faculty and their students are in the unique position of being able to invent platform technologies through the basic research that they carry out. Likewise, many Ph.D. students are looking for entrepreneurial opportunities where they can extend their graduate research into self-directed careers. Together, faculty members and their entrepreneurial students are needed to be "thought leaders" for startups and related business issues for creating jobs and to support the U.S economy in addition to publishing papers.

Critical issues for faculty that need to be dealt with include conflict of interest, time management, an understanding of the business world, and continuing to be professorial and academic. The most important decisions for the company revolve around selection of the management team, when to seek venture funding/creating investor stakeholder value, and which markets to target. Additionally, economic considerations are critical as no product/process will be successful unless it's the lowest cost option on the basis of cost/performance for the customer.

See more of this Session: Chemical Engineering Principles for Nanotechnology (Plenary)
See more of this Group/Topical: Nanoscale Science and Engineering Forum

Carbon Nanotube Networks

Zhenan Bao, Department of Chemical Engineering, Stanford University, Stanford, CA

Single-walled carbon nanotubes (SWNTs) exhibit superior electronic and mechanical properties, in addition to environmental stability, yielding ideal components for flexible electronic devices. While devices based on a single SWNT have yielded interesting findings on their fundamental charge transport, obstacles their device fabrication still inhibit practical implementation. Individual SWNT-based devices are not feasible for large scale integration at this time, however, recently developed random nanotube networks offer an excellent compromise with on/off ratios up to 105 upon optimization of tube chirality ratio and/or device geometry. Still, the nature of SWNTnt randomness in terms of chirality, tube density, and bundling makes device reproducibility poor, greatly inhibiting high performance.

In this talk, I will present our recent work in sorting SWNTs using functionalized surfaces and applications of sorted and aligned SWNTs for sensor applications

See more of this Session: Chemical Engineering Principles for Nanotechnology (Plenary)
See more of this Group/Topical: Nanoscale Science and Engineering Forum

Latex/Vesicle Templated Synthesis of Hollow Inorganic Nanoparticles

Spyros Monastiriotis, Chemical Engineering, The Graduate Center and The City College of the City University of New York, New York, NY

Alex Couzis, Chem. Eng., City College and Graduate Center of the City University of New York, NY, NY

Hollow inorganic nanospheres exhibit catalytic activities and properties from their solid counterparts with the advantages of high surface area and low density. This research focuses on the development of facile scalable methods for the fabrication of porous hollow inorganic spheres using equilibrium catanionic vesicles and amidine functionalized polystyrene latex as templates. Catanionic vesicles [Cetyltrimethylammonium Bromide (CTAB), Sodium Perfluorooctanate (FC7)] and amidine functionalized polystyrene latex with the same size (≈ 90nm) and concentration were the tools in order to study the behavior of acid catalyzed hydrolysis of Tetramethoxysilane (TMOS) and Dihydroxybis(ammonium lactato)titanium(IV) (TALH) as precursors for the formation of the inorganic shell. The vesicle size was tuned by the composition ratio of the two surfactants CTAB and FC7 and characterized by using Cryo-Transmission Electron Microscopy (Cryo-TEM), Small Angle Neutron Scattering (SANS), and Dynamic Light Scattering. By tuning pH, precursor's concentration and template population we can control shell thickness of the inorganic hollow particle and also minimize the gel formation. Multiple inorganic layers were fabricated by surface functionalization of the previous inorganic deposition on the vesicular template. Silica and Titania hollow particles were calcined and characterized with Cryogenic Transmission Electron Microscopy (Cryo-TEM), Transmission Electron Microscopy (TEM), Scanning Electron Microscopy (SEM), Scanning Transmission Electron Microscopy (STEM), Atomic Force Microscopy (AFM), Dynamic Light Scattering (DLS)

See more of this Session: Functional Nanoparticles and Nanocoatings On Particles II
See more of this Group/Topical: Particle Technology Forum

Functionalized Alumina Particles as pH-Responsive Drug Carriers

Bradley R. Gordon, Department of Chemical and Biomolecular Engineering, University of Maryland, College Park, MD

Charles E. Luckett, Department of Chemistry and Biochemistry, University of Maryland, College Park, MD

Daniel D. Lim, Department of Chemistry and Biochemistry, University of Maryland, College Park, MD

Sheryl H. Ehrman, Department of Chemical and Biomolecular Engineering, University of Maryland, College Park, MD

Douglas S. English, Department of Chemistry and Biochemistry, University of Maryland, College Park, MD

Porous alumina particles were functionalized and made pH-responsive using three silane coupling agents (SCAs) with different carbon chain lengths and terminal groups. Release of the anionic dye sulforhodamine 101 (SR101) and the cationic anti-cancer drug Doxorubicin (DOX) from the particles was performed at 37°C and at pH's 5.0 and 7.4 to mimic physiological conditions. The isoelectric points of the functionalized particles were analyzed and are expected to play a crucial role in the pH-response and release characteristics. Charge interactions between the diffusing species and the internal surface of the particles were found to dominate the overall transport for both SR101 and DOX. The effects of silane chain length and layer formation on the release characteristics were also explored and are further discussed in this work.

See more of this Session: Functional Nanoparticles and Nanocoatings On Particles II
See more of this Group/Topical: Particle Technology Forum

Coating Ultra-Thin Microporous/Mesorporous Films On Particle Surface

Xinhua Liang, Department of Chemical and Biological Engineering, University of Colorado, Boulder, CO

Miao Yu, Chamical and Biological Engineering, University of Colorado at Boulder, Boulder, CO

Alan W. Weimer, Department of Chemical and Biological Engineering, University of Colorado, Boulder, CO

Porous thin films/coatings have wide applications in separations, corrosion protective coatings, gas storage, controlled drug delivery, and catalyst encapsulation for controlling reaction selectivity and maintaining catalytic activity. Ultra-thin microporous/mesorporous films/coatings are especially important for some applications. Inorganic films/coatings with few defects, such as zeolite membranes and mesorporous films, have been prepared by hydrothermal synthesis and sol-gel method. While these films/coatings have well defined uniform porous structures, it is difficult to control their thickness with the precision of angstroms. In order to create a porous structure, organic templates must be removed generally by calcination at high temperature. In this paper, we demonstrate a general method to create porous inorganic films/coatings with precisely controlled thickness down to several angstroms from dense organic/inorganic hybrid aluminum alkoxide (alucone) films grown by molecular layer deposition (MLD). The organic constituents in the film can be removed either by mild water etching at room temperature, or calcination in air at elevated temperatures. Because of the layer by layer growth process for MLD, the deposited polymer films have regular structure and the oxidation of MLD polymer films produces uniform interconnected highly porous structures with high surface area. The porous structure has both micropores and mesorpores with BET surface area as high as 1452 m2/g. We expect this porous thin film/coating growth technique, which offers both precise control of the thickness down to several angstroms and well defined porous structure, to have wide applications in catalyst functionalization, controlled drug delivery and nano-filtration.

See more of this Session: Functional Nanoparticles and Nanocoatings On Particles II
See more of this Group/Topical: Particle Technology Forum

The Impact of ATRP Initiator Spacer Length On Grafting Poly(Methyl Methacrylate) From Silica Nanoparticles

David L. Green, Department of Chemical Engineering, University of Virginia, Charlottesville, VA

Chinlun Huang, Merck and Company, Philadelphia

The focus of this work is to quantify the impact of the carbon spacer length (CSL) of immobilized alkoxysilanes initiators on grafting poly(methyl methacrylate) (PMMA) from the surfaces of monodisperse silica nanoparticles. PMMA was grafted using surface-initiated atom transfer radical polymerization (SIATRP), a facile technique to produce well-controlled polymer brushes of high graft densities; the polymerizations were carried out in environmentally friendly 4:1 (v/v) methanol-water solutions at room temperature. Monoethoxysilane initiators of 3, 11, and 15 carbon spacer lengths were synthesized and characterized with 1H NMR and 13C NMR. The initiators were then used to modify the surfaces of monodisperse silica nanoparticles in methyl isobutyl ketone, producing dense initiator monolayers with site densities between 1.80 - 3.64 initiators/nm2. PMMA was subsequently grafted from the functionalized nanoparticles using both CuCl and CuBr catalysts. It was found that polymerizations performed with CuBr were uncontrolled, whereas those with CuCl were controlled. PMMA graft densities ranged between 0.10 - 0.43 polymers/nm2, which increased with the initiator carbon spacer length (CSL). Interestingly, longer CSLs make nanoparticle surfaces hydrophobic, causing nanoparticle aggregation in methanol-water solutions. Our results indicate that surface hydrophobicity is correlated to increases in PMMA graft density through the adsorption of hydrophobic MMA monomers on initiators with longer CSLs. Thus, to augment PMMA graft densities, a subtle balance must be struck between enabling particle stability and increasing MMA adsorption in methanol-water solutions.

See more of this Session: Functional Nanoparticles and Nanocoatings On Particles II
See more of this Group/Topical: Particle Technology Forum

Facile Preparation of Highly-Scattering Metal-Nanoparticle Coated Polystyrene Latex Beads

Jung-Hyun Lee, School of Chemical and Biomolecular Engineering, Georgia Institute of Technology, Atlanta, GA

Mahmoud A. Mahmoud, School of Chemistry and Biochemistry, Georgia Institute of Technology, Atlanta, GA

Valerie Sitterle, Georgia Tech Research Institute, Atlanta, GA

Jeffrey Sitterle, Georgia Tech Research Institute, Atlanta, GA

Carson Meredith, School of Chemical and Biomolecular Engineering, Georgia Institute of Technology, Atlanta, GA

Noble-metal coated polymer composite microspheres and nanoshells have attracted intense interest due to applications in electronics, photonics, medical imaging, drug delivery, surface-enhanced Raman scattering (SERS) and catalysis. By incorporating nanoparticles (NPs) exhibiting preferential scattering characteristics on a polymeric bead surface, the composite microsphere can be used as a scattering contrast agent for medical imaging, and other optical and electronic devices. Here we report a new technique for the facile preparation of highly-scattering metal NP-coated PS latex beads. Different sizes (30, 60 and 80 nm), chemistries (Au and Ag), and shapes (sphere and cube) of NPs were successfully incorporated on commercially available PS beads using this technique. Homogeneous and dense metal coatings were obtained via solvent-controlled swelling and heteroaggregation of polymer capped-metal NPs with the PS beads by the addition and removal of an organic solvent. The organic solvent controlled the heteroaggregation of colloidal particles as well as the swelling/shrinking mechanism of PS beads in this strategy. No prior functionalization of the bead was needed. The fabricated composite beads were stable during long-term water storage. The morphology and coverage of the metal coating on the beads, and thus optical properties, can be effectively controlled by the proper choice of the concentration of solvent and metal NPs. Complete and close packed metal coating on the PS bead can be obtained with smaller and spherical metal AuNPs, while metal surface coverage reached to a certain limitation with larger or cubic metal NPs. The resulting metal coated-PS microspheres exhibited highly enhanced scattering and tunable optical characteristics, which were investigated with the dark-field microscopy and UV-vis absorption spectra. Their performances as SERS substrates were evaluated with Raman Spectroscopy

See more of this Session: Functional Nanoparticles and Nanocoatings On Particles II
See more of this Group/Topical: Particle Technology Forum

Preparation of Quantum Dot-Embedded Polymeric Nanoparticles in a Micromixer

Yanjie Zhang, Chemical & Biological Engineering, Iowa State University, Ames, IA

Aaron R. Clapp, Dept. of Chemical & Biological Engineering, Iowa State University, Ames, IA

We describe a process for stabilizing luminescent quantum dots within a block co-polymer particle. The method uses a staged micromixer which initially forms nanocrystal-homopolymer aggregates in the first mixer, which are then fed into the second mixer where they are stabilized by a suitable block co-polymer. The precipitation is controlled by the nanocrystal and polymer concentrations, flow rate, resident time, and solvent ratio. Under turbulent mixing conditions, the generated particles have a low polydispersity index and relatively small size (~100-500 nm diameter). Below the turbulence threshold, polydispersity increases dramatically. The loading of quantum dots and ultimate composite particle size depends strongly on the efficiency of aggregation obtained within the first mixer and can be controlled by adjusting the homopolymer concentration as well as the capping ligands present on the nanocrystal surface. Careful selection of the block co-polymer introduced in the second mixer provides long-term colloidal stability in aqueous solutions and preserves the luminescence properties of the quantum dots. This method is a rapid (< 1 minute) and effective means for stabilizing quantum dots within a polymeric particle for applications including cell labeling, drug delivery, and particle image velocimetry.

See more of this Session: Functional Nanoparticles and Nanocoatings On Particles II
See more of this Group/Topical: Particle Technology Forum

Enhanced Emission of Gold Nanoparticles Due to Electron Transfer From Surface Bound Molecules and Its Use in pH Sensing

Chang-won Lee,Bioengineering, University of Utah, Salt Lake City, UT

Curtis Takagi, Materials Science and Engineering, University of Utah, Salt Lake City, UT

Agnes Ostafin, Materials Science and Engineering, University of Utah, Salt Lake City, UT

Gold nanoparticles have long been of interest in the development of SPR and SERS chemical and biological sensors. Recently, fluorescence-based sensing has been suggested as a viable alternative sensing method because of the development of emissive gold nanoparticles produced either in organic solution or inside the confines of dendrimers. These have properties of quantum dots, and the effect of environment on their emissions is not well understood. Here we report an all water synthesis of water-soluble gold nanoparticles with atypical near infrared fluorescence believed to originate via absorbance of light energy by a surface bound molecule and transfer of electrons to the metal which subsequently fluoresces. Particles were developed with mercaptooctanoic acid as a capping agent under reduction conditions with sodium borohydride. The product was purified with ethanol trituration for further application or characterization. TEM analysis of the nanoparticle showed its diameter to be 2.2 ± 0.6 nm (Figure 1A). The gold nanoparticle solution shows bright red emission with the excitation on UV plate (Figure 1B). The fluorescence of the gold nanoparticle varies over the physiological pH range and has a possibility to be used as a nanosized pH sensor without the need to conjugate additional organic molecules to the particle surface. In comparison to the widely used quantum dots or fluorescent dye molecules which usually have broad excitation spectra and overlapped region with emission spectra, these particles have a distinct excitation and emission spectra (excitation spectra peak @ 280 nm and emission spectra peak @ 610 nm) with 200 nm gap between them with large Stokes shift. The fluorescence of the gold nanoparticle is more efficient than other reported fluorescent gold nanoparticles with quantum yield (QY) of 0.015 measured with fluorescein dye molecule as a standard (QY=0.95) and the life time of the fluorescence is around 1.7 μs. The distinct excitation (red shifted from that of pure MOA) and emission without significant overlap between them with a 200 nm gap between them is suggestive of excited electron transfer from surface bound molecules to the core gold for the emission. A computational simulation study using Gaussian03 program and b3pw91/lanl2dz functional/basis set was also performed to evaluate the change of the band gap of mercaptooctanoic acid with and without Au-S bond formation. The result showed similar red shift of the band gap. The emission intensity exhibited strong pH sensitivity and supports the proposed mechanism that excitation of surface capping MOA in specific charge states is an important determinant of emission in this system. The distinct excitation (red shifted from that of pure MOA) and emission without significant overlap between them with a 200 nm gap between them is suggestive of excited electron transfer from surface bound molecules to the core gold for the emission. A computational simulation study using Gaussian03 program and b3pw91/lanl2dz functional/basis set was also performed to evaluate the change of the band gap of mercaptooctanoic acid with and without Au-S bond formation. The result showed similar red shift of the band gap. The emission intensity exhibited a strong pH sensitivity and supports the proposed mechanism that excitation of surface capping MOA in specific charge states is an important determinant of emission in this system.

Figure 1. (A) TEM picture of gold nanoparticles, (B) Gold nanoparticle solution without (left) and with (right) UV excitation on a UV transilluminator plate (wavelength 254 nm).

See more of this Session: Nanotechnology and Nanobiotechnology for Sensors I
See more of this Group/Topical: Nanoscale Science and Engineering Forum

Periodic Plasmonic Nanostructures as Efficient SERS Substrates for Biosensing

Tzung-Hua Lin, Cheimcal Engineering, University of Florida, Gainesville, FL

Nicholas Linn, Chemical Engineering, University of Florida, Gainesville, FL

Xuefeng Liu, School of Chemical and Material Engineering, Jiannan University, Wuxi, China

Inkook Jun, Material Science and Engineering, University of Florida, Gainesville, FL

Bin Jiang, Mathematics and Statistics, Portland State University, Portland, OR

Peng Jiang, Chemical Engineering, University of Florida, Gainesville, FL

Surface-enhanced Raman scattering (SERS) is a non-invasive technique that enables the detection and characterization of both small organic and big biological molecules at very low concentrations. To overcome the reproducibility issue of stochastically aggregated colloidal nanoparticles in SERS, various periodic metallic nanostructures created by self-assembly have been exploited. Unfortunately, most of these bottom-up approaches are limited in creating centimeter-sized samples, and are not compatible with standard microfabrication, thereby impeding the cost-efficiency and scale-up of these unconventional methodologies. Here we report numerous scalable templating techniques for fabricating wafer-sized SERS substrates with high and reproducible enhancement factors. Non-closed-packed colloidal crystals prepared by a simple spin-coating technique are used as templates to create periodic arrays of nano-dimples, nano-pyramids, nano-nipples, nanoholes, and nanoflasks, as well as binary nanohole-nanoparticle assemblies. These nanostructures have high-density sharp features (circular edges or nanoscale tips) that function as "hot spots" to concentrate electromagnetic field. We have also developed a finite-element method (FEM) model to simulate the spatial distribution of the electromagnetic field and predict the corresponding SERS enhancement factor from the periodic metallic nanostructures.

See more of this Session: Nanotechnology and Nanobiotechnology for Sensors I
See more of this Group/Topical: Nanoscale Science and Engineering Forum

Composition Effect of Ag-Cu Alloy Nanoparticles On Luminescence Enhancement/Quenching of Vicinal Luminophores

Sanchari Chowdhury, Chemical and Biomedical Engineering, University of South Florida, Tampa, FL

Venkat Bhethanabotla, Department of Chemical and Biomedical Engineering, University of South Florida, Tampa, FL

Rajan Sen, Civil Engineering, University of South Florida, Tampa, FL

Composition effect of Ag-Cu alloy nanoparticles on luminescence enhancement/quenching of vicinal luminophores

The emission of luminophores is significantly influenced at the close proximity of conducting metallic nanostructures. Metal nanoparticles are known to both quench and enhance luminescence depending on the optical properties of nanoparticles, fluorophore-particle separation distance, molecular dipole orientation with respect to particle surface and size of the nanoparticles[1, 2]. Both enhancement and quenching of luminescence due to the proximity of nanoparticles are efficiently utilized for many different applications. Enhanced signal and photostability of luminophores, improved surface immunoassay and DNA detection, enhanced wavelength-ratiometric sensing, and amplified assay detection are few examples of the applications of metal enhanced luminescence[3, 4]. On the other hand, quenching resulting due to metallic nanoparticles is successfully utilized for the improvement of homogeneous and competitive fluorescence immunoassay, optical detection of DNA hybridization, competitive hybridization assay and in optoelectronics[5, 6].

In this work, metal enhanced/quenched luminescence was explored in the vicinity of Ag_xCu_{100-x} alloy nanoparticles at different compositions (x=100, 67, 50, 30 and 0). These Ag-Cu nanoparticles were synthesized using the polyol process. It was observed that the luminescence of dyes was strongly dependent on the Ag-Cu nanoparticle composition. Highest luminescence enhancement of dye Cy 3, commonly used for biological applications, was realized at the vicinity of pure silver nanoparticles. Enhancement effects were found to decrease as the percentage of copper increases in the nanoparticles, leading to the quenching of fluorescence at the pure copper nanoparticle platform. Theoretical calculations based on the corrected Gersten and Nitzan model [7] were carried out to predict the possible enhancement factors for spherical Ag-Cu nanoparticles for a range of compositions. Results of these calculations corroborate the experimental findings. Signal manipulation of luminescent dyes using the composition of alloy nanoparticles will extend applications of metal enhanced luminescence to several fields ranging from optoelectronics to biological research.

References:

(1) Lakowicz, J. R.; Shen, Y.; D'Auria, S.; Malicka, J.; Fang, J.; Gryczynski, Z.; Gryczynski, I. *Anal. Biochem.* **2002**, *301*, 261-277.

(2) Anger, P.; Bharadwaj, P.; Novotny, L. *Phys. Rev. Lett.* **2006**, *96*, 113002-113001-113004.

(3) Zhang, J.; Matveeva, E.; Gryczynski, I.; Leonenko, Z.; Lakowicz, J. R. *J. Phys. Chem. B* **2005**, *109*, 7969-7975.

(4) Lakowicz, J. R. *Anal. Biochem.* **2001**, *298*, 1-24.

(5) Zai-Sheng, W.; Jian-Hui, J.; FU, L.; Guo-Li, S.; Ru-Qin, Y. *Anal. Biochem.* **2006**, *353*, 22-29.

(6) Ao, L.; Gao, F.; Pan, B.; He, R.; Cui, D. *Anal. Chem.* **2006**, *78*, 1104-1106.

(7) Mertens, H.; Koenderink, A. F.; Polman, A. *Phys. Rev. B* **2007**, *76*, 115123-115121-115112.

See more of this Session: Nanotechnology and Nanobiotechnology for Sensors I
See more of this Group/Topical: Nanoscale Science and Engineering Forum

Nanoengineered Transparent Metallic Nanofibrous Membrane and Its Application for Humidity Sensing

Wenzhao Jia, Chemical, Materials and Biomolecular Engineering, University of Connecticut, Storrs, CT

Ying Wang, Chemical, Materials and Biomolecular Engineering, University of Connecticut, Storrs, CT

Joysurya Basu, Chemical, Materials and Biomolecular Engineering, University of Connecticut, Storrs, CT

Timothy Strout, Chemical, Materials and Biomolecular Engineering, University of Connecticut, Storrs, CT

C. Barry Carter, Chemical, Materials and Biomolecular Engineering, University of Connecticut, Storrs, CT

Yu Lei, Chemical, Materials and Biomolecular Engineering, University of Connecticut, Storrs, CT

Transparent metallic nanofibrous membranes of Au, Pd, Pt, Ni, Ag and Au-Pd alloy have been successfully fabricated. The as-prepared samples were characterized by SEM, TEM, FT-IR, and the results reveal that the metallic nanofibers consist of a lot of metal nanoparticles held together. The as-prepared transparent metallic nanofibrous membranes show good mechanical and conductive properties, and offer an excellent sensing platform. Its application for humidity detection is demonstrated and the developed humidity sensor exhibits fast response, excellent recovery behavior, and good sensitivity.

See more of this Session: Nanotechnology and Nanobiotechnology for Sensors I
See more of this Group/Topical: Nanoscale Science and Engineering Forum

DNA Hybridization Detection Using Spectral Changes of Zinc Selenide Nanocrystals

Jun Wang, Chemical Engineering, University of Massachusetts, Amherst, MA

Tracy Heckler, Chemical Engineering, University of Massachusetts, Amherst, MA

Pedro Lei, Chemical and Biological Engineering, SUNY at Buffalo, Amherst, NY

Stelios Andreadis, Chemical and Biological Engineering, SUNY-Buffalo, Buffalo, NY

T. J. Mountziaris, Chemical Engineering, University of Massachusetts, Amherst, MA

Fluorescent labeling of biological molecules is used widely for analytical purposes in biotechnology and bioengineering. It involves the use of an organic dye linked to a moiety that selectively bonds a particular biological molecule, allowing the detection of the latter by the fluorescence of the dye. Typical organic fluorescent dyes have narrow excitation spectra and broad emission spectra that exhibit red tailing.

Semiconductor nanocrystals or quantum dots have emerged as a new class of fluorescent markers with distinct advantages over traditional organic dyes [1-4]. Their attractive properties include a narrow, symmetric, and strong emission that is size-tunable, continuous excitation by any wavelength smaller than the emission wavelength, resistance to photobleaching, as well as excellent optical and chemical stability that allows their use in lengthy experiments, both in vitro and in vivo. The ability to synthesize different populations of quantum dots with narrow emission spectra permits multiplexing, a property that is very important for simultaneous detection of several analytes, that would be very tedious and expensive if done sequentially.

In addition to their use as multi-color "passive" tags for biomolecules, quantum dots have significant potential as "active" sensors, because their optical spectra change when the detection molecules attached to their surface hydridize with complementary ones in solution. The focus of this work is the development of new biological detection and DNA analysis schemes by using the changes in the optical spectra of ZnSe quantum dots.

We have synthesized highly-fluorescent ZnSe quantum dots, modified their surface with functional molecules to render them water-dispersible, and conjugated them with oligonucleotides. Hybridization of complementary oligonucleotides conjugated with ZnSe quantum dots results in a significant increase in the fluorescence intensity of the quantum dots and a measurable red shift in the emission wavelength.

Experiments were performed in which DNA hybridization was studied using ZnSe quantum dots conjugated with oligonucleotides as active sensors and free oligonucleotides in solution as targets. An increase in the emission intensity of the quantum dots was observed upon hybridization of the tagged oligonucleotides with free oligonucleotides. A dependence of the detected fluorescence emission intensity on the length and location of the hybridized part was detected, indicating that spectral changes can be used to identify the DNA stuctures attached to the quantum dots. Ongoing experiments are focusing on determining the sensitivity of this techique for detecting DNA mutations. Applications of these phenomena in the development of novel DNA analysis strategies will be discussed.

References

1. M. Bruchez, et al., Science 281, 2013 (1998).

2. W.C.W. Chan and S. Nie, Science 281, 2016 (1998).

3. M.-Y. Han, et al., Nature Biotechnology 19, 631 (2001).

4. W. C. W. Chan, et al., Current Opinion in Biotechnology, 13, 40 (2002).

See more of this Session: Nanotechnology and Nanobiotechnology for Sensors I
See more of this Group/Topical: Nanoscale Science and Engineering Forum

A Simulation Approach to the Thermodynamics and Dynamics of Self-Assembled Mesophases of Multi-Faceted and Multi-Lobed Particles

Umang Agarwal, Chemical and Biomolecular Engineering, Cornell University, Ithaca, NY

Fernando Escobedo, Chemical and Biomolecular Engineering, Cornell University, Ithaca, NY

Recent advances in colloidal chemistry have made possible synthesis of a

great variety of nanoparticle shapes. Hence, it is of great interest to predict the

various morphologies they can form and to acquire a general understanding of how to control their directed self assembly via changes of geometric features (like shape and

aspect ratio). We have attempted to get a clearer picture of the conditions at which different types of mesophase can form by performing isothermal-isobaric Monte Carlo simulations on particles belonging to different symmetry groups. Furthermore, we studied the rheological properties of ordered phases formed by anisotropic, multi-lobed particles using Non-Equilibrium Molecular Dynamics and microheology simulations and compared these results with those for isotropic phases. These dynamic simulations provide a means to probe the mechanism by which defects motion and flow pattern are

coupled.

See more of this Session: Self and Directed Assembly at the Nanoscale I
See more of this Group/Topical: Nanoscale Science and Engineering Forum

Computer Simulation of Self-Assembly of Dipolar and Quadrupolar Colloid Particles for the Design of Stimuli-Responsive Materials

Amit Goyal, Department of Chemical and Biomolecular Engineering, North Carolina State University, Raleigh, NC

Carol K. Hall, Department of Chemical and Biomolecular Engineering, North Carolina State University, Raleigh, NC

Orlin D. Velev, Department of Chemical and Biomolecular Engineering, North Carolina State University, Raleigh, NC

Colloidal particles with directional interactions that self-assemble into pre-defined microstructures have the potential to serve as the foundation for a new generation of micro- and nano-structures of remarkable complexity and precision. Dipolar colloid particles tend to align end-to-end and self-assemble into variety of structures ranging from co-crystals of novel symmetry, to open networks (gels) of cross-linked chains of particles. Quadrupolar colloid particles also tend to self-assemble in a wide variety of structural motifs including sheets, tubes and shells depending upon external conditions. We use molecular dynamics computer simulation to explore the self-assembly, structure, crystallization and/or gelation of systems of colloid particles with permanent dipole moments or quadrupole moments immersed in a high-dielectric solvent.

Phase diagrams for monodisperse and bidisperse system of dipolar colloid particles in the temperature-packing fraction plane are calculated using discontinuous molecular dynamics simulations. An interesting aspect of these phase diagrams is the appearance of co-crystals containing large and small dipolar colloid particles. Even more interesting is the appearance of two unique bicontinuous gel structures: the first consisting of two single-species interpenetrating networks of cross-linked chains and the second consisting of a network of cross-linked chains of one species coated by particles of the other species.

We also explore the structure formation of systems of colloid particles with permanent quadrupole moment. We introduce a simple quadrupole-quadrupole discontinuous potential model that gives rise to the self-organization of surface and tubular structures. A new type of anisotropic colloid particle is introduced having a displaced quadrupole moment with a unique symmetry; systems comprised of these particles form tubular structures. Our simulations predict optimal conditions for making tubes of specific diameter and length out of quadrupolar colloidal particles; this might eventually be a route to the formation of high quality tubes.

See more of this Session: Self and Directed Assembly at the Nanoscale I
See more of this Group/Topical: Nanoscale Science and Engineering Forum

Effects of Nanoparticle Size Polydispersity On the Tethered Nanoparticle Phase Diagram

Carolyn L. Phillips, Applied Physics, University of Michigan, Ann Arbor, MI

Sharon C. Glotzer, Chemical Engineering and Materials Science and Engineering, University of Michigan, Ann Arbor, MI

Recent simulations predict that aggregating nanospheres functionalized with polymer "tethers" can self-assemble to form a cylinder, perforated lamallae, lamallae, and even the double gyroid phase also seen in block copolymer and surfactant systems. We study the impact of nanoparticle size polydispersity on the properties of the phase diagram. We show that in the portions of the phase diagram characterized by an icosahedral packing motif, a low amount of polydispersity lowers the energy and a large amount of polydispersity raises the potential energy of the system by disrupting the icosahedral packing. In the portions of the phase diagram characterized by crystalline packing, polydispersity raises the energy of the system and induces a phase transition from crystalline to liquid within the nanosphere packing of the microphase.

See more of this Session: Self and Directed Assembly at the Nanoscale I
See more of this Group/Topical: Nanoscale Science and Engineering Forum

Computer Simulation of Nanoparticle Solutions in Confinement

Chen Wang, Department of Chemical and Biological Engineering, Missouri University of Science and Technology, Rolla, MO

Ramesh Chembeti, Department of Chemical and Biological Engineering, Missouri University of Science and Technology, Rolla, MO

Jee-Ching Wang, Department of Chemical and Biological Engineering, Missouri University of Science and Technology, Rolla, MO

Nanometer-sized particles possess extraordinary properties that are not available in either isolated molecules or bulk materials. To fully realize their potential, nanoparticles need to be fabricated into predesigned nanostructures or multidimensional ordered arrays. Significant progress has been made recently in synthesizing surfactant-capped nanoparticles that are dissolvable in typical non-polar solutions and capable of self-assembling into ordered arrays when deposited on a smooth surface or at a water-air interface. We study the feasibility of a new assembly route, namely using nanoscale confinement to promote the assembly of surfactant-capped nanoparticles. We first use the molecular dynamics simulation technique and atomistic models to investigate the interactions and behaviors of nanoparticle monomers and oligomers dissolved in hexane confined between two solid surfaces. We show that by applying proper perpendicular and parallel pressures on the confined nanoparticle solutions, nanoparticles form ordered aggregates. The nanoparticle interactions and the simulation system are then coarse grained to perform Brownian Dynamics simulations to study nanoparticle assembly facilitated by nanoconfinement over larger time and length scales. We conclude that nanoscale confinement with proper pressure control could be a new useful strategy for promoting nanoparticle self-assembly.

See more of this Session: Self and Directed Assembly at the Nanoscale I
See more of this Group/Topical: Nanoscale Science and Engineering Forum

Molecular Dynamics Simulation of the Oriented Attachment of Gold Nanoparticles in Liquid n-Hexane

Leonidas Gergidis, Chemical Engineering, PENNSYLVANIA STATE UNIVERSITY, State College, PA

Kristen Fichthorn, Chemical Engineering, PENNSYLVANIA STATE UNIVERSITY, State College, PA

Parallel molecular dynamics (MD) simulations were used to probe the oriented attachment of two gold nanoparticles immersed in liquid n-hexane solvent. For the free moving and for nanoparticles with fixed orientation at different centers of mass separation distances, we have calculated quantitative and qualitative characteristics of the nanocrystal orientations. Preferable vertex to vertex and facet to facet orientations have been observed similar with previous experimental studies of silver nanoparticles in N,N dimethylformamide (Giersig et al. J. Mater. Chem., 2004, 14,607) . Direct computation of the densities of the surrounding fluid exhibited a preferential localization of n-hexane molecules on the facets of the gold nanocrystals avoiding the edges and vertices of the nanocrystals. In addition, visualization of the confined fluid in the interparticle region have shown very interesting ordering and layering effects.

See more of this Session: Self and Directed Assembly at the Nanoscale I
See more of this Group/Topical: Nanoscale Science and Engineering Forum

Monte-Carlo and Brownian Dynamic Simulaitons of Self-Assembly and Gelation of Magnetic Particles IN the PRESENCE of A Magnetic Field

Marco Lattuada, Institute for Chemical and Bioengineering, Department of Chemistry and Applied Biosciences, ETH Zurich, Zurich, 8093, Switzerland

Marco Furlan, Institute for Chemical and Bioengineering, Department of Chemistry and Applied Biosciences, ETH Zurich, Zurich, Switzerland

Massimo Morbidelli, Department of Chemistry and Applied Biosciences, Institute for Chemical and Bioengineering, Zurich, Switzerland

The behavior of dispersions of magnetic particles in the presence and in the absence of a magnetic field has been the subject of numerous experimental studies and theoretical investigations. The goal of these studies was to elucidate the effect of dipolar interactions on the spontaneous formation of clusters and their orientation in the presence of a magnetic field, as well as the existence of such clusters in the absence of a field. However, only a few experimental studies have tried to focus on the behavior of concentrated suspensions of magnetic particles, while the simulation works have only covered the behavior of relatively diluted suspensions. In addition, so far all of these studies have been carried out in the case of colloidally stable magnetic particles.

On the other hand, the self-assembly of destabilized non-magnetic colloidal particles is a very well understood phenomenon, leading to the formation of well know fractal clusters with a fractal dimension ranging between 1.8 and 2.1 depending upon the absence or presence of repulsive interactions, respectively. It is also well know that colloidal dispersions can undergo gelation when fractal clusters reach sufficient size to percolate into an infinite network.

With the purpose of bridging this existing gap, and of utilizing magnetic particles for the preparation of novel porous materials with controlled properties, we set out to understand the self-assembly behavior of either completely or partially destabilized superparamagnetic colloidal particles in the presence of a magnetic field. We have carried out Monte-Carlo simulations to investigate the structure and morphology of clusters and gels made of superparamagnetic colloids in the presence of magnetic field, as a function of particle size and magnitude of repulsive interactions. All the inter-particles interactions, i.e., dipolar, Van der Waals, and electrostatic are rigorously accounted for. Our primary objective is to analyze the anisotropy in the structure of gels obtained with magnetic fields applied at different stages during the gelation process. Additionally, we have also carried out Brownian dynamic simulations on the same system to better understand the effect of magnetic fields on the kinetics of clusters and gel formation. Comparison with experimental data from our lab is performed.

See more of this Session: Self and Directed Assembly at the Nanoscale I
See more of this Group/Topical: Nanoscale Science and Engineering Forum

Peptide Helix Stabilization by Oligo(ethylene oxide) Conjugation

Amit S. Jain, Department of Chemical and Biomolecular Engineering, Tulane University, New Orleans, LA

Hank Ashbaugh, Department of Chemical and Biomolecular Engineering, Tulane University, New Orleans, LA

Synthetic polymer-protein conjugates can improve biomolecule function over natural proteins. In the case of helix forming polypeptides, these hybrid biomacromolecules offer the potential stabilization of peptide helicies to enhance their receptor binding, improve protease resistance, and cell membrane permeability. To gain insights into protein stabilization via polymer conjugation, we have performed replica exchange molecular dynamics simulations of a specific helix forming peptide (1CW) experimentally known to be stabilized by oligo(ethylene glycol). Ethylene glycol is found to preferentially interact with positively charged lysine and thereby stabilize hydrogen-bonding with neighboring amino acids. Side chain interference of lysine with its neighbors is reduced by ethylene glycol, with a consequential increase in helicity. Simulation of polyalanine with lysine mutations shows similar results. We therefore conclude the interaction of lysine with ethylene oxide contributes to helix stability.

See more of this Session: Self and Directed Assembly at the Nanoscale I
See more of this Group/Topical: Nanoscale Science and Engineering Forum

Activation by and Processing of Polyanhydride Nanospheres by Antigen Presenting Cells

Bret D. Ulery, Chemical and Biological Engineering, Iowa State University, Ames, IA

S. Avanti Sarkar, Veterinary Microbiology and Preventative Medicine, Iowa State University, Ames, IA

Bryan H. Bellaire, Veterinary Microbiology and Preventative Medicine, Iowa State University, Ames, IA

Michael J. Wannemuehler, Veterinary Microbiology and Preventative Medicine, Iowa State University, Ames, IA

Balaji Narasimhan, Chemical and Biological Engineering, Iowa State University, Ames, IA

Improving vaccine efficacy is a major area of recent research. It has been shown that biodegradable polyanhydride particles possess the capacity to function both as superb adjuvants and as single dose delivery vehicles. Currently, it is unknown how polymer particles facilitate both antigen presenting cell (APC) activation and particle internalization and processing. The method by which encapsulated antigen goes from being internalized within a particle to being processed and effectively presented on the surface of APCs plays a crucial role in the initiation of an immune response. By observing how polyanhydride nanospheres are uptaken and processed by APCs using microscopy and evaluating activation of these cells by cell surface marker expression and cytokine secretion, we can better understand the potential of these materials as vaccine delivery vehicles. In addition, polyanhydride chemistry effects on APC uptake mechanisms will provide valuable information assisting in the rational design of vaccine delivery platforms. Copolymers of 1,6-bis(p-carboxyphenoxy)hexane and sebacic acid and copolymers of 1,8-bis(p-carboxyphenoxy)-3,6-dioxaoctane and 1,6-bis(p-carboxyphenoxy)hexane were synthesized and fabricated into nanospheres by anti-solvent precipitation. Murine bone-marrow derived dendritic cells (BMDCs) were differentiated and incubated with polyanhydride nanospheres in vitro to allow for particle internalization and cellular activation. Laser scanning confocal microscopy was utilized to visualize particle uptake and processing by BMDCs. Flow cytometry was employed to investigate the capacity for polyanhydride nanospheres to increase cell surface marker expression on these cells. A LUMINEX bead assay was used to measure the secretion of multiple cytokines from APCs. As polymer hydrophobicity decreased, cell surface marker expression was enhanced and more polyanhydride nanospheres were internalized and processed. Since cellular processing is dependent on polymer chemistry and particle size, knowledge of how these variables can be exploited to best design delivery systems that facilitate desired immune responses for a variety of vaccine candidates is highly desirable.

See more of this Session: Bionanotechnology for Gene and Drug Delivery III
See more of this Group/Topical: Nanoscale Science and Engineering Forum

PAMAM Dendrimers for Brain Delivery of Therapeutics for the Treatment of Cerebral Palsy: Chemistry, In Vivo Efficacy and Imaging

Rangaramanujam M. Kannan, Chemical Engineering and Materials Science/NICHD Perinatology Research Branch, Wayne State University, Detroit, MI

Raghavendra Navath, Chemical Engineering/NICHD PRB, Wayne State University, Detroit, MI

Yunus Emre Kurtoglu, Chemical Engineering/NICHD PRB, Wayne State University, Detroit, MI

Hui Dai, Pediatrics, NICHD PRB, Wayne State University, Detroit, MI

Bing Wang, Pediatrics/NICHD PRB, Wayne State University, Detroit, MI

Sujatha Kannan, Pediatrics/Children's Hospital of Michigan, Wayne State University, Detroit, MI

Roberto Romero, Eunice Kennedy Shriver National Institute of Child Health and Human Development, NICHD, NIH, DHHS, Detroit, MI

Maternal intrauterine inflammation resulting in microglial activation has been implicated in the development of periventricular leukomalacia and cerebral palsy. N-acetyl cysteine (NAC) is a drug that is currently being explored for the treatment of neuroinflammation in neonatal and perinatal applications. However, plasma binding of NAC significantly reduces the bioavailability requiring very high doses (100-300 mg/kg in animal models). Our translational research effort is focused on the preparation, in vitro, and in vivo evaluation of PAMAM dendrimer-based nanodevices for the delivery of therapeutics in general, and NAC in particular.

Nanodevice preparation and characterization: We have developed neutral PAMAM dendrimer-based nanodevices where a disulfide linker is used to link the drug to the dendrimer. The conjugates were characterized, by 1H NMR, MALDI-TOF, and HPLC. Each molecule of the dendrimer had approximately 16 molecules of NAC, and the conjugate exhibited high water solubility. The presence of the disulfide linker enables tailored intracellular release of the drug in a manner sensitive to the glutathione levels (low in blood circulation, high inside the cells).

In vivo imaging: FITC-labeled dendrimers were administered in neuroinflammation-associated pregnant rabbit models of cerebral palsy, through subdural and intravenous injections. The histological studies were correlated with microPET studies on 64Cu-labeled dendrimers. Our results are the first to show that dendrimers are able to cross the BBB in a neuroinflammation model. Our studies suggested that dendrimers localize in the activated microglial cells and astrocytes in the corpus callosum, hippocampus, corona radiata, internal capsula, external capsula and around the lateral ventricles in the endotoxin kits. In normal animals, which do not have appreciable activated microglia or astrocyte proliferation in these regions, no appreciable uptake was found. This paves the way for potentially useful drug delivery approaches using dendrimers for the treatment of neuroinflammation, even without specific targeting moieties. When this is combined with the ability of dendrimers to selectively localize in activated microglial cells, significant improvements in vivo performance can be achieved.

In vivo efficacy: The nanodevices were evaluated extensively in a rabbit model of cerebral palsy. The biodistribution and efficacy of intravenously administered dendrimer-drug conjugates are compared with those of the free drug using a combination of tools. The efficacy is evaluated using a combination of neurobehavioral analysis, assessment of brain tissue level inflammatory cytokine analysis. Our studies show that dendrimer-drug conjugates are 10-100 times more efficacious that free drug, suggesting that these conjugates (~ 18000 Da) are able to cross the blood brain barrier and deliver the drug significantly better than free drug. Detailed mechanistic and efficacy studies will be presented.

See more of this Session: Bionanotechnology for Gene and Drug Delivery III
See more of this Group/Topical: Nanoscale Science and Engineering Forum

Understanding Lipid Membrane Interactions with Carbon Nanotubes

Vamshi K. Gangupomu, Chemical and Biological Engineering, Drexel University, Philadelphia, PA

Franco Capaldi, Mechanical Engineering and Mechanics, Drexel University, Philadelphia, PA

Carbon Nanotubes are currently being suggested as nano-needles and nanovectors for delivery of proteins, peptides and genetic materials across cell membranes. The small dimensions of nanotubes and high aspect ratio facilitate delivery with minimal membrane disruption. However, interactions of carbon nanotubes with complex biological structures such as lipid membranes are not completely understood. A fundamental understanding of the mechanisms involved during injection and the magnitude of forces involved in the interaction of membrane and nanotubes is required. We study the forces of interactions between nanotubes and lipid bilayers and also observe the mechanism of internalization of carbon nanotubes by a lipid membrane using Molecular Dynamics (MD). The observed forces provide an insight into the interactions between the nanotube and the membrane lipids and suggest the use of carbon nanotubes as a direct tool to measure the 3D pressure profiles in lipid bilayers. The mechanism of internalization of carbon nanotube by the lipid bilayer was studied by following the free-energy profile, produced using the Jarzynski's equation, as the nanotube traverses the membrane. The obtained free-energy profiles suggested a two stage internalization mechanism that involves both diffusion and an energy dependent process.

See more of this Session: Bionanotechnology for Gene and Drug Delivery III
See more of this Group/Topical: Nanoscale Science and Engineering Forum

Direct Delivery of Non-Viral Vectors of Bone Morphogenic Proteins From Polyelectrolyte Multilayer Thin Films

Raymond E. Samuel, Chemical Engineering, Massachusetts Institute of Technology, Cambridge, MA

Paula T. Hammond, Chemical Engineering, Massachusetts Institute of Technology, Cambridge, MA

The transfection of osteoprogenitor cells in or near the bone fracture site by genes coding for bone morphogenic proteins (BMP) accelerates the fracture healing process. Improvements of the efficacy of non-viral gene transfer methods for the direct delivery of these beneficial BMP genes at fracture repair sites will herald the application of gene transfer technology to this vast clinical condition. We have utilized polyelectrolyte mutilayer (PEM) thin films composed of plasmid DNA vectors encoding human BMP cDNAs and hydrolytically degradable poly (â-amino) ester to directly transfect MC3T3-E1 mouse preosteoblast cells grown on the surface of PEM thin films. MC3T3-E1 cells were able to form stable attachments to the PEM thin films, uptake the plasmid DNA vectors encoding BMP genes, and express the transgene encoded in the plasmid DNA vectors. The transfection efficiency was significantly higher for MC3T3-E1 cells growing in direct contact with the PEM thin film than for those cells growing at distal sites in the monolayer cultures. Local transfection efficiency were improved by increasing the plasmid DNA load within the PEM via increasing the number of deposited layers and also by inclusion of natural polycations (protamine and poly-L-arginine which are known to enhance DNA condensation) within the architecture of the PEM thin films. These results suggest that non-viral gene transfer of BMP genes from PEM thin films may potentially enhance the differentiation of stem cells and increase bone formation at in vivo fracture healing sites.

See more of this Session: Bionanotechnology for Gene and Drug Delivery III
See more of this Group/Topical: Nanoscale Science and Engineering Forum

Rose Bengal Conjugated with Carbon Nanotube for Photodynamic and Hyperthermic Cancer Phototherapy

Ramasamy Anbarasan, Chemical Engineering, National Taiwan University, Taipei, Taiwan

Chung-Hao Wang, Chemical Engineering, National Taiwan University, Taipei, Taiwan

Ching-An Peng, Chemical Engineering, Michigan Technological University, Houghton, MI

Rose Bengal (RB) is a water-soluble xanthene dye, known to produce singlet oxygen with a high absorption coefficient in the visible region of the spectrum. Although RB has the potential as a photosensitizer for photodynamic therapy (PDT), the clinical application of RB in PDT has been hampered due to insufficient lipophilicity and lack of tumor-accumulating property. Carbon nanotube (CNT) has recently been considered as a thermal-coupling agent for localized hypothermic therapy under near-infrared (NIR) laser exposure. In this study, RB and folic acid were coupled with CNT with the aid of Sb2O3 as a catalyst. UV-Vis, FT-IR, and XPS spectral analysis confirmed the chemical binding of RB and folic acid on CNT. RB-tagged CNTs functionalized with folate moiety were incubated with MCF-7 breast cancer cells for 6 and 24 hours, respectively. After replaced with fresh culture media, cell culture plates were first exposed with 540-nm LED light (4 J/cm2) and then followed with 808-nm NIR laser (600 mW/cm2) for 3 min. The amount of singlet oxygen triggered by RB detected by sensing probe Singlet Oxygen Sensor Green was increased significantly for MCF-7 cells first challenged with FA-CNT-RB for 24 hours and then sequentially exposed with LED light and NIR laser. The cell viability determined by MTT assay revealed 90% eradiation rate for cells treated with RB-CNT-FA which is much better than cells treated with CNT-FA (60% killing rate) and RB-CNT (40% killing rate).

See more of this Session: Bionanotechnology for Gene and Drug Delivery III
See more of this Group/Topical: Nanoscale Science and Engineering Forum

Use of Nanochannel Membranes to Regulate the Release of Cytokines From Biodegradable Reservoirs

Hongyan He, BioLOC LLC., Columbus, OH

Chi Yen, Department of Chemical and Biomolecular Engineering, The Ohio State University, Columbus, OH

W.S. Winston Ho, William G. Lowrie Department of Chemical and Biomolecular Engineering, The Ohio State University, Columbus, OH

L. James Lee, Department of Chemical and Biomolecular Engineering, The Ohio State University, Columbus, OH

The management of locally recurrent malignant melanoma continues to pose a significant challenge. These lesions are typically very painful, and currently available treatments such as repeated intratumoral injections of interferon-alpha (IFN-á) are costly and inconvenient. A particulate-like nanoporous miniature device (NMD) based on biodegradable polymer was developed for controlling local delivery of immunological agents to the tumor microenvironment. The main part of the device consists of a nanoporous controlled gate and a fabricated drug reservoir loaded with IFN-á. To improve the biocompatibility of the devices, a hydrophilic poly(ethylene glycol) monoacrylate was grafted onto the device. Microscopic visualization of the nanoporous controlled gate from in vitro experiments exhibited stability over the desired drug delivery period of two months. In vitro release experiments demonstrated linear release of IFN-á. The IFN-á released was biologically functional, as demonstrated by flow cytometric analysis of peripheral blood mononuclear cells for P-STAT1 (phosphorylated-signal transducer and activator of transcription 1) following exposure to the supernatants from the release experiments. We demonstrated that subcutaneous implantation of the NMDs was well tolerated and associated with an anti-tumor effect in a murine model of human melanoma. There was no evidence of significant inflammatory response or host encapsulation. Thus, these experiments show that our NMD can be fabricated and employed in vivo as a versatile drug delivery platform.

See more of this Session: Bionanotechnology for Gene and Drug Delivery III
See more of this Group/Topical: Nanoscale Science and Engineering Forum

Rapid Analysis of Weak Protein Interactions Using Self-Interaction Nanoparticle Spectroscopy

Aditya N. Bengali, Dept of Chemical & Biological Engineering, Rensselaer Polytechnic Institute, Troy, NY

Joseph J. Grimaldi, Dept of Chemical & Biological Engineering, Rensselaer Polytechnic Institute, Troy, NY

Peter M. Tessier, Dept of Chemical & Biological Engineering, Rensselaer Polytechnic Institute, Troy, NY

Weak protein interactions regulate diverse protein condensation behavior (e.g., crystallization and aggregation), yet they are notoriously difficult to characterize, especially in a high-throughput manner. We are developing a nanoparticle-based platform, namely self-interaction nanoparticle spectroscopy (SINS), to detect protein self-interactions in a high-throughput manner (Tessier et al., *J Am Chem Soc*, 2008). In this presentation we will discuss different approaches of modifying gold nanoparticles for efficient, stable immobilization of proteins with diverse physical properties. Moreover, we will demonstrate that protein self-interactions can be characterized using SINS in good agreement with measurements of osmotic second virial coefficients. Finally, we will discuss our use of SINS for rapidly analyzing the self-association behavior of a library of homologous antibody fragments with the goal of improving the design and selection of aggregation-resistant antibodies.

See more of this Session: Nanotechnology and Nanobiotechnology for Sensors II
See more of this Group/Topical: Nanoscale Science and Engineering Forum

Ultrafast, Highly Sensitive Label-Free Pathogen Detection Via Chemically Modified Graphene (CMG) Sensors

Ashvin Nagaraja, Electrical Engineering, Kansas State University, Manhattan, KS

Nihar Mohanty, Chemical Engineering, Kansas State University, Manhattan, KS

Vikas Berry, Chemical Engineering, Kansas State University, Manhattan, KS

In this presentation, we demonstrate an ultra-fast, highly sensitive detection of a defined genetic sequence via reversible attachment onto surface modified graphene (SMG), leading to pertinent modification of the inherent electronic density of states, which in turn could be evaluated by I-V measurements. Further, we demonstrate the label free detection of specific DNA and RNA markers of pathogenicity via hybridization with immobilized oligonucleotides on the SMG substrate. Current methods of pathogen detection require hours to days for precise results while the presented method accomplishes close to 90 % accuracy in under a minute. By bypassing the traditional PCR based protracted techniques, this work would enable the development of a standalone, label free, miniaturized, highly sensitive, and re-usable device, for utilization in a variety of molecular diagnostic applications.

See more of this Session: Nanotechnology and Nanobiotechnology for Sensors II
See more of this Group/Topical: Nanoscale Science and Engineering Forum

Layer-by-Layer Assembly of Multiwall Carbon Nanotube Ultrathin Films for Biosensing Applications

Saroja Mantha, Materials Engineering, Auburn University, Auburn, AL

Valber A. Pedrosa, Materials Research and Education Center, Auburn University, Auburn, AL

Virginia A. Davis, Department of Chemical Engineering, Auburn University, Auburn, AL

Aleksandr L. Simonian, Materials Research and Education Center, Auburn University, Auburn, AL

The unique mechanical, electrical, and optical properties of multi-walled carbon nanotubes (MWNTs) make them very attractive for the fabrication of new advanced materials. The layer-by-layer (LBL) deposition method involves the alternate adsorption of oppositely charged macromolecules such as polymers and biomacromolecules. This approach provides a simple and flexible method to produce high-strength multifunctional thin films from MWNTs and a variety of enzymes and polyions. Organophosphorus hydrolase (OPH) is an enzyme that exhibits the ability to hydrolyze a large variety of organophosphorus compounds by producing less toxic products such as p-nitrophenol and diethyl phosphate. OPH was immobilized on the surface functionalized MWNT. The LBL assembly was constructed starting with the well dispersed aqueous solutions of positively charged MWNT-Polyethyleneimine (PEI) and negatively charged MWNT-DNA. The bilayers of polyions was held together by electrostatic interaction. Furthermore, an alternate layer of positively charged MWNT-OPH and MWNT-DNA was adsorbed on five bilayers of MWNT-PEI/MWNT-DNA system. FT-IR shows the surface functionalization of MWNTs. TEM and SEM images indicating highly dispersed solutions. This system was used to detect a model organophosphate, paraoxon, with high sensitivity and stability. Absorption spectroscopy and electrochemical flow injection analysis was used as a transduction platforms.

Keywords: Organophosphorus hydrolase, paraoxon, polyethyleneimine

See more of this Session: Nanotechnology and Nanobiotechnology for Sensors II
See more of this Group/Topical: Nanoscale Science and Engineering Forum

Investigating Enzyme Kinetics Using Nanofluidic Devices

Edgar D. Goluch, Kavli Institute of Nanoscience, Section Molecular Biophysics, Delft University of Technology, 2628 CJ Delft, Netherlands

Marcel A.G. Zevenbergen, Kavli Institute of Nanoscience, Section Molecular Biophysics, Delft University of Technology, 2628 CJ Delft, Netherlands

Bernhard Wolfrum, Ibn-2, Forschungszentrum Jülich GmbH, JARA-FIT, 52428 Jülich, Germany

Pradyumna S. Singh, Kavli Institute of Nanoscience, Section Molecular Biophysics, Delft University of Technology, 2628 CJ Delft, Netherlands

Armand W.J.W. Tepper, Leiden Institute of Chemistry, Leiden University, 2333 CC Leiden, Netherlands

Hendrik A. Heering, Leiden Institute of Chemistry, Leiden University, 2333 CC Leiden, Netherlands

Gerard W. Canters, Leiden Institute of Chemistry, Leiden University, 2333 CC Leiden, Netherlands

Serge G. Lemay, Kavli Institute of Nanoscience, Section Molecular Biophysics, Delft University of Technology, 2628 CJ Delft, Netherlands

We have developed a nanofluidic device consisting of two parallel electrodes in a nanometer-scale cavity with openings at both ends. Electrochemically active molecules that enter the cavity transfer charge between the electrodes when a potential difference is applied. Typically, electrochemical reactions only involve one or a few electrons per molecule, rendering direct detection of a single molecule in solution virtually impossible. This limitation is overcome, in this situation, by redox cycling, where each molecule can transfer, on average, thousands of electrons by repeatedly traveling between the electrodes before escaping back out into the bulk. Further, the devices are created using standard microfabrication technology, and can therefore be produced reliably in large quantities.

Numerous sensing applications can be envisioned with such nanofluidic devices, ranging from water purity to drug screening. This talk will focus on employing these devices for determining enzyme kinetics. The reaction kinetics can be monitored directly by the device if the enzyme is capable of converting an electrochemically inactive substrate into an active product. Since the nanochannel is open to the bulk, the composition of the solution inside the device is a representative sample of the bulk solution. Therefore, as the concentration of product in the bulk solution increases, the average number of product molecules in the nanochannel also goes up leading to a larger signal. Our model enzyme is tyrosinase, which converts electrochemically inactive monophenols into active diphenols and quinones.

See more of this Session: Nanotechnology and Nanobiotechnology for Sensors II
See more of this Group/Topical: Nanoscale Science and Engineering Forum

Single Molecule Aptamer-Target Interactions for Sensor Applications

Xiaojuan Zhang, Chemical and Life Science Engineering, Virginia Commonwealth University, Richmond, VA

Vamsi K. Yadavalli, Chemical and Life Science Engineering, Virginia Commonwealth University, Richmond, VA

Highly specific interactions between oligonucleotides and proteins are ubiquitous in nature. Short synthetic oligonucleotide (DNA/RNA) sequences known as aptamers are increasingly being used for applications such as biodiagnostics and biosensors to directly exploit their exquisite molecular recognition abilities. As aptamers emerge as extremely valuable tools in the development of therapies and diagnostic devices, understanding the modes of interaction of aptamers and their targets is of vital importance in the rational design of such devices. Force spectroscopy using an atomic force microscope (AFM) is a valuable tool for studying such interactions and molecular biorecognition at a single molecule level. Here we present studies of the interactions of highly specific aptamers (RNA and DNA) with target proteins on a mixed self-assembled monolayer (SAM) platform. In this study, mixed SAMs consisting of N-hydroxysuccinimide (NHS) and oligoethylene glycol (OEG)-terminated thiols on ultraflat gold surfaces were used to covalently immobilize proteins. By the optimization of attachment sites via lysine-NHS linkages on a protein-resistant layer of the OEG SAM, it is possible to isolate single proteins for study in a controlled fashion. In particular, we present data on the specific interaction profiles, rupture forces and environmental determinants for the binding of specific aptamers with the clinically relevant biomarker vascular endothelial growth factor (VEGF). It is envisioned that this technology will enable insight into the fundamental biophysical interactions between aptamers and their target proteins for various sensor applications.

See more of this Session: Nanotechnology and Nanobiotechnology for Sensors II
See more of this Group/Topical: Nanoscale Science and Engineering Forum

Graphene Sheets-Oil Nanocomposites: Equilibrium and Transport Properties From Molecular Simulation

Deepthi Konatham, School of Chemical Biological and Materials Engineering, The University of Oklahoma, Norman, OK

Alberto Striolo, School of Chemical Biological and Materials Engineering, The University of Oklahoma, Norman, OK

Graphene sheets (GSs), one-atom-thick layers of carbon atoms, are receiving enormous scientific attention because of extraordinary electronic and mechanical properties. These intrinsic properties will lead to innovative nano-composite materials that could be used to produce novel transistors and thermally-conductive polymeric materials. Such materials could be obtained by dispersing thermally conductive nanoparticles within polymers.

We conducted molecular dynamics simulations of functionalized graphene sheets dispersed in liquid organic linear alkanes (oils) at room conditions. We observed that as the GS volume fraction in GS-oil dispersion increases, the system manifests different morphologies. At low GS volume fraction, the GSs appear uniformly dispersed throughout the matrix. At high GS volume fraction, the GS yield a system with no positional order, but with long range orientational order. This is reminiscent of a lyotropic phase transition from a disordered to a nematic phase which was characterized using the order parameter approach. The phase boundaries are likely to depend on the GS size and functionalization. Nematic phases may yield to anisotropic heat transfer properties.

We conducted non-equilibrium molecular dynamics simulations for GSs dispersed in octane to characterize heat-transfer properties of the GSs. It was found that the resistance to heat transfer offered by the interface between the GSs and the oil, called the Kapitza Resistance (RT), is less than that found for carbon nanotubes dispersions.

See more of this Session: Self and Directed Assembly at the Nanoscale II
See more of this Group/Topical: Nanoscale Science and Engineering Forum

Self Assembly of DNA Segments On Graphene and Carbon Nanotube Surfaces in Aqueous Environment: a Molecular Simulation Study

Xiongce Zhao, Center for Nanophase Materials Sciences, Oak Ridge National Laboratory, Oak Ridge, TN

Molecular dynamics simulations were performed to study the interaction of double stranded DNA segments with the surfaces of graphene and carbon nanotubes in aqueous solution. Three different kinds of self assembly phenomena were observed. Firstly, it is found that a DNA segment can `stand up' on the carbon surfaces with its helix axis perpendicular to the surfaces of graphene or nanotube. Secondly, a DNA segment can also lie on the carbon surface with its axis parallel to the surface, if both of its ends can manage to interact with the carbon surfaces. In this case, the ending basepairs of the DNA are broken due to severe deformations. Thirdly, it is observed that short DNA segments can concatenate to each other to form a longer DNA when they are placed in the grooves of nanotube bundles. The self assembly usually happens in less than 10 nanoseconds. The very fast self assembly of DNA segments on carbonaceous surfaces emphasizes the dominant role of hydrophobic forces in DNA-carbon interaction in aqueous environment.

See more of this Session: Self and Directed Assembly at the Nanoscale II
See more of this Group/Topical: Nanoscale Science and Engineering Forum

Quantifying Nanoparticle Self-Assembly : A QCM-D Study

Archana Jaiswal, Q-Sense, Glen Burnie, MD

Mark A. Poggi, Q-Sense, Glen Burnie, MD

Bartosz A. Grzybowski, Department of Chemical and Biological Engineering and Department of Chemistry, Northwestern University, Evanston, IL

Stoyan K. Smoukov, Department of Chemical and Biomolecular Engineering, North Carolina State University, Raleigh, NC

There has been an increasing demand for analytical tools that can quantify the interaction and/or reaction of nano-scale particles, polymers as well as bio-molecules, with various surfaces. In-situ monitoring of changes of nano-scale materials, is of critical importance. If scientists and engineers have a more complete understanding of the behavior of materials at the nano-scale, new materials can be built from the nano-scale up.

Quartz Crystal Microbalance with Dissipation Monitoring (QCM-D), a nanomechanical acoustic-based analytical technique, provides in situ, real-time characterization of the interactions of various polymeric and bio-materials at variety of interfaces. With QCM-D, simultaneous measurement of resonance frequency change (ÄF) and energy dissipation change(ÄD) is performed by periodically switching off the driving power of oscillation of the sensor crystal and recording the decay of damped oscillation as the adsorption and/or structural changes takes place at sensor crystal surface (Figure 1). While change in frequency provides information about mass changes, changes in dissipation (D) provides structural information about the viscoelastic properties of adsorbed films in real time.

This presentation will first review work that has previously been done utilizing QCM-D to quantify self-assembly processes. Then we will discuss recent results that have shown that we can follow (in real-time) the grafting of chemically-modified nanoparticles onto surfaces.

See more of this Session: Self and Directed Assembly at the Nanoscale II
See more of this Group/Topical: Nanoscale Science and Engineering Forum

Self-assembling of Noncovalently Connected Micelles between P (AA-co-St) and PVP in Aqueous Media

Yun FANG[*], *Jing SHA, Fangmin LIU, Zhongyu LAI*

School of Chemical & Material Engineering, Jiangnan University, Wuxi, Jiangsu 214122, China

Abstract

A series of amphiphilic random copolymers poly (acrylic acid-co-styrene) (P (AA-co-St)) with different ratio of acrylic acid to styrene were synthesized by free radical copolymerization. The copolymers were characterized by means of IR, GPC, DSC and ^1H NMR. The self-assembling of P (AA-co-St) with hydrophilic macromolecule poly (vinylpyrrolidone) (PVP) was carried out in aqueous media and the micellization was studied by means of DLS and TEM. The results of DLS showed that hydrokinetic radiuses of micelles rises with initial concentrations of P (AA-co-St) and PVP and molecular weight of PVP while rising with reducing contents of AA, separately. As the concentration of PVP is kept constant and that of P(AA-co-St)s is in the rang of 1.0~1.5 mg/ml, AA content of P (AA-co-St) and molecular weight of PVP have a significant effect on micelle size. Besides, several factors such as different adding speed in the micellization, pH, and the concentrations of NaCl have effects on micelle size. TEM results demonstrated that the micelles were regular spheres with nucleus and shell. Special shape micelles appeared such as about 40nm vesicles when PVP k-90 (1.0 mg/ml) assembled with P(AA-co-St) (n_{AA} : n_{St}=1:3) (1.0 mg/ml).

Introduction

During the traditional micellization of macromolecule, block or graft copolymers were chosen as the most universal object for study as their high regular structure[1-5] and they were studied extensively. As block copolymers were synthesized with difficulties and their price were expensive, the signification of their feasibility explores and theory research was more important than that of application research, so their actual application was limited. During the last years, people started to choose random copolymers which were simple and cheap. Amphiphilic random copolymers can realize micellization in selective solvent to form aggregates of sphere[6], vesicle[7], wheel[8] and bowl[9]. Either block copolymers or random copolymers, the micellization are conducted in selective solvent which involved organic solvent and their applications in life were limited. The core-shell structure of micelles which was formed by self-assembling of block copolymers and random copolymers was covalently connected. "Block-copolymers-free strategy"[10] which was the more simple strategy for micellization was developed

[*] Corresponding author: yunfang@126.com. Supported by the National Natural Science Foundation of China (20871059)

by Jiangming research group in fudan university. The core-shell structure of micelles formed in this method was connected with the most universal and the simplest noncovalent bond which was differ from that of the micelles self-assembled of block copolymers and random copolymers.

In this paper, a series of amphiphilic random copolymers poly (acrylic acid-co-styrene) (P (AA-co-St)) with different compositions were synthesized by free radical copolymerization in the solution of dioxane. The synthesized P (AA-co-St) were characterized by means of infrared spectrum (IR), gel permeation chromatography (GPC), differential scanning calorimetry (DSC) and nuclear magnetic resonance (NMR); The synthesized P (AA-co-St) were assembled with polyvinylpyrrolidone (PVP) into micelles in aqueous media. DLS and TEM were used to study the micellization of the system.

Experiments

Materials Acrylic acid (AA): AR, used after decompress distillation, purchased from Chinese medicine Shanghai chemical reagent company. Styrene (St): AR, used after decompress distillation, purchased from Chinese medicine Shanghai chemical reagent company. Dioxane: AR, purchased from Chinese medicine Shanghai chemical reagent company. Polyvinylpyrrolidone (PVP): AR, purchased from ISP. Tetrahydrofuran (THF): AR, purchased from Chinese medicine Shanghai chemical reagent company.

P (AA-co-St) preparation Monomer AA, St and initiator AIBN were quantified in 100ml rockered flask. Dioxane was added. Pour nitrogen to remove oxygen for 10min. The rockered flask was put in oil and milled for 12h at 60 ~ 70□. After the reaction, the sample was deposited and was purified for 3 times. The sample was dried at 30□.

Micelle preparation P (AA-co-St)/PVP micelle solutions were prepared by adding 1mL of P (AA-co-St) /THF solutions into 20 mL of PVP/H_2O solution dropwise under ultrasonic. The resultant solutions were transparent and bluish. The concentrations of P (AA-co-St)/THF ranged from 0.1 to 1.5 mg/ml and PVP/H_2O ranged from 0.01 to 1.0 mg/ml.

Character of P (AA-co-St) The actual composition of P (AA-co-St) was confirmed by ^{1}H NMR using DMSO as solvent by a Bruker NMR. The infrared spectrum was confirmed by a Canada ABB Bomem FTLA2000-104. Molecular weight and polydispersity (M_W/Mn) were determined by gel permeation chromatography (GPC) using monodispersed polystyrene as standard. Glass transition temperature (Tg) of P (AA-co-St) was determined by differential scanning calorimetry (DSC).

Dynamic Light Scattering The micelle solutions were clarified by filtering thro ugh 0.45μm Millipore filter. The measurements were performed in German ALV-5022F dynamic light scattering. All the DLS were measured at 25±0.1℃ and at a scattering angle of 90°.

Transmission Electron Microscopy TEM observations were conducted on a Philips Tecnai 12 transmission electron microscopy at an accelerating voltage of 120 kv. The specimens were prepared by spraying a drop of the micelle solution on a carbon-coated copper grid followed by water evaporation.

Results and Discussion

Character of P (AA-co-St) Five sorts of P (AA-co-St) were synthesized that were 1:3, 1:2, 1:1, 2:1 and 3:1(molar ratio of Acrylic acid to Styrene). IR spectrum demonstrated that we had synthesized random copolymers P(AA-co-St) which we needed (see Figure 1).The peak locations were similar. The actual composition, Molecular weight and polydispersity M_W/Mn and Glass transition temperature Tg of P (AA-co-St) were listed in Table 1.

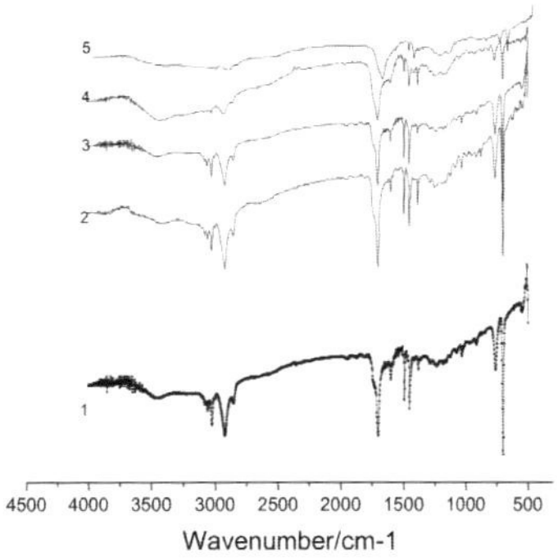

Figure 1 Infrared absorbtion spectrum of P (AA-co-St) with different feed composition

Curve 1: n_{AA}□n_{St} = 1:3; Curve2: n_{AA}□n_{St} = 1:2; Curve3: n_{AA}□n_{St}=1:1 Curve4: n_{AA}□n_{St}= 2:1; Curve5: n_{AA}□n_{St} = 3:1

Table 1 Properties of P (AA-co-St) with different composition

n_{AA}: n_{St} a)	n_{AA}: n_{St} b)	Mn×10^4	Mw×10^4	Mw/Mn	Tg /^{o}C
1:3	1 : 2.27	2.23	2.99	1.34	127.3
1:2	1 : 1.6	2.39	2.95	1.23	132.3
1:1	1 : 0.92	2.39	2.94	1.23	140.7
2:1	1.81 : 1	2.55	2.94	1.35	146.6
3:1	2.2 : 1	2.55	3.46	1.36	145.5

a):Feed composition b): actual composition of P(AA-co-St) determined by means of ^{1}H NMR.

Mn: number-average relative molecular mass; Mw: weight-average relative molecular mass;

Tg: glass transition temperature.

Effect of initial concentrations on hydrokinetic radiuses of micelles THF is the fine solvent of P (AA-co-St) while water is the fine solvent of PAA and the precipitator. When P (AA-co-St)/THF solutions were added into abundant PVP/H_2O solution dropwise, there were no precipitation but the solution was transparent and bluish for the proton giver of PAA and the proton acceptor of PVP can form hydrogen bond. There were six groups in the research. In the group 1 to 5, the initial concentration of PVP k-30 was kept constant and the initial concentration of P (AA-co-St) of five different molar ratio of Acrylic acid to Styrene was changed. Group 1 to 5 corresponded to $n_{AA} : n_{St}$ =1:3, 1:2, 1:1, 2:1 and 3:1,separately. In the group six, the initial concentration of P(AA-co-St) ($n_{AA} : n_{St}$ =1:3) was kept constant and the initial concentration of PVP k-30 was changed. The hydrokinetic radiuses of micelles were measured by DLS. For example, the R_h distributing of micelles in the first group was in Figure 2. The DLS results of group 1 to 5 were in Figure 3. In Figure 2, the curves of R_h distributing were single peaks and the hydrokinetic radiuses R_h increased when the initial concentration of P (AA-co-St) ($n_{AA} : n_{St}$ =1:3) increased. High concentration of P (AA-co-St) induced large nucleus which induced large micelles. In groups 2 to 5, the size of micelles had the similar changing trade (see Fig.3). When the content of AA in P (AA-co-St) reduced, the size of micelles increased. Low content of AA meant high content St which induced large nucleus. What more, when the initial concentration of P (AA-co-St) was low (0.1 mg/ml and 0.5 mg/ml), the content of AA had a little effect on the size while the initial concentration of P (AA-co-St) was above 1.0 mg/ml, the content of AA had a significant effect on the size.

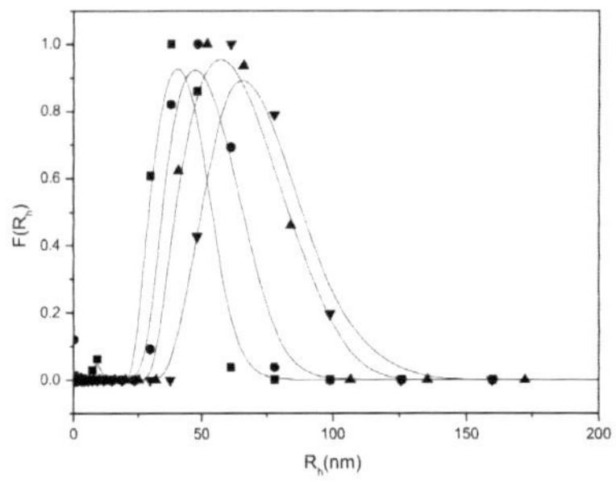

Figure 2 The hydrodynamic radius distributing of micelles formed with 0.1 mg/mlPVP k-30/H_2O and P (AA-co-St) ($n_{AA} : n_{St}$ =1:3)/THF with different concentration (■, 0.1 mg/ml; •, 0.5 mg/ml; ▲, 1.0 mg/ml; ▼, 1.5 mg/ml)

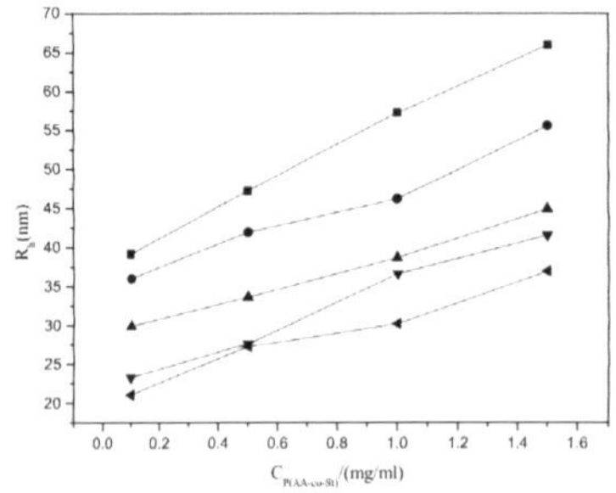

Figure 3 Effect of initial concentration of P (AA-co-St)/THF with different composition (■ $n_{AA} : n_{St}$=1:3; • $n_{AA} : n_{St}$=1:2; ▲ $n_{AA} : n_{St}$=1:1; ▼ $n_{AA} : n_{St}$=2:1; ◄ $n_{AA} : n_{St}$=3:1) on the hydrodynamic radius R_h of micelles

The initial concentration of PVP also had an effect on the size of micelles (see Fig.4). When keep the concentration of P (AA-co-St) (n_{AA} : n_{St} =1:3) 1.0 mg/ml constant, R_h increased with the initial concentration of PVP. When initial concentration of P (AA-co-St) was kept constant, high concentration of PVP meant that more PVP massed to the nucleus which induced large micelles.

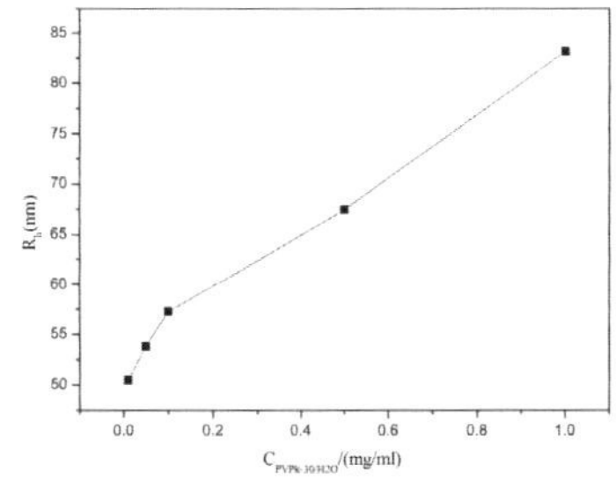

Figure 4 Effect of initial concentration of PVP k-30/H_2O on the hydrodynamic radius R_h of micelles

Effect of molecular weight of PVP on hydrokinetic radiuses of micelles To study the effect of molecular weight of PVP on the size of micelles, PVP k-90 and k-30 were chosen to assembled with P (AA-co-St) (n_{AA} : n_{St} =1:3) (see Fig.5).

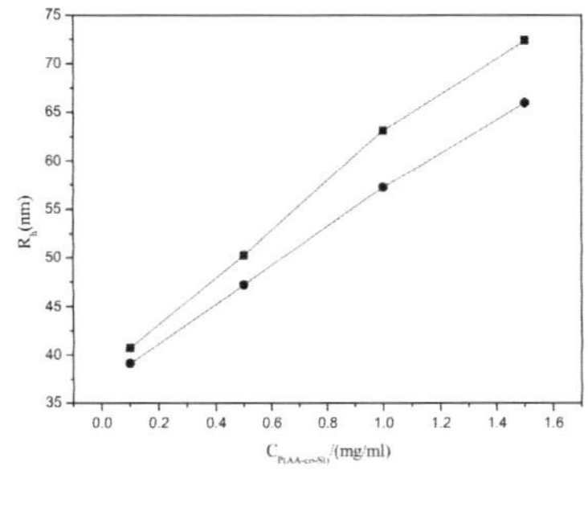

Figure 5 Effect of molecular weight of PVP on the hydrodynamic radius R_h of micelles, ■ PVP k-90; ● PVP k-30

When molecular weight of PVP increased, the size of micelles increased. When the initial concentration of P (AA-co-St) was low (0.1 mg/ml and 0.5mg/ml),the molecular weight of PVP had a little effect on the size while the initial concentration of P (AA-co-St) was above 1.0 mg/ml, the molecular weight of PVP had a significant effect on the size. It is similar to that of the content of AA in P (AA-co-St).Increasing molecular weight of PVP induced thicker shell which induced large micelles. When the initial concentration of P (AA-co-St) was low, the nucleus was small and less PVP massed to the nucleus which could explain that the molecular weight of PVP had a little effect on the size. When the initial concentration of P (AA-co-St) was high, the nucleus was large and more PVP massed to the nucleus. That is why the molecular weight of PVP had a significant effect on the size.

Effect of several factors on hydrokinetic radiuses of micelles The micellization was kinetics controlled .So the size of micelles could be affected by many factors such as different adding speed in the micellization, pH, and the concentrations of NaCl. Effect of different adding speed in the micellization could get different size of micelles (see Fig.6).When adding interval was 0.5 min/d, the size was larger. Then when adding interval was1, 3, 4 and 5 min/d, the size increased with the interval increased. When adding interval was 7 min/d, the size decreased. The cause of the phenomenon was that when P (AA-co-St)/THF was added to PVP/H_2O, the solvent environment worsened at the moment and PS aggregated. When adding interval was 0.5 min/d, as the speed was fast, the increasing velocity was far larger than its shrinking velocity. So the size was large when adding interval was1, 3, 4 and 5 min/d, there was enough time for PS to adjust frame. The pH had a significant effect on the size (see Fig 7). In acid condition, the size was larger than that of the alkali condition. In acid condition, the size had little changes. In alkali condition, the size decreased with pH increased. When the condition was acid, PAA had many carboxyls which caused more and more PVP aggregated to PAA. Then the size was large. The carboxyls got saturation at the moment in acid condition which caused the size changed little. When the condition was alkali, PAA ionization and the degree of ionization increased with pH increased which caused less and less PVP aggregated to PAA .Then the size was small. The concentration of NaCl also had an effect on the size (see Fig 8).The size increased with the concentration of NaCl increased. The cause of the phenomenon was that Na^+ had a static screen effect on AA^- which weakened repellency of PAA. Then more PAA attended the assembling of the micelles which caused more PVP massed to PAA. With the increasing of the concentration of NaCl, effect of screen became more and more strong and repellency became more and more weak. The size then increased.

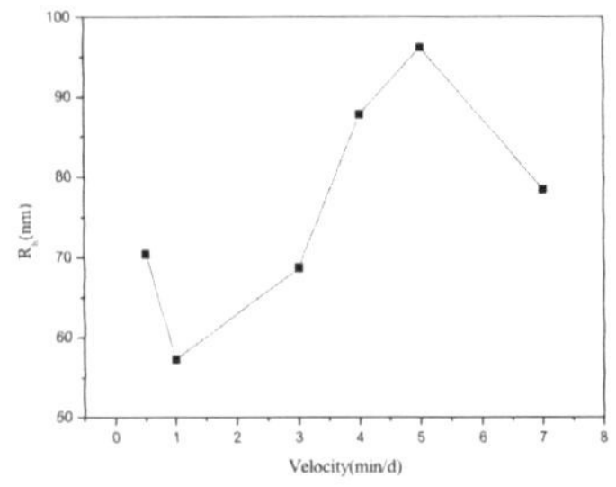

Figure 6 Effect of different adding velocity during micellization on the hydrodynamic radius R_h

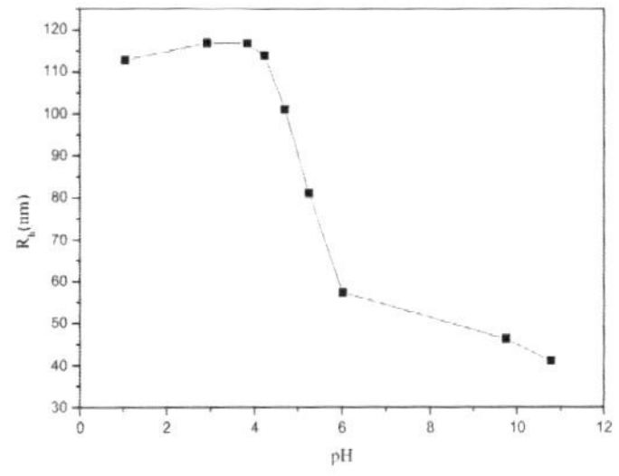

Figure 7 Effect of pH on the hydrodynamic radius R_h

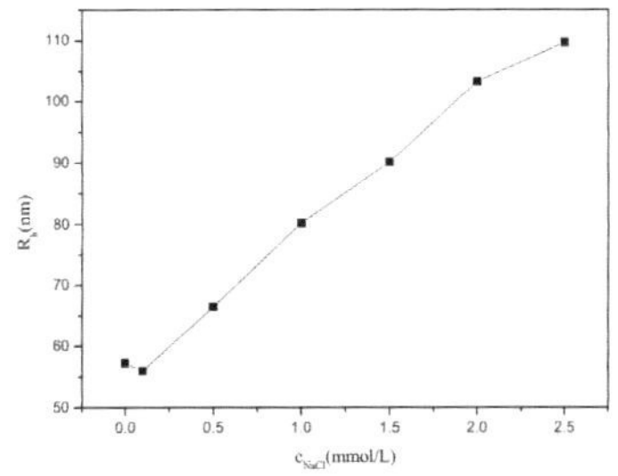

Figure 8 Effect of concentration of NaCl on the hydrodynamic radius R_h

TEM images of micelles TEM was used to observe the morphology of micelles. The TEM images of micelles assembled by $PVP_{k\text{-}30}$ and P(AA-co-St) (n_{AA} : n_{St} =1:3) were seen in Fig.9. The micelles were regular spheres. The micelles had two parts. They were fuscous nucleus and tint periphery which represent nucleus and shell of the micelles, separately. The nucleus was consisting of hydrophobic PS and PVP composed the shell. PAA was on the bonder of nucleus and shell. What's more, there was special shape micelles appeared in the system. When PVP k-90 and P (AA-co-St) (n_{AA} : n_{St} =1:3) assembled, vesicles were formed, about 40 nm (see Fig.10).

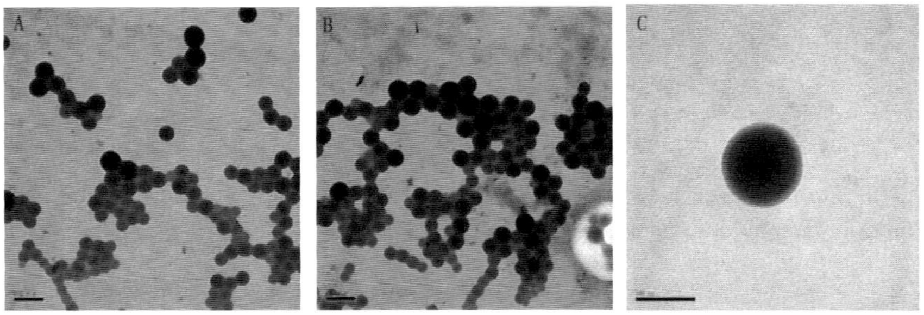

Figure 9 TEM images of the micelles assembled by P (AA-co-St) (n_{AA} : n_{St} =1:3) (1.0 mg/ml) and PVP k-30 (0.1 mg/ml)

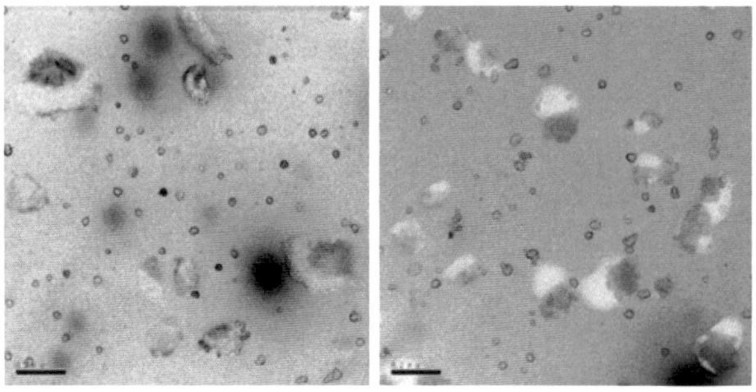

Figure 10 TEM images of the micelles made of P (AA-co-St) (n_{AA} : n_{St}=1:3) (1.0 mg/ml) and PVP k-90 (1.0 mg/ml)

In conclusion, the self-assembling of amphiphilic random copolymers P (AA-co-St) with PVP was carried out in aqueous media. DLS results showed that hydrokinetic radiuses of micelles rises with initial concentrations of P (AA-co-St) and PVP and molecular weight of PVP while rising with reducing contents of AA, separately. As the concentration of PVP is kept constant and that of P (AA-co-St) is in rang of 1.0~1.5 mg/ml, AA content of P (AA-co-St) and molecular weight of PVP have a significant effect on micelle size. Micelle size was also affected by several factors such as different adding speed in the micellization, pH and the concentrations of NaCl. TEM results demonstrated that the micelles were regular spheres with nucleus and shell.

References

1. Lifeng zhang, Kui Yu, Adi Eisenberg (1996), "Ion-Induced Morphological Changes in "Crew-Cut" Aggregates of Amphiphilic Block Copolymers", Science, June, pp.1777-1779.
2. Kabanov AV, Bronich TK, Kabanov VA, Yu K, Eisenberg A (1998), "Spontaneous Formation of Vesicles from Complexesof Block Ionomers and Surfactants", J. Am. Chem. Soc, September, pp. 9941-9942.
3. Izabel Cristina Riegel, Adi Eisenberg (2002), "Novel Bowl-Shaped Morphology of Crew-Cut Aggregates from Amphiphilic Block Copolymers of Styrene and 5-(*N*,*N*-Diethylamino)isoprene", Langmuir, March, pp. 3358-3363.
4. Hamley IW, Ansari IA, Castelletto V, Nuhn H, Rosler A, Klok HA (2005), "Solution Self-Assembly of Hybrid Block Copolymers Containing Poly(ethylene glycol) and Amphiphilic *â*-Strand Peptide Sequences", Biomacromolecules, February, pp. 1310-1315.
5. Li Yan Qiua, You Han (2007), Self-assembled polyethylenimine-graft-poly(e-caprolactone) micelles as potential dual carriers of genes and anticancer drugs", Biomaterials, June, pp. 4132-4142.
6. Lifang Zhu, Chongfeng Ma, Xiaojie Li, Mingqing Chen, Cheng Yang, Xiaoya Liu (2007), Oil Chemistry, January, pp. 388-392.

7. Faysal Ilhan, Trent H.Galow, Mark Gray, Gilles Clavier, Vincent M.Rotello (2000), "Giant Vesicle Formation through Self-Assembly of Complementary Random Copolymers", J. Am. Chem. Soc, June, pp. 5895-5896.
8. Lifeng Yan, Guangzao Zhang, Guangming Liu, Jing Ji, Wan Li(2006) , "Emulsifier-Free Synthesis and Self-Assembly of an Amphiphilic Poly(styrene-*co*-acrylic acid) Copolymer", J. Applied Polymer Science, February, pp. 3718-3726.
9. Xiaoya Liu, Joon-Seop Kim, Jun Wu, Adi Eisenberg (2005), "Bowl-Shaped Aggregates from the Self-Assembly of an Amphiphilic Random Copolymer of Poly(styrene-*co*-methacrylic acid) ", Macromolecules, August, pp. 6749-6751.
10. Min Wang, Ming Jiang, Fanglin Ning, Daoyong Chen, Shiyong Liu, Hongwei Duan (2002), "Block-Copolymer-Free Strategy for Preparing Micelles and Hollow Spheres: Self-Assembly of Poly(4-vinylpyridine) and Modified Polystyrene", Macromolecules, June, pp.5980-5989.

Synthesis of Hollow Cu_2O Submicrospheres by SDS-PVP Necklace-like Soft Clusters

Ye FAN, Yun FANG, Yi GUO, Lei KE*

School of Chemical and Material Engineering, Jiangnan University, Wuxi, Jiangsu, China

Abstract

The nano-/microsized Cu_2O with special morphologies were prepared from copper sulfate ($CuSO_4$) reduced by hydrazine ($N_2H_4 \cdot H_2O$) in aqueous solution of sodium dodecyl sulfate (SDS)-polyvinylpyrrolidone (PVP). XRD proof indicates that the products are pure cubic Cu_2O crystals, and grown preferentially along (111) lattice plane. SEM and TEM images show that the products are hollow Cu_2O submicrospheres with regular morphology and controlled size, and confirm that the hollow submicrospheres are composed of primary Cu_2O nanocrystals with diameters of ca. 10 nm. The N_2 adsorption isotherm indicates that the products are mesoporous shells. The SDS-PVP necklace-like clusters act as dual soft-template and induced an unique template-induced secondary assembly. As a photocatalyst, the hollow Cu_2O submicrospheres showed excellent catalytic degradation ability for methyl orange.

Introduction

The nano-/microsized materials have superior physicochemical properties compared to macroscopic materials. The preparation of nano-/microsized particles has received considerable attention. As a p-type semiconductor, the band gap of cuprous oxide (Cu_2O) is 2.2 eV, it is a promising material with potential applications in catalysis, sensing, antifouling coatings, electronic component and solar energy conversion etc. Many groups have prepared Cu_2O nanocrystals with different morphology by various methods. In the recent years, soft template attracted a wide spread attention, especially the surfactant and polymer system. Such as, Luo et al reduced Cu_2O nanoflowers from $Cu(CH_3COO)_2 \cdot H_2O$ by $NaBH_4$ in the CTAB-DMF aqueous solution[1], and Kuo et al fabricated truncated rhombic dodecahedral Cu_2O nanocages and nanoframes by particle aggregation and acidic etching in SDS aqueous solution[2], Li's group synthesized Cu_2O nanoflowers by surfactant–assisted electrochemical deposition[3], Zhu and co-works prepared Cu_2O hollow microspheres in the present of PVP and investigated their application in DNA biosensor of hepatitis B virus[4], Xu et al prepared single-crystalline Cu_2O hollow nanocubes with PEG-assisted[5].

---

* Corresponding author: yunfang@126.com. Supported by the National Natural Science Foundation of China (20871059).

Our group prepared silver nanoparticles[6], gold nanorings[7], multipod-like gold nanoparticles[8] and nickel nanoparticles[9]. The research indicated that surfactant–polymer template not only played an important role in controlling the size and morphology, but also obtained more research results compared to single template. In this paper, The hollow Cu_2O submicrospheres were prepared from copper sulfate ($CuSO_4$) reduced by hydrazine ($N_2H_4{\cdot}H_2O$) in the soft template of sodium dodecyl sulfate (SDS)-polyvinylpyrrolidone (PVP) necklace–like soft cluster, preliminary discussed the template-induced secondary assembly synthesis of hollow Cu_2O submicrospheres in virtue of the dual soft-template provided by SDS-PVP soft clusters and investigated the photocatalytic ability of hollow Cu_2O submicrospheres.

Experimental

Synthesis All reagents were of analytical grade, used without further purification. In a typical experiment, the aqueous solution of $CuSO_4$-SDS-PVP was adjusted pH to 10 ± 0.5, $N_2H_4{\cdot}H_2O$ was added into it and well–mixed, the reaction mixture quickly turns to be a light yellow suspension. After reacted 55 min in (40±1.0) °C water bath, the reaction mixture was cooled naturally to room temperature to obtain Cu_2O sol. The Cu_2O sol centrifugated (12,000 rpm) and the resultant precipitates were washed with distilled water and anhydrous ethanol.

Photocatalysis A 0.06 g amount of hollow Cu_2O submicrospheres was dispersed in a 100.0 mL methyl orange (5 mg/L) under stirring in the visible light. The dispersion was centrifuged to separate the solid particles, and the methyl orange (MO) concentration was determined using UV–Vis spectrometer.

Characterization The morphology and microstructure of the products were characterized using a transmission electron microscope (TEM, Tecnai G^2 20) with an accelerating voltage of 200 kV. X-ray diffraction (XRD) patterns collected on a D8Advance diffractometer with a graphite monochromator and Cu Kα radiation (λ = 0.15418 nm) were used to analyze the crystal structure of the products. N_2 adsorption - desorption isotherms were measured on a QuadraSorb SI apparatus at 77 K. The pore size distribution was calculated from the adsorption branch using the Barett-Joyner-Halenda (BJH) theory.

Results and Discussion

Characterization of structure and morphology The products were identified as the cubic symmetry of cuprite (Cu_2O, JCPDS no. 05-0667) by X-ray diffractometry (Figure 1A). The mean size of the primary Cu_2O nanocrystals is ca. 8.0 nm on the basis of the Debye–Schrrer equation from the 111 reflection. The SEM image (Figure 1B) shows that the products are regular submicrospheres with rough surface, and the hollow structure was preliminary presumed according to the significant switch and shade contrast. The TEM image (Figure 1C) clearly shows the hollow submicrospheres were assembled from a lot of nanocrystals.

The N_2 adsorption–desorption isotherm of the hollow Cu_2O submicrospheres shown in Figure 2A exhibits an acute uptake and a hysteresis loop in the relative pressure range of 0.8 – 1.0, indicating the presence of the mesopores. The corresponding pore size distribution curve (Figure 2B) calculated by the BJH method based on the adsorption branch displays two peaks. The first peak in the range of 1 to 5 nm is attributed to the aggregation of the nanocrystals within the shells of the hollow Cu_2O submicrospheres. The peak in the range of 8 – 60 nm further confirms the hollow structure of the Cu_2O submicrospheres.

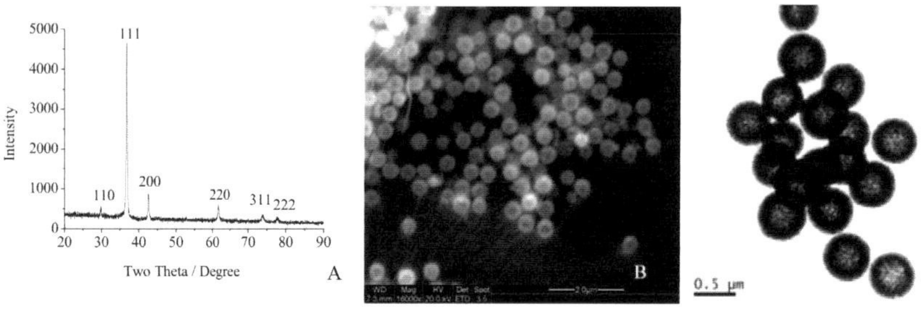

Figure 1 Characterization by means of (A) XRD, (B) SEM, (C)TEM of hollow Cu_2O submicrospheres which were reduced by $N_2H_4{\cdot}H_2O$ (20 $mmol{\cdot}L^{-1}$) in the system of $CuSO_4$ (1 $mmol{\cdot}L^{-1}$)-PVP K30 (0.25 $mmol{\cdot}L^{-1}$)-SDS (35 $mmol{\cdot}L^{-1}$)

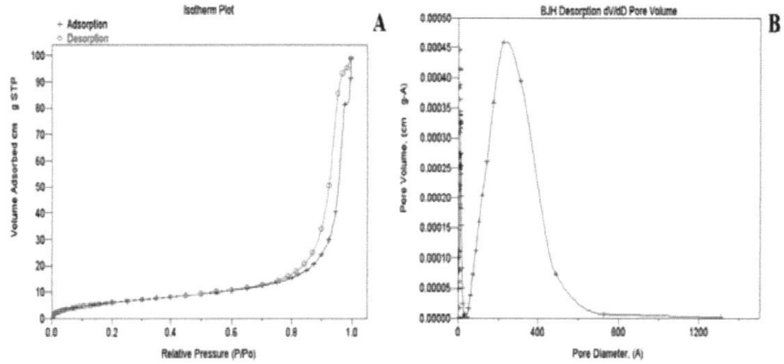

Figure 2 (A) N_2 adsorption-desorption isotherm and (B) pore size distribution curve of Cu_2O hollow submicrospheres which were obtained from the system of $CuSO_4$ (10 $mmol{\cdot}L^{-1}$)-SDS (9 $mmol{\cdot}L^{-1}$)-PVP K30 (0.25 $mmol{\cdot}L^{-1}$) reduced by $N_2H_4{\cdot}H_2O$ (25 $mmol{\cdot}L^{-1}$)

The growth process of hollow Cu_2O submicrospheres The electron microscope images of the products in different reaction stages were shown in Figure 3. Figure 3A shows that the primary Cu_2O nanocrystals with the size of ca. 10 nm were mainly produced in the reaction system at 30 s. After 2 min, the Cu_2O nanocrystals aggregated orderly to form spheres with the size of ca. 100 nm. Figure 3C is a TEM image of the sample collected after 10 min, the size of most spheres distributed in the range of 100~150 nm, the middle of the sphere was loose and with a open structure. With increasing the reaction time to 55 min (SEM in Figure 3D, TEM in Figure 3E), the obtained spheres were uniform and presented obvious hollow structure, the mean size of the hollow Cu_2O submicrospheres was 200 nm.

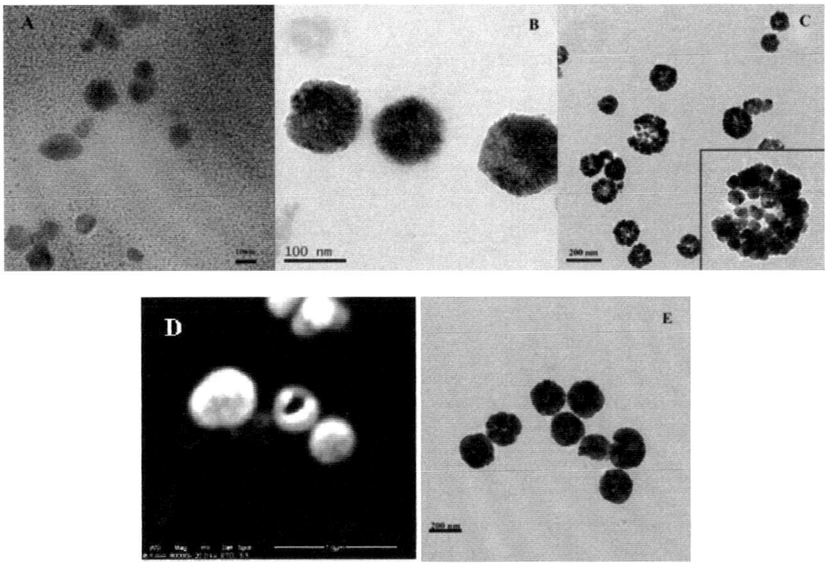

Figure 3 Electron microscopic images of the hollow Cu_2O submicrospheres at different reaction stages obtained from the system of $CuSO_4$ (10 mmol·L^{-1})-SDS(9 mmol·L^{-1})-PVP (0.25 mmol·L^{-1}) reduced by $N_2H_4·H_2O$ (25 mmol·L^{-1})

(A) TEM at 30 s (B) TEM at 2 min (C) TEM at 10 min (D) SEM at 55 min (E) TEM at 55 min

Unique template-induced secondary assembly of SDS–PVP soft clusters According to the characteristics of SDS–PVP necklace-like soft clusters[10, 11], the assembling mechanism of hollow Cu_2O submicrospheres might be that, in the synthesis process of hollow Cu_2O submicrospheres, SDS – PVP soft clusters played a role of dual soft–template and induced an unique template-induced secondary assembly to synthesis hollow Cu_2O submicrospheres. In this process, SDS bound micelle as the first soft-template, induced the primary Cu_2O nanocrystals with diameters of ca. 10 nm; the PVP linked on SDS bound micelle as the secondary soft–template, induced the primary Cu_2O nanocrystals to assemble to hollow Cu_2O submicrospheres.

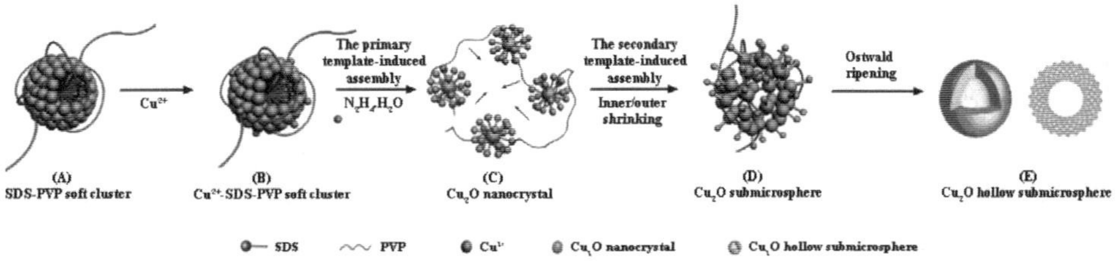

Figure 4 Schematic illustration of the template-induced secondary assembly synthesis of hollow Cu_2O submicrospheres in virtue of the dual soft-template provided by SDS-PVP soft clusters

Photocatalysis The UV-vis spectra of methyl orange (MO) solutions in the presence of hollow Cu_2O submicrospheres were taken over time. When 0.06 g of hollow Cu_2O submicrospheres was added into 100.0 mL MO solution (5 mg/L), the photocatalytic degradation rate of MO was 92.2 % within 2 h and

97.7% of MO was removed (Figure 5) after 5 h. The results indicate that the hollow Cu_2O submicrospheres exhibited superior catalytic degradation ability for methyl orange, because of their hollow nanostructures and high specific surface area.

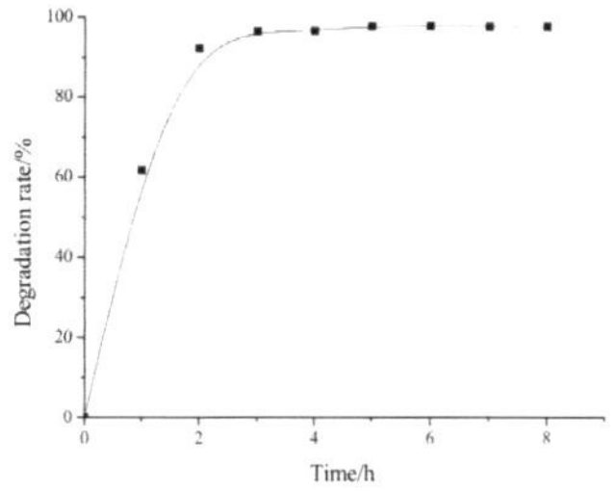

Figure 5 The effect of hollow Cu_2O submicrospheres on the photocatalytic degradation rate

In conclusion, hollow Cu_2O submicrospheres were aggregated from primary Cu_2O nanocrystals which reduced from $CuSO_4$ in soft - template of SDS–PVP cluster by $N_2H_4{\cdot}H_2O$. The SDS–PVP necklace-like clusters act as dual soft-template and induced an unique template-induced secondary assembly.

References

1. Luo YS, Li SQ, Ren QF, Liu JP, Xing LL, Wang Y, Yu Y, Jia ZJ, and Li JL. Facile Synthesis of Flowerlike Cu_2O Nanoarchitectures by a Solution Phase Route. *Crystal Growth & Design*. 2007; 7: 87-92.
2. Kuo CH, Huang MH. Fabrication of Truncated Rhombic Dodecahedral Cu_2O Nanocages and Nanoframes by Particle Aggregation and Acidic Etching. *J. Am. Chem. Soc.* 2008; 130: 12815–12820.
3. Li J, Shi Y, Cai Q, Sun QY, Li HD, Chen XH, Wang XP, Yan YJ and Vrieling E. G. Patterning of Nanostructured Cuprous Oxide by Surfactant-Assisted Electrochemical Deposition. *Crystal Growth & Design*. 2008; 8: 2652-2659.
4. Zhu HT, Wang JX, Xu GY. Fast Synthesis of Cu_2O Hollow Microspheres and Their Application in DNA Biosensor of Hepatitis B Virus. *Crystal Growth & Design*. 2009; 9: 633-638.
5. Xu YY, Jiao XL, Chen DR. PEG-Assisted Preparation of Single-Crystalline Cu_2O Hollow Nanocubes. *J. Phys. Chem. C*. 2008; 112: 16769–16773.
6. Huang SY. Aggregation Properties of Inorganic Electrolyte/SDS/PEG Ternary System. *Master Dissertation: Jiangnan University, China*, 2005.
7. Wang CR, Fang Y, Feng JW. Synthesis of Au Nanorings Using SDS-PEG Cluster as the Soft Template under Mild Conditions. *Acta Chimica Sinica*. 2007; 65: 1177-1180.
8. Wang CR, Fang Y, Li B. Self-Assembly of Multipod-like Gold Nanoparticles in the Soft Template of SDS-PVP Clusters. *Acta Physico-Chimica Sinica*. 2008; 24: 183-186.
9. Ruan XY, Fang Y, Fan Y. Preparation of Ultrafine Nickel Powders in SDS-PVP Aqueous Solution.

Acta Physico-Chimica Sinica. 2008; 24: 1513-1518.

10. Fang Y, Liu XF, Xia YM, Zong LY. Characterization of Clusters Interaction Sites at Cluster of Sodium Dodecyl Sulfate and Water Soluble Nonionic Macromolecules Investigated by ^{1}H NMR、^{13}C NMR and 2D NMR. *Chemical Journal of Chinese Universities*. 2006; 27: 731-734.
11. Chen FB. Studies on Soft Cluster Template in the Synthesis of Cu-containing Nano/Submicro Particles. *Master Dissertation: Jiangnan University, China*, 2007.

Expanding On the In Vivo Capabilities of the Vesosome, a Novel Lipid-Based Drug Delivery Vehicle

Benjamin J. Wong, Department of Chemical Engineering, University of California, Santa Barbara, Santa Barbara, CA

Jason Schmidt, Chemical Engineering, University of California, Santa Barbara, Santa Barbara, CA

Joseph Zasadzinski, Chemical Engineering, University of California, Santa Barbara, Santa Barbara, CA

Although an optimal drug delivery vehicle should have several properties that are universally desired, such as the ability to localize at the site of disease, deliver contents at a rate appropriate for maximum therapeutic benefit and retain its contents over this timeframe, other characteristics, such as loading capacity and circulation lifetime, may be more or less desirable, based on the type of treatment in question. Liposome-based drug delivery systems have been widely used as drug carriers, both in industry and research, and have proven to be adequate in handling the task even though content retention and controlled release remain problematic. Our group has demonstrated the in vivo viability of the vesosome - a large lipid bilayer enclosing many smaller liposomes - and shown that it is the most suitable candidate for addressing these issues. The external lipid bilayer offers a second barrier of protection for interior components and also serves as the anchor for active targeting components. Furthermore, internal compartmentalization permits customization of separate environments for multiple therapeutics and release triggers, highlighting the vesosome's potential as a single site, single dose, multiple component drug treatment.

Prior work on the vesosome centered around modifying it for in vivo active targeting, a phenomena that requires less than a half hour circulation half-life to be effective. The approximate two-hour half life of the vesosome is more than sufficient for this task, but current the current synthesis has proven problematic for purification of the final product. Since the processes of vesosome formation and functionalization are based on the phase behavior exhibited by homogeneous lipid mixtures, slight changes in lipid composition, such as the addition of PEG-lipids or fluorescently-labeled lipids for active targeting and in vivo imaging, have had a notable impact on the structures that form. However, vesosome formation has still been achieved using heterogeneous lipid mixtures. A variety of microscopy techniques, including freeze-fracture and cryo transmission electron microscopy as well as fluorescence and confocal microscopy, are used to characterize these functionalized drug carriers.

Our current work is focused on maintaining a suitable half-life for the vesosome while modifying the synthesis to permit for both increased purity and improved in vivo function. Preliminary in vivo active targeting experiments have shown the viability of the vesosome in directed aggregation at disease sites and formulation changes have shown a new synthesis to maintain a comparable half-life while improving upon product purity. Continuing work on the vesosome focuses on quantifying the effectiveness of active targeting and the implementation of multi-drug encapsulation and delivery.

See more of this Session: Self and Directed Assembly at the Nanoscale II
See more of this Group/Topical: Nanoscale Science and Engineering Forum

A mesoporous material is a material containing pores with diameters between 2 and 50 nm. They have high surface areas, providing a vast number of sites where sorption processes can occur. These materials have numerous applications in catalysis, separation and many other fields. It is possible to prepare numerous new materials with a regular and controlled morphology on the nano and micro scale by simply adjusting the morphology of the template material. The nature of organic templates is one of the most important factors for designing and synthesizing mesoporous materials. In general, surfactants with cationic, anionic, and neutral charges and amphiphilic block copolymers have been utilized as organic templating agents. One of the current scientific challenges is the creation of mesoporous crystalline metal oxides. The reason for this is that crystalline oxides are chemically, thermally, and mechanically robust. For example, mesoporous, crystalline TiO_2 is of interest for several applications including solid-state solar cells or photocatalysis, where organized porosity would be helpful to improve the contact between the oxide and the other species (e.g. the anode in the case of solar cells).

One important strategy of nanostructure synthesis is self-assembly using soft chemistry, which is inexpensive and easy to scale-up. The advantage of this method is that the template allows nanomaterials with a variety of morphologies to be prepared, while the disadvantage is that this technique requires high calcination temperatures to remove the template, potentially resulting in a collapse of the tubular structure in the product.

Recently, amphiphilic block copolymers have been increasingly used to organize mesostructured composite solids, because the architectures of these copolymers can be used to control the morphology and interactions between the inorganic and organic species.

In this study, for the first time PMMA-b-PAA, an amphiphilic diblock copolymer, was synthesized by reversible fragmentation chain-transfer (RAFT) polymerization through one step to use as a self-assembling agent to synthesize of mesoporous TiO_2.

Results and Discussion:

Table 1 summarizes how the M_n of PMMA and PMMA-b-PAA change with conversion (or time), with the conversions being determined using *in-situ* FTIR. It is clear that M_n increases directly with time, along with having low PDIs, further proving the livingness of this polymerization.

Table 1. Molecular weight, PDIs, and conversions of PMMA and PMMA-b-PAA at different reaction times

Samples	Time (h)	M_n (g/mol)	PDI	Conversion (%)
PMMA	1	5,760	1.3	15
PMMA	5	9,200	1.2	60
PMMA-b-PAA	1	10,200	1.3	7
PMMA-b-PAA	2	14,000	1.4	20
PMMA-b-PAA	3	16,000	1.3	30

To investigate the morphology of the block copolymer of PMMA-b-PAA at the investigated polymerization times, and subsequent PAA chain lengths, SEM and TEM analysis were employed. The SEM micrographs are given in Figure 1. Figure 1a shows SEM images of PMMA while Figure 1b-d illustrates that of PMMA-b-PAA at different times, with the resultant increase in chain length of PAA. After one hour, the cluster of PAA started growing from the PMMA backbone, moreover the cluster size as well as cluster density of PAA segments increases with copolymerization time which indicates the livingness of the diblock copolymerization.

Figure 1. SEM images of (a) PMMA, and PMMA-b-PAA after (b) 1, (c) 2, and (d) 3 hours

In order to obtain more detailed structural information, TEM analysis was performed for the different copolymer samples. To prepare the TEM samples, the investigated PMMA-b-PAA samples were dissolved in methanol, a good solvent for PAA but a poor solvent for PMMA. As a result of the preferential affinity of the solvent for the PAA block, the solution will contain micelles consisting of a PMMA core surrounded by a PAA corona. Figure 2a shows that a small micelle was formed in the PMMA-b-PAA_1 (diblock copolymer after one hour) sample, which can be attributed to the small length of the PAA block. As the length of the PAA block increased, PMMA-b-PAA_2, and PMMA-b-PAA_3 (diblock copolymer after two and three hours, respectively), the micelles are growing, to make cross-linked "pearl necklace" networks or Dendrimer-like Star (Figure 2 b-c). The hydrogen bonding between PAA chains in the corona facilitates network self-assembly containing PMMA within the core of the micelles.

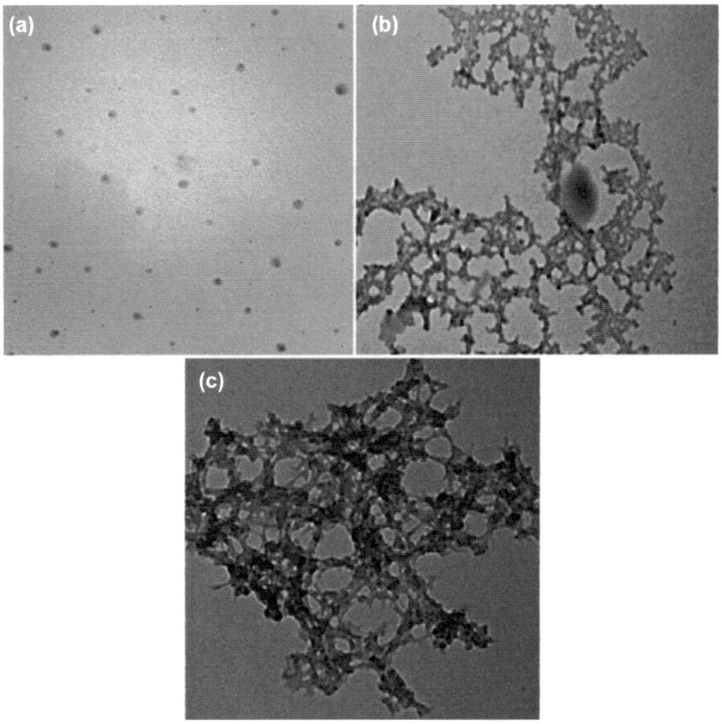

Figure 2. TEM images of (a) PMMA-b-PAA$_1$, (b) PMMA-b-PAA$_2$, and (c) PMMA-b-PAA$_3$

The crystal structure of the TiO_2 synthesized at 65 °C and calcined at 600°C was examined by XRD, with the spectra given in Figure 3. In the wide-angle powder XRD pattern, five crystal peaks at high 2θ values of 25.28°, 37.80°, 48°, 53.89°, and 62.7° appear, which is consistent with the anatase phase of titanium dioxide materials.

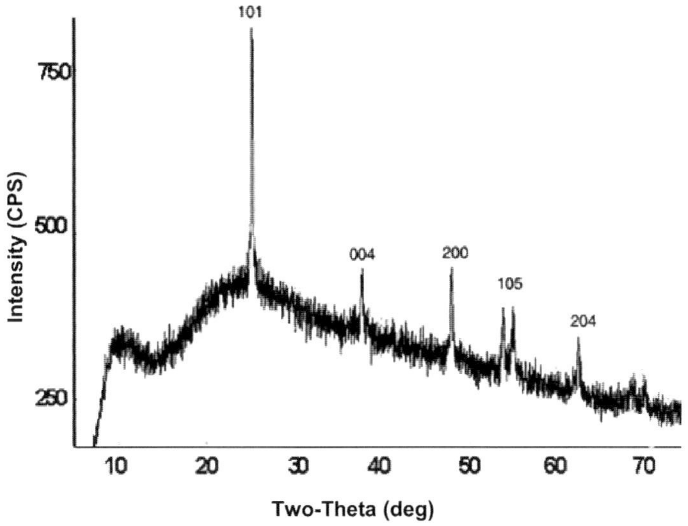

Figure 3. X-ray diffraction (XRD) patterns of mesoporous TiO_2 after calcination at 600°C

The isotherms provided in Figure 4 show that the TiO_2 materials exhibit a type IV isotherm with a hysteresis loop. At PMMA-b-PAA_3 the hysteresis loop is positioned at the top, giving a higher surface area whereas PMMA-b-PAA_1 and PMMA-b-PAA_2 have lower values (T_1, T_2, and T_3 exhibit PMMA-b-PAA_1, PMMA-b-PAA_2, and PMMA-b-PAA_3 as templates, respectively).

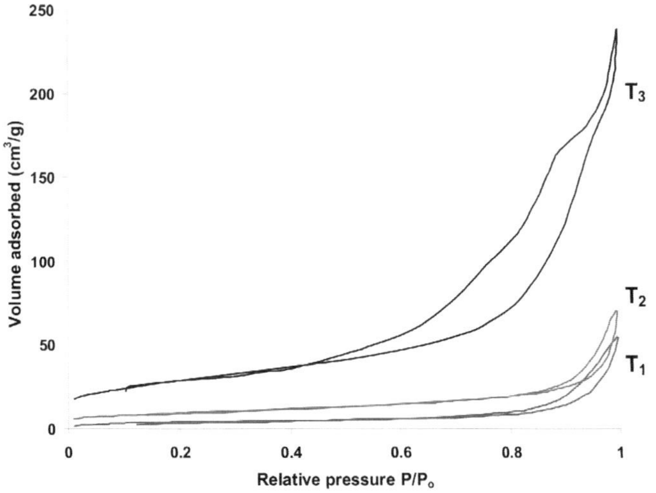

Figure 4. Isotherm plots of mesoporous TiO_2 using (a) T_1 (b) T_2 (c) T_3 as templates

The calculated pore volumes, pore diameter, and surface areas are given in Table 2.

Table 2. Surface area, pore volume and average pore diameter of calcined TiO_2 using different chain length templates

Template	BET (m^2/g)	Pore Vol. (cm^3/g)	Pore Dia (nm)	Morphology
T_1	15	0.085	10.2	Spherical
T_2	28	0.14	10.8	Rod/Cylinder
T_3	105	0.37	13.4	Hollow Core/ mesoporous shell

The size of the synthesized mesoporous TiO_2 was investigated by dynamic light scattering (DLS) measurements. The samples of mesoporous TiO_2 calcined at 600 °C for 2 hours were dispersed in THF using sonication. Figure 5 shows particle size evolution of these particles with respect to the chain length of PAA in amphiphilic copolymer of PMMA-b-PAA. As the chains of PAA are increasing, while other reaction conditions remained unchanged, the particle size of mesoporous TiO_2 was increased.

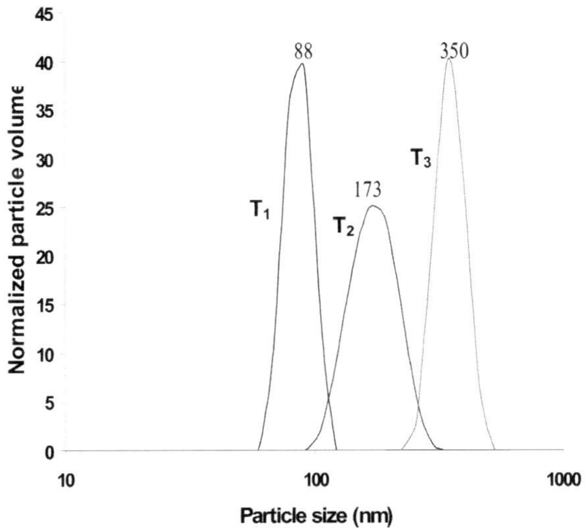

Figure 5. Particle size distributions of the mesoporousTiO_2 using different templates

SEM was used to study the morphology of mesoporous TiO_2 after calcination at 600°C. Samples with various morphologies were obtained when the chain length of PAA in PMMA-b-PAA was increased. As observed in these micrographs, the short PAA in diblock copolymer template (T_1) produced the agglomerated spherical TiO_2 particles. (Figure 6a-b) With increasing the chain length of PAA (T_2), the morphology of TiO_2 has been changed from spherical to rod/cylinder with diameter and length ranging from 40-300 nm and 200-4000 nm, respectively (Figure 6c-d). It can be attributed to the morphology of the block copolymer template, cross-linked pearl necklace. Further increasing the PAA length (T_3), the TiO_2 particles assembles into hollow core, having the wall thickness ~ 110 nm. (Figure 6e-f)

Figure 6. SEM images of mesoporous TiO_2 using T_1 (a-b), T_2 (c-d), and T_3 (e-f) as templates

Figure 7 shows the TEM image of mesoporous TiO_2 calcined at 600 °C for 2 hours. Using the short chain length of PAA (T_1) (Figure 7a-b) both the crystalline and amorphous phases can be observed. Selected area electron diffraction patterns recorded on mesoporous TiO_2 show the walls of these materials are comprised of nano crystalline oxides with characteristic diffuse electron diffraction rings (Figure 7a inset), which is consistent with the XRD measurements (Figure 5). Upon increasing the PAA chain length resulting in changing the T_2 template morphology, the rod/cylinder mesostructure TiO_2 with short-range order can be observed. (Figure 7c-d) The TEM image in Figure 7e-f indicates that the morphology of synthesized TiO_2 after increasing the chain length of PAA in diblock copolymer template (T_3) was changed to hollow core/mesoporous shell. However, because of the thickness of the shell wall ca. ~110 nm and the agglomeration of TiO_2 particles, electron beams can't pass the shell which results the dark image of core. The mosoporosity of the shell can be observed from Figure 7e.

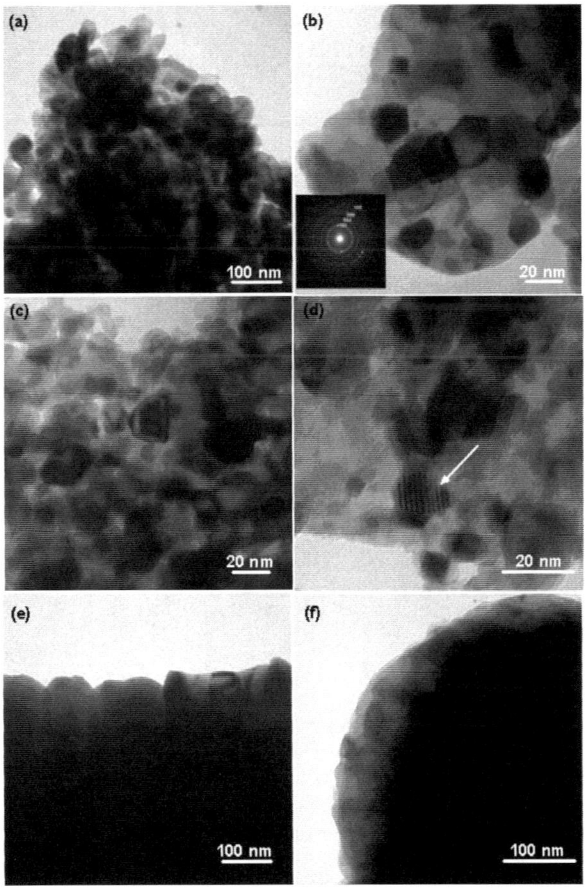

Figure 7. TEM images of mesoporous TiO_2 using (a-b) T_1, (c-d) T_2, and (e-f) T_3 as templates

Effect of Different Catalyst Supports On the (n,m) Selective Growth of Single-Walled Carbon Nanotube From Co-Mo Catalyst

Yuan Chen, School of Chemical and Biomedical Engineering, Nanyang Technological University, Singapore, Singapore

Bo Wang, School of Chemical and Biomedical Engineering, Nanyang Technological University, Singapore, Singapore

Yanhui Yang, School of Chemical and Biomedical Engineering, Nanyang Technological University, Singapore, Singapore

Co-Mo catalysts supported on four different high surface area oxides (SiO2, Al2O3, MgO, and TiO2) were evaluated to investigate the (n,m) selectivity control in single-walled carbon nanotube (SWCNT) synthesis. Results from Raman spectroscopy and thermogravimetric analysis showed that Co-Mo catalysts supported on SiO2 and MgO possessed good selectivity toward SWCNTs, while photoluminescence and ultraviolet-visible-near infrared spectroscopy results indicated that these two catalyst supports induced the same (n,m) selectivity to near-armchair tubes, such as (6,5) and (7,5) tubes. Catalysts supported on TiO2 produced a mixture of multi-walled carbon nanotubes (MWCNTs) and SWCNTs, whereas catalysts supported on Al2O3 mainly grew MWCNTs. Characterization of catalysts by ultraviolet-visible diffuse reflectance spectroscopy suggested that the surface morphology of metal clusters over different supports was not directly responsible for the (n,m) selectivity. Analysis of monometallic (Co or Mo) and bimetallic (Co-Mo) catalysts using temperature program reduction demonstrated that catalyst supports changed the reducibility of metal species. The interaction between supports and Co/Mo metals perturbed the synergistic effect between Co and Mo, leading to the formation of different metal species that are responsible for the observed distinction in SWCNT synthesis.

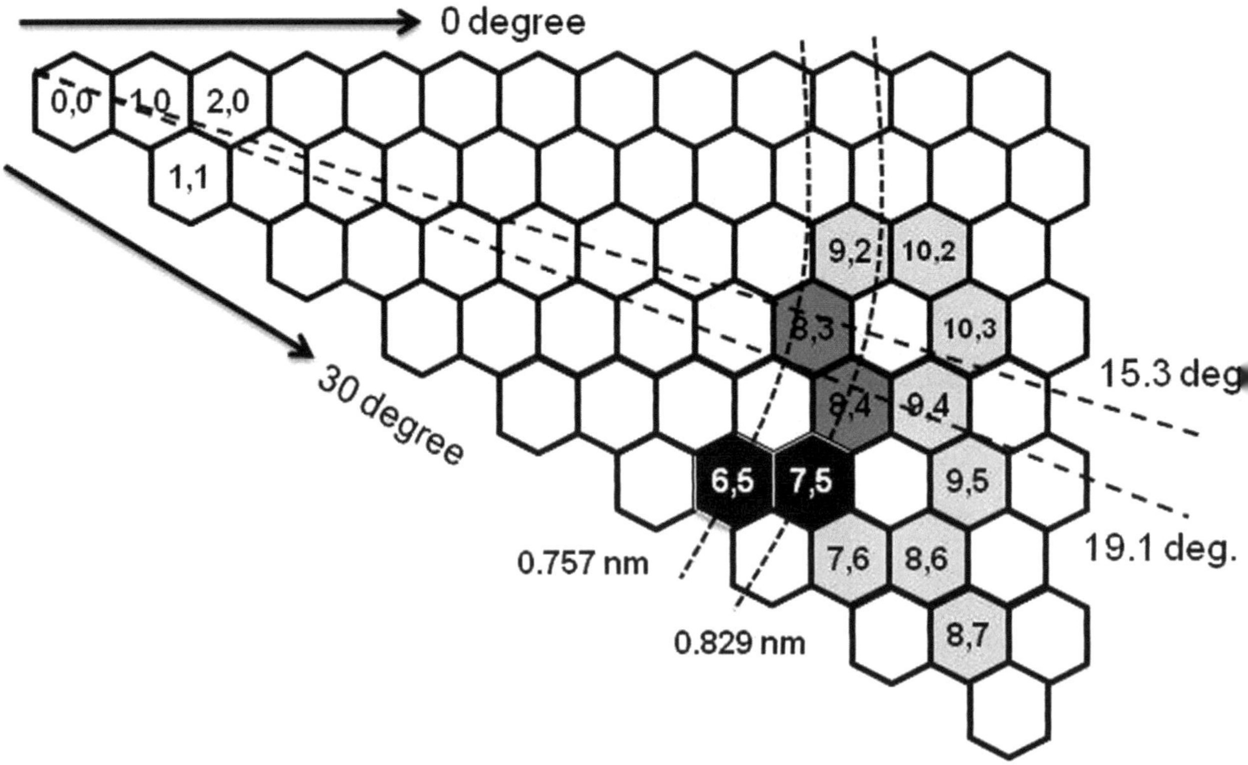

See more of this Session: Carbon Nanotubes I: Synthesis
See more of this Group/Topical: Nanoscale Science and Engineering Forum

Sol-Gel Synthesis and Characterization of Co-Mo/Silica Catalysts for Single-Walled Carbon Nanotube Production

Veronica M. Irurzun, M. Pilar Ruiz, Yongqiang Tan, and Daniel E. Resasco*
School of Chemical Engineering and Materials Science, University of Oklahoma, Norman, Oklahoma 73019 (USA)
*resasco@ou.edu

Introduction

Previous studies by our group [1-3] have demonstrated that Co-Mo/silica catalysts exhibit high selectivity to the growth of single-walled carbon nanotubes (SWNT) by CO disproportionation. The effect of parameters such as the extent of interaction of different Co-Mo species with the support, metal loading, Co:Mo molar ratio, and reaction conditions on the yield and structural properties of SWNT were analyzed. In this context, the present study focuses on the potential effects in the Co:Mo interaction using different supports.

The sol-gel method has been extensively used in the preparation of supported metal catalysts because it typically results in homogeneous materials with a high metal dispersion [1,2]. In particular, this method has been used to obtain active phases of Ni-Mo and Co-Mo [3] bimetallic catalysts supported on silica for the oxidative dehydrogenation of alkanes. It has been observed that catalysts prepared by the sol-gel method can stabilize the bimetallic active phases much more efficiently than those prepared by other methods. Therefore, sol-gel appears as a suitable method to produce Co-Mo/SiO_2 catalysts, which are highly selective towards for single-walled nanotubes (SWNT) when the gaseous feed is CO [4]. In this study, a series of silica-supported Co-Mo catalysts prepared by the sol-gel method has been prepared, varying the ratio of the added ammonium hydroxide to silica precursor (NH_4/TEOS). The concentration of ammonium hydroxide has an important effect on the rate of silica polymerization [5] and consequently, it is expected that changes in the NH_4/TEOS ratio will affect the resulting morphology and, consequently, the yield and structure of the SWNT that each catalyst produces. The carbon yield and structural characteristics of the nanotubes that result by varying the NH_4/TEOS ratio in the sol-gel process have been compared with the intention of drawing relationships between the catalyst structure and the catalytic performance.

As a comparison, a number of silica supports synthesized by different methods (precipitation, flame hydrolysis, etc.) with varying surface properties have been used to prepare catalysts by the incipient wetness impregnation. It is believed that silica supports with different surface properties (surface area, point of zero charge, residual impurities, particle size, crystallinity, etc.) may have effects on the yield and quality of the nanotube product. The carbon yield and structural characteristics of the resulting nanotubes from using the various silica supports have been compared with the aim of obtaining structure/properties relationships for this interesting application of heterogeneous catalysts.

Materials and Methods

1. Sol gel catalysts

In all the Co-Mo/SiO_2 catalysts prepared, the metal loading and Co/Mo molar ratio were kept constant at 9 wt% and 1:3, respectively. An aqueous solution of $Co(NO_3)_2.6H_2O$ was combined with an isopropanol solution of $MoCl_5$; to this mixture, tetraethyl orthosilicate (TEOS) and ammonium hydroxide (28 %) solutions were added in different ratios. After aging for 20 hr, the gels were filtered and dried in air at 200 °C for 30 min, and further calcined at 650 °C for 150 min. Before the nanotube synthesis, the catalysts were pre-reduced at 500 °C for 30 min under H_2 (125 sccm) in a ½ inch vertical quartz reactor. When the reaction temperature (750 °C) was reached under He flow, the CO feed (250 sccm) started. After a growth period of 30 min, the system was cooled down in He flow. The catalysts and products were analyzed by Raman scattering, optical absorption, X-ray diffraction, temperature programmed oxidation and reduction, thermo gravimetric analysis, and transmission electron microscopy.

2. Impregnated catalysts

Bimetallic Co-Mo catalysts supported on different silicas were prepared by incipient wetness co-impregnation of aqueous $(NH_4)_6Mo_7O_{24}.4H_2O$ and $Co(NO_3)_2.6H_2O$ solutions, keeping a Co:Mo molar ratio of 1:3. Before the SWNT synthesis, the catalysts were calcined in air and pre-reduced in H_2. When the reaction temperature (750 °C) was reached under He flow, the CO feed (250 sccm) was started. After a growth period of 30 min, the system was cooled down in He. The catalysts and products were analyzed by Raman spectroscopy, optical absorption, TEM, TGA, TPO and X-Ray diffraction.

Results and Discussion

Table 1 summarizes the yield and SWNT selectivity data obtained over the different Co-Mo catalysts synthesized by varying the NH_4/TEOS ratio. As previously demonstrated the selectivity towards SWNT can be semi-quantitatively estimated by the intensity ratio of the D and G bands in the Raman spectrum, the former due to disordered (sp^3) carbon and the latter due to ordered graphene (sp^2) in the SWNT structure. It is clear that the NH_4/TEOS ratio used in the preparation has an optimum at intermediate values. Two factors seem to affect the quality and yield of SWNT, Co-Mo particle size and fraction of metal exposed (Figure 1). At high NH_4/TEOS ratios, the metal particle size is large and the fraction of metal exposed is low. On the other hand, at low NH_4/TEOS ratios, the metal particle size is large but the fraction of metal exposed is high.

Table 1. Carbon yield (%) and quality (D/G)

NH_4/TEOS ratio	**1/6**	**2/6**	**4/6**	**6/6**
Carbon yield (%)	4.2	13.5	9.6	3.4
D/G	0.081	0.046	0.049	0.068

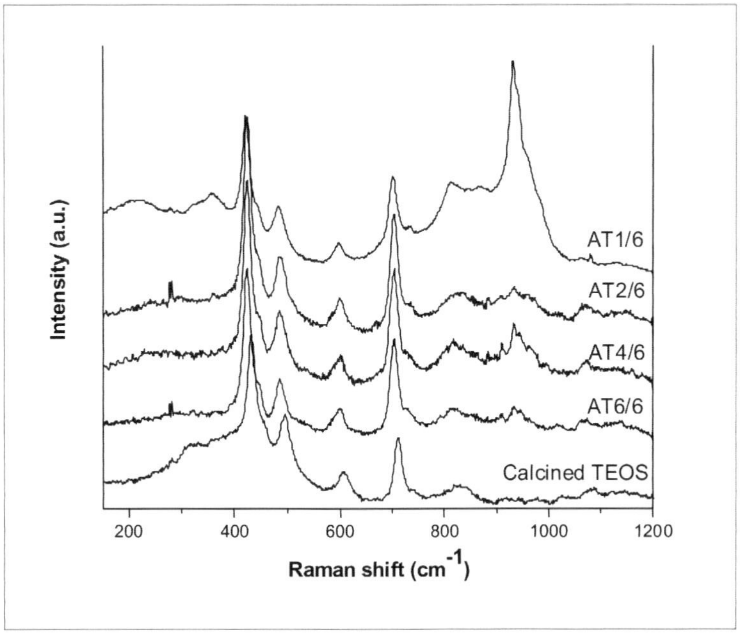

Figure 1. Raman scattering of the Co-Mo calcined catalysts with different NH_4/TEOS ratios (AT). The spectrum of the calcined support is also included (TEOS).

In addition to Raman, the four samples have been investigated with XPS, UV-Vis, TEM, and XRD to identify the types of $CoMoO_x$ species stabilized in each case, and find a relationship between the different species that are present after the calcination step and the resulting SWNT product that is obtained (i.e. D/G ratio, and (n,m) nanotube type – chirality distribution).

As illustrated in Table 2, significant variations in carbon yield and SWNT selectivity are also observed when different silica supports are used. The carbon yield is measured by TPO. The SWNT selectivity is assessed from Raman and TGA analysis. Among the four silicas included in Table 1, the precipitated HiSil 210 results in the highest yield and highest selectivity in the product (i.e. lowest D/G Raman ratio, whereas the D band corresponds to disordered carbon (sp^3) and the G band to ordered graphene (sp^2)). It is apparent that the nature of the silica support has an effect on the type and domain size of the Co-Mo species stabilized during the calcination step. Depending on the nature and state of dispersion of these species, the carbon yield and SWNT selectivity vary widely. Interestingly, while UV-Vis spectroscopy (see Fig.1, right) of the calcined catalysts indicate that all samples have at least a fraction of Co in the octahedral configuration, which is typical of Co molybdate, almost equally good samples (e.g. HS210 and HS915) exhibit very different fractions of Co in octahedral and tetrahedral configurations. On the other hand, Raman analyses (see Fig. 1, left) of the calcined samples suggest that the domain size of the molybdate is the parameter that determines the performance of the catalyst. For example, absence of bands in the 930-970 cm^{-1} region (e.g. HS190), indicative of a very high dispersion results in low SWNT yield. At the other extreme, large molybdate crystallites are not desirable either. Samples such as that supported on silica gel displayed high intensities in this range and high intensity ratio between the 948 cm^{-1} and 958 cm^{-1} bands ($CoMoO_4$ to Mo oxides) [8], indicating the presence of large $CoMoO_4$ crystallites. Such samples have low activity and low selectivity, consistent with the poor performance of unsupported $CoMoO_4$.

Table 2. Carbon yield and SWNT selectivity of the nanotube product

Catalyst	Silica gel	Hisil 190	Hisil 210	Hisil 915
Carbon yield (%)	1.21	0.78	3.33	1.17
D/G Raman ratio	0.225	0.100	0.040	0.100

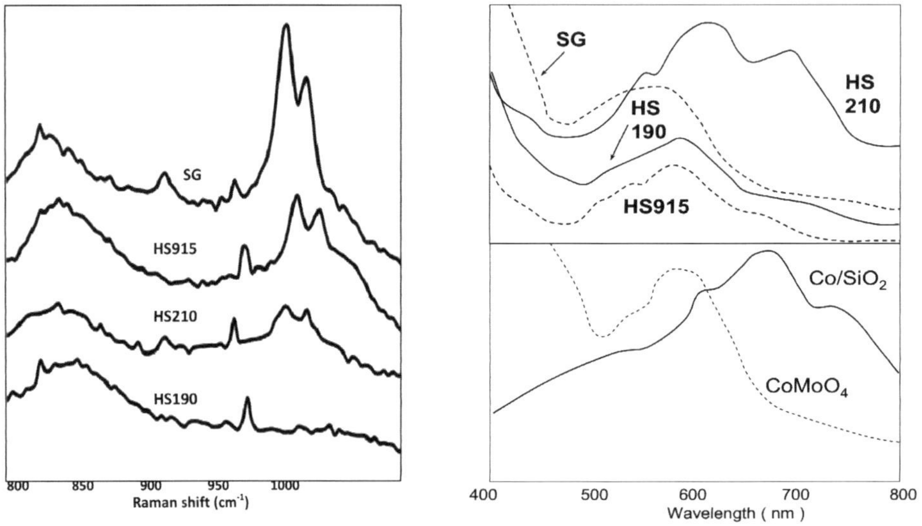

Figure 2. Raman (left) and UV-VIS (right) spectra of calcined Co-Mo catalysts supported on different silicas, as well as two reference compounds, monometallic Co/SiO_2 and $CoMoO_4$.

References

1. Meille, V., *Appl Catal A*, 315, 1 (2006).
2. Samanta, S., Laha, S.C., Mal, N.K., and Bhaumik, K. *J. Mol. Catal. A*. 222, 235 (2004).
3. Maione, A., and Devillers, M., *J. Sol. State Chem*. 177, 2339 (2004).
4. Herrera, J.E., and Resasco, D.E., *J. Catal*. 221, 354 (2004).

Brinker, C.J., and Sherer, G.W. in "Sol-Gel science: The physics and chemistry of sol-gel processing" (Academic press Ed.), 1989.

5. Alvarez, W.E, Kitiyanan, B., Borgna, A., and Resasco, D.E., *Carbon* 39, 547 (2001)
6. Herrera, J.E., Balzano, L., Borgna, A., Alvarez, W.E, and Resasco, D.E., *J. Catal* 204, 129 (2001)
7. Walter, W.E., Pompeo, F., Herrera, J.E., Balzano, L., and Resasco, D.E., *Chem. Mater.* 14, 1853 (2002)
8. Maione A. , and Devillers M. *J. Sol. St. Chem*. 177, 2339 (2004)

Intrinsic and Extrinsic Factors Which Affects the Optical Properties of Individual Single-Walled Carbon Nanotubes

Juan G. Duque, Department of Chemical and Biomolecular Engineering, Rice University, Houston, TX

Stephen Doorn, Physical Chemistry and Applied Spectroscopy, Los Alamos National Laboratory, Los Alamos, NM

Matteo Pasquali, Department of Chemical and Biomolecular Engineering, Rice University, Houston, TX

Multiple studies have shown that the optical properties of single-walled carbon nanotubes (SWNTs) are excitonic in nature.(1-3) However, a complete understanding of how these excitonic states shape the optical properties of SWNTs is lacking. Comparing the results of different published studies is difficult; the luminescence properties of SWNTs are affected by various extrinsic factors originating from synthesis and processing methods which affect the structure, sidewall quality, and environment of SWNTs. Fundamental and applied reports are hindered by poor control over the synthesis and post-processing methods. This is highlighted by the range of reported fluorescence quantum yields and luminescence lifetimes measured from ensembles or single SWNTs - they range between 10-4 to 7 % (4-8) and from 5 to 200 ps,(5,8-10) respectively, depending on the report.

Here we use single molecule imaging and time resolved spectroscopy, photoluminescence maps, UV-vis absorbance, and Raman spectroscopy at multiple excitation wavelengths to identify different extrinsic factors which deeply influence the optical properties of SWNT.11,12 Single molecular studies of individual (6,5) SWNTs showed that SWNT luminescence depends strongly on extrinsic and intrinsic factors such as sample preparation, sample environment, sidewall defects, and tube synthesis methods (HiPco or CoMoCat).(11) More precisely, we showed that surfactant selection and sample medium are crucial factors to obtaining SWNTs with optimum optical properties.

Ensemble measurements showed compelling spectroscopic evidence of substantial differences in chirality and diameter distribution which alter the luminescence properties within different HiPco batches.(12) These variations also affect SWNT solubility both in surfactant suspensions and superacids.

These findings underline the necessity for more standard production and post-processing procedures such that correlations between different reports in the literature can be drawn, and we begin to understand the complex excitonic structure of SWNTs.

References

1. Wang, F.; et. al Science 2005, 308, (5723), 838.

2. Perebeinos, et. al. PRL 2004, 92, (25 I), 257402.

3. Shaver, J.; et.al. Nano Lett. 2007, 7, (7), 1851.

4. O'Connell, M. J.; et.al. Science 2002, 297, (5581), 593.

5. Wang, F.; et.al. Nano Lett. 2007, 7, (12), 3698-3703.

7. Tsyboulski, D. A.; et.al. Nano Lett. 2007, 7, (10), 3080.

8. Berciaud, S.; et.al. PRL 2008, 101, (7), 077402.

9. Jones, M.; et.al. Phys. Rev. B 2005, 71, 115426.

10. Hirori, H.; et.al. PRL 2006, 97, (25), 257401.

11. Duque, J. G.; et.al. Nano Lett.; Submitted.

12. Duque, J. G.; et.al. In Preparation.

See more of this Session: Carbon Nanotubes I: Synthesis
See more of this Group/Topical: Nanoscale Science and Engineering Forum

Production of MWCNT by a FBCCVD (Fluidized Bed Catalyzed Chemical Vapor Deposition) Process

Alina Carmen Tito, Chimica, Materiali e Ingegneria Chimica "G.Natta", Politecnico di Milano, Milano, Italy

Carlo Vittorio Mazzocchia, Politecnico di Milano, Milano, Italy

Pierrot nunga Shambosenge, Politecnico di Milano, Milano, Italy

Massimiliano Bestetti, Politecnico di Milano, Milano, Italy

Among the different techniques that have been applied for the processing of multiwall carbon nanotubes (MWCNT), catalytic chemical vapor deposition (CCVD) in a fluidized bed reactor appears promising in principle due to its relatively low cost, the possibility of its working in a continuous way and the potential high yield production. This technique also assures a good homogeneity of the carbonaceous deposit on the catalyst surface, a good contact between the solid phase (supported catalyst) and the gaseous one (the source of carbon), a good thermal homogeneity into the reactor and it allows a quick mass and heat transfer, avoiding a heat gradient along the bed and a reactant concentration gradient. Multi walled carbon nanotubes have been produced from H2-C2H4 mixtures on Fe-Al2O3 catalyst by a fluidized bed catalytic chemical vapour deposition process. Various parameters such as the temperature, the H2:C2H4 ratio, the amount of metal deposited on the support, the amount of the used catalyst, the reaction time, the typology of catalyst and the reproducibility of the process have been examined. The influence of these parameters on the deposited carbon yield is reported, together with observations of the produced material. The nature of the deposited carbon and the quality of the MWCNT obtained were investigated by SEM, TEM, TG-DTA and XRD analyses. The evolution of the catalyst in time and along the reactor was monitored by SEM and ICP analyses.

See more of this Session: Carbon Nanotubes I: Synthesis
See more of this Group/Topical: Nanoscale Science and Engineering Forum

Vapor-Phase Functionalization of Vertically Aligned Carbon Nanotubes

Yu Mao, Biosystems Engineering, Oklahoma State University, Stillwater, OK

Yumin Ye, Biosystems Engineering, Oklahoma State University, Stillwater, OK

Vertically aligned carbon nanotubes (CNTs) have important applications in biosensing owing to their excellent electronic properties and pre-defined ordered structure. Current research mainly focuses on functionalization of non-aligned nanotubes using solution-based protocols. Functionalization of aligned CNTs is still unexplored due to the difficulty of preserving the desirable alignment. We report a vapor-phase method to functionalize aligned CNTs with the nanotube alignment retained. The process involves flowing of functional monomer vapors around nanotubes and reaction of the monomer molecules to deposit ultra-thin nanocoatings around each individual nanotube. We demonstrated the functionalization of both multi-walled and singled-walled aligned carbon nanotubes using this vapor deposition method. A variety of surface chemistry, including hydrogel and epoxy chemistry, can be incorporated onto the nanotubes. We also demonstrated the control to covalently and non-covalently attach the functional layers to the nanotubes. Enzyme activity on the nanotubes is significantly improved as a result of the vapor-phase functionalization.

See more of this Session: Carbon Nanotubes I: Synthesis
See more of this Group/Topical: Nanoscale Science and Engineering Forum

Conical Carbon Nanotube Arrays: Large Area Synthesis, Field Emission Characteristics

Santosh Rupa Dumpala, Chemical Engineering, University of Louisville, Louisville, KY

Abdelilah Safir, ElectroOptics Research Institute and Nanotechnology Center, University of Louisville, Louisville, KY

David Mudd, Physics, University of louisville, Louisville, KY

Robert W. Cohn, University of Louisville, Louisville, KY

Gamini U. Sumanasekera, Physics, University of Louisville, Louisville, KY

Mahendra K. Sunkara, Department of Chemical Engineering, University of Louisville, Louisville, KY

ABSTRACT

Conical carbon nanotube (CCNT) is a new morphological manifestation of carbon nanaotube with a unique structure, tapering from micron-sized base to nanometer scale tips with central hallow core consisting of a multiwalled carbon nanotube.1. The surfaces of these conical structures are composed of high density of graphene edge planes.2 Also, due to the conical geometry and high aspect ratio, the arrays of the conical carbon nanotubes contain their tips spaced from each other. So, these conical carbon tube arrays are expected to yield dramatically different behavior in terms of both electrochemical and field emission characteristics. However, for all the applications, large area (> 1 cm2 area) synthesis of these arrays on flat substrates has been challenging.

Here, we report our recent success with synthesizing these conical carbon tubular arrays over graphite and other metallic foils using microwave plasma enhanced chemical vapor deposition (MWCVD) reactor. Cylindrically rolled foils immersed vertically into the plasma, resulted in conical carbon tubes up to as long as 30 microns. The field emission characteristics for a CCNT array sample with a tip radius of 5 nm, density of 108/cm2 and having the highest aspect ratio exhibited a low turn-on electric field (< 0.7 V/μm) and a high field enhancement factor (β > 7,500). Other samples with lower emission characteristics were attributed either to the presence of field screening effect resulting from higher CCNT density or due to the corresponding tip and wall characteristics.

References:

1. R. C. Mani, X. Li, M. K. Sunkara, K. Rajan, Nano Lett. 2003, 3, 671

2. B. Chernomordik, S. Dumpala, Z. Chen and M.K. Sunkara, Chem. Vapor Depos. 2008, 14, 1-7.

See more of this Session: Carbon Nanotubes I: Synthesis
See more of this Group/Topical: Nanoscale Science and Engineering Forum

High-Throughput, Ultrafast Synthesis of Solution Dispersed High-Quality Graphene Via a Novel Hydride Chemistry: Raman Spectra and Electrical Gating

Nihar Mohanty, Chemical Engineering, Kansas State University, Manhattan, KS

Ashvin Nagaraja, Electrical Engineering, Kansas State University, Manhattan, KS

Jose Armesto, Chemical Engineering, Kansas State University, Manhattan, KS

Vikas Berry, Chemical Engineering, Kansas State University, Manhattan, KS

In this talk, we will present a study of a novel process for synthesis of organic-phase stabilized graphene by instantaneous reduction of graphene-oxide (GO) in hydrides at high-throughput. Synthesis of such solution-dispersed graphene is important for large-scale production of graphene and for its applications in building graphene-films, FETs, etc. The current dispersion-production-processes employ toxic chemicals, are low in yield or take several days to complete. The route presented in this talk is ultrafast (few seconds), uses commonly available, low-toxicity chemicals, has a high yield of ~ 68 % and shows a significant reduction in the sp3 bonds as confirmed by Raman spectra. We will also demonstrate that thus-produced single graphene sheets exhibit high carrier mobilities of 50-200 cm2/V/s. The high-throughput processing time-scale, high stability of graphene dispersion, benignity of the reagents and low-cost will enable expedited research and incorporation of graphene into next-generation graphene technologies & applications.

See more of this Session: Carbon Nanotubes I: Synthesis
See more of this Group/Topical: Nanoscale Science and Engineering Forum

Synthesis, Colloidal Stability, and Magnetic Properties of Nanoparticles for Magnetic Fluid Hyperthermia

Adriana P. Herrera, Chemical Engineering, University of Puerto Rico, Mayaguez, PR

Carola Barrera, Chemical Engineering, University of Puerto Rico, Mayaguez, PR

Vanessa Ayala, Chemical Engineering, University of Puerto Rico, Mayaguez, PR

Hector L. Rodriguez, Chemical Engineering, University of Puerto Rico, Mayaguez, PR

Magda Latorre, Chemical Engineering, University of Puerto Rico, Mayaguez, PR

Madeline Torres-Lugo, Chemical Engineering, University of Puerto Rico, Mayaguez, PR

Carlos Rinaldi, Chemical Engineering, University of Puerto Rico, Mayaguez, PR

In vitro and in vivo experiments in which cells are in contact with magnetic nanoparticles and subjected to high frequency oscillating magnetic fields have shown that the particles may induce cell death. Owing to the biocompatibility of superparamagnetic iron oxide nanoparticles, these observations have garnered considerable interest for the treatment of cancer, in which the nanoparticles could be directly injected into a tumor or functionalized to selectively target cancer cells. The subsequent application of an oscillating magnetic field would result in destruction of the cancer, without many of the deleterious side effects common to radio- and chemotherapy. These observations have been termed Magnetic Fluid Hyperthermia (MFH). This talk will discuss criteria for designing nanoparticles for use in cancer treatment, the mechanisms and limitations of magnetic fluid hyperthermia, research on development and characterization of magnetic nanoparticles coated with biocompatible and thermoresponsive polymers, and biological assays being used to quantify the efficacy of the treatment and determine the mechanisms through which cell death occurs.

See more of this Session: Magnetic Nanoparticles in Biotechnology and Biomedicine I
See more of this Group/Topical: Nanoscale Science and Engineering Forum

The Design of Well-Defined Polymer-Magnetite Complexes for Biomedical Applications

William C. Miles, Chemical Engineering, Virginia Tech, Blacksburg, VA

P.P. Huffstetler, Macromolecular Science and Engineering, Virginia Tech, Blacksburg, VA

J.D. Goff, Macromolecular Science and Engineering, Virginia Tech, Blacksburg, VA

C.M. Reinholz, Macromolecular Science and Engineering, Virginia Tech, Blacksburg, VA

J.S. Riffle, Macromolecular Science and Engineering, Virginia Tech, Blacksburg, VA

R.M. Davis, Chemical Engineering, Virginia Tech, Blacksburg, VA

Magnetite nanoparticles have many uses, in particular as contrast agents for magnetic resonance imaging (MRI) and for incorporation into magnetic ferrofluids.[1,2] Magnetite particles have relatively low cytotoxicities, and their surfaces are amenable to the adsorption of biocompatible macromolecules.[3] Thus, specific functionalities can be attached to magnetite nanoparticles in order to optimize a specific property for appropriate applications.[4] Rapid improvement in biomaterials would be greatly enhanced by starting with a well-defined system with a controlled, narrow particle size distribution rather than using systems where aggregation is not controlled well and the particle size distribution is poorly defined, as is the case with some commercial products. Thus, the goal of our work has been to develop a model to distinguish between slight aggregates and single polymer-particle systems and to allow for design of complexes with controlled size and colloidal stability.

In this work, we examined two distinctly different polymer-magnetite systems. The first consisted of poly(dimethylsiloxane) (PDMS) stabilized magnetite nanoparticles. The magnetite was synthesized using two different methods. The first method was the coprecipitation of iron chlorides with ammonium hydroxide. Through adaptation of a density distribution model and comparison to dynamic light scattering (DLS) results, it was determined that there were slight aggregates formed from this method. Following extensive magnetic separation, a single polymer-particle system could be obtained. However, approximately 80 percent of the material must be separated to achieve this well-defined system, and so a better synthetic route was needed. Thus, a second magnetite synthesis was adapted from Pinna et al.[5], which consists of the high temperature decomposition of iron acetylacetonate at high temperature in benzyl alcohol. Again, through comparison of the model to DLS, it was determined that the complexes were not aggregated. The model matched the experimentally obtained numbers to approximately seven percent, which is in line with experimental error.

The second polymer-magnetite system utilized the high temperature magnetite synthesis, but used a polyether as the polymer stabilizer. The polyether stabilizers used were poly(ethylene oxide) (PEO) homopolymers and amphiphilic poly(propylene oxide)-b-poly(ethylene oxide) (PPO-b-PEO) block copolymers. The same methodology was used to model the complexes as was used for the PDMS system. Again, the model matched within eight percent of experimental measurements obtained from DLS, indicating a well-defined particle size distribution. The extended Derjaguin-Landau-Verwey-Overbeek (DLVO) theory was used to predict the colloidal stability of these complexes. The model was able to predict when the onset of aggregation would occur, thus allowing us to determine when a single-particle system began to cluster.

Thus, we now have a tested tool for designing the size of polymer-magnetite complexes to within eight percent of a desired value. This tool also allows for the determination of the onset of aggregation, allowing the optimization of the polymer loading. Finally, because there are no adjustable parameters in the model, this method should be applicable to various polymer-particle systems, and allow for the rapid development of well-defined systems for biomedical applications.

1. Mefford, et al. Chemistry of Materials, 2008, 20(6), 2184-2191.

2. Huffstetler, P. P, et al. Polymer Preprint, 2008, 49(2), 1103-1104.

3. Harris, L., et al. Chemistry of Materials 2003, 15(6), 1367-1377.

4. Miles, W.C., et al. Langmuir 2009, 25(2), 803-813.

5. Pinna, N., et al. Chemistry of Materials 2005, 17, 3044-3049.

See more of this Session: Magnetic Nanoparticles in Biotechnology and Biomedicine I
See more of this Group/Topical: Nanoscale Science and Engineering Forum

Formation and Characterization of Bilayer-Decorated Magnetoliposomes

Yanjing Chen, Chemical Engineering, University of Rhode Island, Kingston, RI

Geoffrey D. Bothun, Chemical Engineering, University of Rhode Island, Kingston, RI

Bilayer-decorated magnetoliposomes have been designed with hydrophobic superparamagnetic iron oxide (SPIOs) nanoparticles using reverse phase evaporation. The ability to embed the nanoparticles (5 nm, oleic acid surface coating) within dipalmitoylphosphatidylcholine (DPPC) bilayers was examined as a function of the lipid to nanoparticle ratio. This ratio varied up to 5000:1, which corresponded to 5000 lipid molecules per nanoparticle. Magnetoliposome structure was observed by cryogenic transmission electron microscopy (cryo-TEM), and the presence of SPIO nanoparticles confirmed by energy dispersive spectroscopy (EDS). Under cryogenic preservation, electromagnetic radiation heated the embedded nanoparticles and holes were burnt within the lipid bilayers, further verifying decoration. Lipid phase behavior was examined by differential scanning microcalorimetry (Nano DSC) and showed that the nanoparticles eliminated the gel to rippled gel pretransition and increased and broadened the rippled gel to fluid melting temperature. Carboxyfluorescein leakage experiments were used to demonstrate the ability to control magnetoliposome bilayer permeability using external radio frequency (RF) inductive heating. Our results show that SPIO nanoparticles can be bound to lipid bilayers and externally heated to induce and control leakage of encapsulated agents by adjusting the lipid to nanoparticle ratio. Results will also be presented on the design of cationic magnetoliposomes selectively decorated with anionic SPIO nanoparticles bound at the lipid headgroup/water interface.

See more of this Session: Magnetic Nanoparticles in Biotechnology and Biomedicine I
See more of this Group/Topical: Nanoscale Science and Engineering Forum

Synthesis of Nanoparticles of Magnetite and Drug-Polymer Complexes Using a Multi-Inlet Vortex Mixer

Raquel Mejia, Macromolecular Science & Engineering Program, Virginia Tech, Blacksburg, VA

Oguzhan Celebi, Macromolecular Science & Engineering Program, Virginia Tech, Blacksburg, VA

Nikorn Pothayee, Macromolecular Science & Engineering Program, Virginia Tech, Blacksburg, VA

William Miles, Chemical Engineering, Virginia Tech, Blacksburg, VA

Judy Riffle, Chemistry, Virginia Tech, Blacksburg, VA

Richey M. Davis, Chemical Engineering, Virginia Tech, Blacksburg, VA

Functional nanoparticles that have well-defined size distributions and are colloidally stable in aqueous media are important for biomedical applications. Clusters of superparamagnetic particles are particularly interesting as magnetic resonance imaging (MRI) contrast agents. For drug nanoparticles, control of the particle size distribution is very important in biological systems for cellular uptake and targeted delivery. Thus, development of a method to control the size of sterically stabilized nanoparticles is needed. We have used a multi-inlet vortex mixer (MIVM)[1,2] to create clusters of superparamagnetic particles of magnetite and to create drug-polymer complexes. In the MIVM, particle formation occurs by the rapid mixing of an organic active and a stabilizing amphiphilic polymer, dissolved in an organic solvent, with an antisolvent to create solutions with high supersaturation values. To produce nanoparticles of homogeneous particle size, the mixing time has to be lower than the nucleation and growth time scale.

The MIVM process was used to make clusters comprised of hydrophobically modified magnetite nanoparticles (diameter ~10 nm) dispersed in Tetrahydrofuran (THF) stabilized with an amphiphilic polymer, Pluronics F-127 (a triblock of Poly (ethylene oxide-b-propylene oxide-b-ethylene oxide)). Clusters of these magnetite nanoparticles were formed with hydrodynamic diameters ranging from 60-230 nm in aqueous media. The control of the composition of these clusters and the response of the cluster colloidal stability to an external magnetic field will be discussed.

The MIVM process was also used to encapsulate the antibiotic doxycycline using an amphiphilic polymer designed to achieve intracellular delivery of the antibiotic and to increase the drug loading content. Doxycycline is water-soluble and so it was complexed first with cholic acid to form the water-insoluble doxycycline-cholate salt that was soluble in THF. Particles of this salt were precipitated in water and encapsulated with a pentablock polymer (Poly(caprolactone)-poly(ethylene oxide)-poly(propylene oxide)-poly(ethylene oxide)-poly(caprolactone)). After the nanoparticles were formed, they were dialyzed to remove organic solvent and unimer chains and were then characterized by dynamic light scattering and electron microscopy. Nanoparticles with number average hydrodynamic diameters in the range 50-200 nm were formed. The control of drug loading in these particles will be discussed along with cell uptake studies and the pharmacokinetics of antibiotic release.

References:

1.Liu, Y; Cheng, C; Liu, Y; Prud'homme, RK; Fox, RO. Mixing in a multi-inlet vortex mixer (MIVM) for flash nano-precipitation. Chemical Engineering Science 2008, 63, 2829-2842.

2.Johnson, BK; Prud'homme, RK. Mechanism for rapid self-assembly of block copolymer nanoparticles. Phys. Rev. Lett. 2003, 91, 118302:1-4.

See more of this Session: Magnetic Nanoparticles in Biotechnology and Biomedicine I
See more of this Group/Topical: Nanoscale Science and Engineering Forum

Hermetically Coated Superparamagnetic Fe_2O_3 Particles with SiO_2 Nanofilms

Alexandra Teleki, Department of Mechanical and Process Engineering, Swiss Federal Institute of Technology Zurich, Zurich, Switzerland

Marcel Suter, Department of Mechanical and Process Engineering, Swiss Federal Institute of Technology Zurich, Micro- and Nanosystems, Zurich, Switzerland

Piran R. Kidambi, Department of Mechanical and Process Engineering, Swiss Federal Institute of Technology Zurich, Zurich, Switzerland

Olgac Ergeneman, Department of Mechanical and Process Engineering, ETH Zurich, Institute of Robotics and Intelligent Systems, Zurich, Switzerland

Frank Krumeich, Department of Mechanical and Process Engineering, Swiss Federal Institute of Technology Zurich, Zurich, Switzerland

Bradley J. Nelson, Department of Mechanical and Process Engineering, ETH Zurich, Institute of Robotics and Intelligent Systems, Zurich, Switzerland

Sotiris E. Pratsinis, Department of Mechanical and Process Engineering, Swiss Federal Institute of Technology Zurich, Zurich, Switzerland

Magnetic nanoparticles are frequently coated with SiO_2 to improve their functionality and bio-compatibility in a range of biomedical and polymer nanocomposite applications. Here, a scalable flame aerosol technology is used to produce highly dispersible, superparamagnetic iron oxide nanoparticles hermetically coated with silica to retain full magnetization performance.[1] Iron oxide particles were produced by flame spray pyrolysis (FSP) of iron acetylacetonate in xylene/acetonitrile solutions and the resulting aerosol was *in-situ* coated with silicon dioxide by oxidation of swirling hexamethlydisiloxane vapor.[2] The process allows independent control of the core Fe_2O_3 (maghemite) particle properties and the thickness of their silica coating film. This ensures that the non-magnetic SiO_2 layer can be closely controlled and minimized. The optimal SiO_2 content for complete (hermetic) encapsulation of the magnetic core particles was determined by isopropanol chemisorption.[3] The magnetization of Fe_2O_3 coated with about 2 nm thin SiO_2 layers was nearly identical to that of uncoated, pure Fe_2O_3 nanoparticles.

[1] Teleki, A., M. Suter, P. R. Kidambi, O. Ergeneman, F. Krumeich, B. J. Nelson, and S. E. Pratsinis, Hermetically-coated superparamagnetic Fe_2O_3 particles with SiO_2 nanofilms, *Chem. Mat.* **in press**, (2009).

[2] Teleki, A., M. C. Heine, F. Krumeich, M. K. Akhtar, and S. E. Pratsinis, In situ coating of flame-made TiO_2 particles with nanothin SiO_2 films, *Langmuir* **24**, 12553 (2008).

[3] Teleki, A., M. K. Akhtar, and S. E. Pratsinis, The quality of SiO_2 coatings on flame-made TiO_2-based nanoparticles, *J. Mater. Chem.* **18**, 3547 (2008).

See more of this Session: Magnetic Nanoparticles in Biotechnology and Biomedicine I
See more of this Group/Topical: Nanoscale Science and Engineering Forum

Enhanced Magnetic Properties in Functionalized Iron Oxide Nanoparticles for Biomedical Applications

Dattari Nagesha, Physics, Northeastern University, Boston, MA

Brian D. Plouffe, Chemical Engineering, Northeastern University, Boston, MA

Minh Phan, Chemical Engineering, Northeastern University, Boston, MA

Laura H. Lewis, Chemical Engineering, Northeastern University, Boston, MA

Srinivas Sridhar, Physics, Northeastern University, Boston, MA

Shashi K. Murthy, Chemical Engineering, Northeastern University, Boston, MA

Iron oxide nanoparticles are essential components of next-generation drug delivery systems, MRI contrast agents, and cell separation systems. All of these applications involve some form of surface functionalization of the nanoparticles to either render them hydrophilic or to give them targeting capabilities. The impact of such surface functionalization techniques on the magnetic properties of these nanoparticles is, however, not well-understood. This presentation will describe changes in magnetic properties in Fe3O4 nanoparticles originally synthesized with oleic acid acting as a hydrophobic surfactant when functionalized with dopamine. These changes include a significant increase in the remanence, saturation magnetization, and blocking temperature after dopamine functionalization, even though TEM and XRD studies revealed no change in the particles' size and/or structure. The results are consistent with an increase in the magnetic size of the nanoparticle core induced by the dopamine ligand exchange process. Potential mechanisms for this change will be discussed along with implications for the use of these nanoparticles in magnet-activated microfluidic cell separation.

See more of this Session: Magnetic Nanoparticles in Biotechnology and Biomedicine I
See more of this Group/Topical: Nanoscale Science and Engineering Forum

Comparative Analysis of Functionalizing Magnetite Nanoparticles with Hydrophilic Polymers Using ATRP

Reynolds A. Frimpong, Chemical and Materials Engineering, University of Kentucky, Lexington, KY

Adriana P. Herrera, Chemical Engineering, University of Puerto Rico, Mayaguez Campus, Mayaguez, PR

Carlos Rinaldi, Chemical Engineering, University of Puerto Rico, Mayaguez Campus, Mayaguez, PR

J. Zach Hilt, Chemical and Materials Engineering, University of Kentucky, Lexington, KY

Recent advances in colloidal science have led to various improved techniques of synthesizing magnetic nanoparticles. Thermal decomposition, for example, has been used lately to produce magnetic nanoparticles with very narrow size distribution in comparison with the co-precipitation method. The magnetic nanoparticles produced by the thermal decomposition method, however, are generally hydrophobic and require post processing to functionalize them with hydrophilic molecules particularly where biomedical applications are envisaged. Various post processing approaches have therefore been developed to exchange the hydrophobic stabilizers for hydrophilic ones.

In this work, a ligand exchange method has been used to attach initiation sites on the surface of magnetite nanoparticles. The initiator was subsequently used to functionalize the magnetite nanoparticles with different hydrophilic polymers using atom transfer radical polymerization (ATRP). The success of the ATRP was dependent on the density of the initiating sites available on the nanoparticle surface. Particularly, we compared the ATRP reaction for iron oxide particles synthesized by the thermal decomposition and co-precipitation methods. The surface properties of the nanoparticles were examined to provide an insight into the factors that promote better ligand exchange and ATRP reaction for the co-precipitation iron oxide nanoparticles.

See more of this Session: Magnetic Nanoparticles in Biotechnology and Biomedicine I
See more of this Group/Topical: Nanoscale Science and Engineering Forum

Effect of Core Material and Graft Molecular Weight On the Energy Dissipation Rate of Magnetic Nanoparticles in Oscillating Magnetic Fields

Adriana P. Herrera, Chemical Engineering, University of Puerto Rico, Mayaguez Campus, Mayaguez, PR

Carola Barrera, Chemical Engineering, University of Puerto Rico, Mayaguez Campus, Mayaguez, PR

Yashira Zayas, Chemical Engineering, University of Puerto Rico, Mayaguez Campus, Mayaguez, PR

Nayla Bezares, Chemical Engineering, University of Puerto Rico, Mayaguez Campus, Mayaguez, PR

Carlos Rinaldi, Chemical Engineering, University of Puerto Rico, Mayaguez Campus, Mayaguez, PR

Magnetic nanoparticles subjected to high frequency oscillating magnetic fields dissipate energy in the form of heat, inducing an increase in the temperature of the surrounding medium. This effect forms the basis of emerging applications of magnetic nanoparticles in cancer treatment and in magnetically actuated drug delivery. With the objective of improving the fundamental understanding of this phenomenon, the energy dissipation from magnetite and cobalt ferrite nanoparticles with narrow particle size distribution and surface modified with grafted poly (ethylene glycol) (PEG) of different molecular weights was studied. Specific absorption rates (SARS) in water were measured for all particles and correlated to physical and magnetic nanoparticle properties, such as core size, hydrodynamic diameter, saturation magnetization, and blocking temperature.

See more of this Session: Magnetic Nanoparticles in Biotechnology and Biomedicine I
See more of this Group/Topical: Nanoscale Science and Engineering Forum

Polymer Nanosphere-Based Vaccination Strategies against Yersinia Pestis

Bret D. Ulery, Chemical and Biological Engineering, Iowa State University, Ames, IA

Michael J. Wannemuehler, Veterinary Microbiology and Preventative Medicine, Iowa State University, Ames, IA

Jing Yu, Veterinary Microbiology and Preventative Medicine, Iowa State University, Ames, IA

F. Chris Minion, Veterinary Microbiology and Preventative Medicine, Iowa State University, Ames, IA

Devender Kumar, Center for Immunology and Microbial Disease, Albany Medical College, Albany, NY

Dennis W. Metzger, Center for Immunology and Microbial Disease, Albany Medical College, Albany, NY

Balaji Narasimhan, Chemical and Biological Engineering, Iowa State University, Ames, IA

Yersinia pestis, the causative agent of bubonic and pneumonic plague, is a Category A bioterrorism agent for which only moderately effective multi-dose whole-killed or live-attenuated vaccines currently exist. Subunit vaccines have been recently studied, but current strategies have shown only short-lived protection and require multiple injections. Our work is focused on the design of a recombinant protein-based (rF1p-V10) vaccine, delivered with polyanhydride nanosphere adjuvants as an inoculation strategy that will lead to a single-dose plague vaccine. For this study, random copolymers of 1,6-bis(p-carboxyphenoxy)hexane and sebacic acid and copolymers of 1,8-bis(p-carboxyphenoxy)-3,6-dioxaoctane and 1,6-bis(p-carboxyphenoxy)hexane were synthesized and fabricated into rF1pV10-loaded nanospheres by anti-solvent precipitation. Nanospheres were delivered subcutaneously or intranasally to C57BL/6 mice. Antigen-specific immune responses were assessed by antibody titers using an ELISA and lymphocyte proliferation was evaluated with 3H-thymidine incorporation. Survival after intranasal CO92 strain Y. pestis challenge was monitored for 14 days post-infection to determine protective capability. The results show that rF1p-V10 delivered with an adjuvant can protect mice against challenge and polyanhydride nanospheres act as adjuvants for sustained delivery conveying more robust immunity. These studies provide evidence that polyanhydride nanosphere vehicles hold great potential to be used in a wide variety of vaccination strategies.

See more of this Session: Nanotechnology for Biotechnology and Pharmaceuticals I
See more of this Group/Topical: Nanoscale Science and Engineering Forum

Drug Nanoparticle Isolation by Hydrogen Bonding Coacervation

Robert K. Prud'homme, Chemical Engineering, Princeton University, Princeton, NJ

Mustafa Akbulut, Chemical Engineering, Princeton University, Princeton, NJ

Suzanne D'Addio, Chemical Engineering, Princeton Unversity, Princeton, NJ

Concepcion Kafka, Chemical Engineering, Princeton Unversity, Princeton, NJ

To ensure long term stability most nanoparticle dispersions need to be prepared in dry form. The drying step is either costly as in the case of freeze drying, or results in aggregation as in the case of pan drying. We present a novel drying process that involves the unique hydrogen bonding complex between polyethylene glycol and poly carboxylic acid compounds. These interactions, triggered at low pH result in rapid floccualation of the nanoparticle dispersion. The resulting precipitate can be dried and redispersed back to the original particle size. The redisolution is triggered by breaking the complex via ionization of the acid groups by neutralizing the solution to a neutral pH.

See more of this Session: Nanotechnology for Biotechnology and Pharmaceuticals I
See more of this Group/Topical: Nanoscale Science and Engineering Forum

Plasmon Resonant Nanoparticles for Lipid-Based Drug Delivery

Guohui Wu, Department of Chemical Engineering, University of California, Santa Barbara, Santa Barbara, CA

Joseph Zasadzinski, Chemical Engineering Department, University of California, Santa Barbara, Santa Barbara, CA

Optimal drug carriers for intravenous drug delivery of chemotherapeutic drugs need to have the following properties: (1) good retention; (2) specific targeting; and (3) spatially and temporally controlled release. Vesosomes are multi-compartment lipid-based nanoparticles consisting of multiple smaller liposomes encapsulated within bigger liposomes. Due to the double-bagged structure, vesosomes have superior drug retention and stability. We used small angle x-ray scattering to characterize the lipid bilayer phase transformations among interdigitated phase, gel-phase, and fluid-phase. With thorough understanding of the vesosome formation, we successfully incorporated tri-block copolymers into vesosomes to shrink the nanoparticle size to minimize immunogenicity and prevent non-specific adsorption of proteins and peptides on the vesosome surface.

Plasmon resonant particles with strong optical absorption at the near-infrared (NIR) light enable development of novel biomedical application. They can act as molecular contrast agents and their localized heating enables thermal ablation. Recently, we showed Plasmon resonant gold nanoshells can generate transient cavitations if irradiated with fs-NIR laser. This allows their novel application in targeting and controlled release from lipid-based drug carriers where the optically triggered cavitations can temporarily rupture lipid membranes with a mechanism similar to that of ultrasound which is known to induced controlled drug delivery.

See more of this Session: Nanotechnology for Biotechnology and Pharmaceuticals I
See more of this Group/Topical: Nanoscale Science and Engineering Forum

Voltage Gated Carbon Nanotube Membranes for Programmed Transdermal Drug Delivery

Ji Wu, Materials, University of Kentucky, Lexington KY, KY

Caroline Strasinger, Pharmacy, University of Kentucky, Lexington KY, KY

Audra Stinchcomb, Pharmacy, University of Kentucky, Lexington KY, KY

Bruce J. Hinds, Chemical and Materials Eng., University of Kentucky, Lexington KY, KY

Carbon nanotubes membranes have been successfully used as the active element in a programmed transdermal delivery system of clonidine. Membranes composed of aligned CNTs passing across a polymer film are chemically modified through diazonium electrochemical grafting for a high anionic surface charge density at CNT tips and along CNT cores. Electro-phoresis/osmosis is observed by measured cationic and neutral fluxes through the CNT membrane as a function of applied bias. High electro-osmotic flows of 0.05 cm/s at -300mV bias are seen. In-vitro studies using diffusion cells with human skin samples show modulated transdermal delivery. An initial flux (0V applied) was 2.1 nm/hr-cm2 while the 'on' state (-300 mV bias) the flux increases to 12 nm/hr-cm2, a 5.7 fold increase. Therapeutically useful fluxes are between 1.7 and 5.4 nmol/hr-cm2. Low power programmable transdermal delivery devices can have important application in pain management and addiction treatment.

See more of this Session: Nanotechnology for Biotechnology and Pharmaceuticals I
See more of this Group/Topical: Nanoscale Science and Engineering Forum

Mucus-Penetrating Nanoparticles for Transmucosal Drug/Gene Delivery

Samuel K. Lai, Department of Chemical and Biomolecular Engineering, The Johns Hopkins University, Baltimore, MD

Ying-Ying Wang, Department of Biomedical Engineering, The Johns Hopkins University, Baltimore, MD

Ming Yang, Department of Biomedical Engineering, The Johns Hopkins University, Baltimore, MD

Jung Soo Suk, Department of Biomedical Engienering, The Johns Hopkins University, Baltimore, MD

Justin Hanes, Chemical and Biomolecular Engineering, The Johns Hopkins University, Baltimore, MD

Viscoelastic human mucus gels protect the lungs, GI tract, reproductive tract, eye and more from infectious pathogens and foreign particulates. The human mucus barrier was recently reported as impermeable to polymer nanoparticles as small as 59 nm. Particles that cannot move through the outer layers of mucus are cleared from the mucosal tissue within seconds to at most a few hours. This reality has strongly limited development of long-lasting controlled release systems to mucus-covered tissues, and likely has played a major role in thwarting gene therapy of mucosal tissues as well, especially in cystic fibrosis.

We hypothesized that adhesion to mucus was a critical rate-limiting barrier to nanoparticle transport through mucus layers. We sought to mimic the hydrophilic and net-neutral surface properties of viruses capable of rapidly moving through human mucus. Our initial search for a candidate material that could endow these surface properties on synthetic particles led us to poly(ethylene-glycol), or PEG. Paradoxically, PEG had a considerable history of use as a muco-adhesive.

This talk will describe our recent discovery that coating synthetic nanoparticles with high densities of low molecular weight PEG allows particles with sizes of at least 500 nm in diameter to rapidly transport through undiluted human mucus nearly as fast as they move through pure water. In contrast, high molecular weight PEG coating makes them even more adhesive to mucus than without coatings. We show that PEG density is especially critical as particle size diminishes from 500- to 100-nm. We also discovered that the spacings (pores) within the human mucus mesh are much larger than previously appreciated, which provides a significant opportunity for controlled drug delivery using large nanoparticles. We will describe the development of next generation mucus-penetrating particles, composed entirely of GRAS materials, that offers sustained drug release over days. We will also discuss the development of particles that rapidly penetrate the highly viscoelastic sputum expectorated from cystic fibrosis patients as well as sinus mucus collected from chronic rhinosinusitis patients.

See more of this Session: Nanotechnology for Biotechnology and Pharmaceuticals I
See more of this Group/Topical: Nanoscale Science and Engineering Forum

Drug Delivery Carriers Masquerade as ²self² to Avoid Body Clearance

Pia L. Rodriguez, Chemical and Biomolecular Engineering, University of Pennsylvania, Philadelphia, PA

Richard Tsai, University of Pennsylvania, Philadelphia, PA

Dennis E. Discher, Chemical & Biomolecular Engineering, University of Pennsylvania, Philadelphia, PA

Particle injected or surface implanted in the body are susceptible to recognition by the immune cells such as macrophage, which are responsible for the clearance of invading pathogens. However macrophage at the same time leave our own ²self² cells alone. The foreign versus self difference does not reside by steric repulsion by glycosylated proteins that have been mimicked using polymer PEG in drug delivery carriers. We have sought to address how macrophages specifically recognize self and to define its limits in systems from nano to micron scale systems by focusing on a cell-surface protein found on all of our cells. In the first stage of macrophage engulfment, the target objects are recognized with the deposition of antibodies or plasma complement proteins that binds to both foreign and self cells. However prior to macrophage engulfing its target, it checks for an identification of self cells, the protein CD47. Initially we demonstrate that 100 aa extracellular component of CD47 when coupled to synthetic particles is sufficient in reducing uptake. The potency of CD47 is not limited to micron sized particles but applies to nano-sized particle because of the inhibitory effect when CD47 is recognized by the receptor SIRPα on macrophages. By exploiting the bodies own system of distinguishing ²self² from foreign using the latest polymer PEG drug delivery carriers coupled with CD47 lead to effective delivery of drug and genetic material to its intended targets by avoiding the immune system.

See more of this Session: Nanotechnology for Biotechnology and Pharmaceuticals I
See more of this Group/Topical: Nanoscale Science and Engineering Forum

Growth of Unidirectional Indium Nitride (InN) Nanorods On Silicon Substrate Using Hydride Metal-Organic Vapor Phase Epitaxy

Vaibhav Chaudhari, Chemical Engineering, University of Florida, Gainesville, FL

Rangarajan Krishnan, Chemical Engineering, University of Florida, Gainesville, FL

David Wood, Chemical Engineering, University of Florida, Gainesville, FL

Tim J. Anderson, Chemical Engineering, University of Florida, Gainesville, FL

In recent years, lots of efforts have been directed towards growth of one dimensional (1D) electronic materials such as nanorods and nanowires. Indium nitride (InN) in particular has become material of great interest due to narrow direct bandgap (0.7-0.9 eV) as well as superior electron transport properties. InN is promising candidate for IR emitters, detectors as well as solar cell applications. There is also a lot of interest in integrating III-V technology to mature Si technology for ease in scale up as well as economic benefits. In this study, InN nanorods are grown by novel Hydride Metal Organic Vapor Phase Epitaxy (H-MOVPE) technique. Using H-MOVPE technique, InN nanorods (with 40nm -100nm diameter) are grown on Si(100) and Si(111) substrates at 560 °C to 600 °C. It was previously observed that intermediate GaN layer can assist vertical nanorod growth in case of Sapphire substrate. In this study, it has been demonstrated that use of GaN layer can help to produce almost vertically aligned nanorods in <0002> direction on silicon substrates as well. It has been observed that silicon substrate orientation has some effect on nanorod direction distribution. In addition, GaN layer growth mode (MOVPE or H-MOVPE), nitridation of substrate, nitridation temperature have also been observed to play important role in final InN naorods orientations. The near unidirectional growth of InN nanorods has been verified by techniques such as X-ray diffraction (XRD), Scanning Electron Microscopy (SEM) and Transmission Electron Microscopy (TEM). InN nanorods have been further characterized by room temperature Photoluminescence (PL) for bandgap and Energy Dispersive Spectroscopy (EDS).

See more of this Session: Nanowires A: Synthesis, Modeling and Applications
See more of this Group/Topical: Nanoscale Science and Engineering Forum

Controlling the Nucleation and Growth of III-V Semiconductor Nanowires

Chandrashekhar Pendyala, Chemical Engineering, University of Louisville, Louisville, KY

Jeong H Kim, Chemical Engineeering, University of Louisville, Louisville, KY

Jacek B Jasinski, Institute for Advanced Materials and Renewable Energy, University of Louisville, Louisville, KY

Zhiqiang Chen, Institute for Advanced Materials and Renewable Energy, University of Louisville, Louisville, KY

Mahendra K Sunkara, Chemical Engineering, University of Louisville, Louisville, KY

Self catalytic synthesis of GaSb nanowires with antimony tip-led growth and bulk nucleation of GaSb nanowires out of a gallium droplet is demonstrated. We propose a concept based on the equilibrium solubility of the constituents and the critical nuclei size at the synthesis conditions to explain the observed growth modes. The proposed framework can be used to predict the synthesis conditions that will lead to either tip-led or bulk nucleation. We consolidate our argument with the synthesis of gallium tip led GaSb nanowires at pre-determined conditions and extend the applicability of the same by the synthesis of antimony tip-led Indium antimonide nanowires. The ability to grow nanowires with control over the material at the tip can be used to develop strategies for self-catalytic synthesis of ternary nanowires and nanowire based hetero-structures.

See more of this Session: Nanowires A: Synthesis, Modeling and Applications
See more of this Group/Topical: Nanoscale Science and Engineering Forum

Self-Catalyzed Epitaxial Growth of Dislocation-Free Indium Phosphide Nanowires On Silicon

Li Gao, Department of Chemical and Biomolecular Engineering, University of California, Los Angeles, Los Angeles, CA

Robyn L. Woo, Department of Chemical and Biomolecular Engineering, University of California, Los Angeles, Los Angeles, CA

Baolai Liang, Department of Electrical Engineering, University of California, Los Angeles, Los Angeles, CA

Marta Pozuelo, Department of Materials Science and Engineering, University of California, Los Angeles, Los Angeles, CA

Sergey Prikhodko, Department of Materials Science and Engineering, University of California, Los Angeles, Los Angeles, CA

Mike Jackson, Department of Materials Science and Engineering, University of California, Los Angeles, Los Angeles, CA

Niti Goel, Intel Corporation, Santa Clara, CA

Mantu K. Hudait, Intel Corporation, Hillsboro, OR

Diana L. Huffaker, Department of Electrical Engineering, University of California, Los Angeles, Los Angeles, CA

Mark S. Goorsky, Department of Materials Science and Engineering, University of California, Los Angeles, Los Angeles, CA

Suneel Kodambaka, Department of Materials Science and Engineering, University of California, Los Angeles, Los Angeles, CA

Robert F. Hicks, Department of Chemical and Biomolecular Engineering, University of California, Los Angeles, Los Angeles, CA

The successful integration of III-V materials onto silicon would make it possible to fabricate an array of new electronic and photonic devices with greater functionality and at a lower cost. However, growing compound semiconductors on silicon is extremely challenging due to the polar/non-polar interface and the large lattice mismatch between the materials. One method of managing the high density of defects generated at the interface is to deposit pseudomorphic buffer layers up to 5.0 microns thick before making the active device structure. We have investigated an alternative approach to this problem: Free standing nanowires of indium phosphide have been grown epitaxially on Si(100) and Si(111) by metalorganic vapor-phase epitaxy. Liquid indium droplets were used to catalyze crystal nucleation. High-resolution transmission electron micrographs revealed that the nanostructures fully accommodated the strain due to the lattice mismatch and formed interfaces with the silicon that were free of dislocations. The preferred growth orientation was along <111> crystal directions. On Si(111), 100% vertical nanowires were achieved at 370 °C and a V/III ratio equal to 200. The wire density was 1.0×10^9 cm^{-2}, while the average dimensions were 3.9 μm in height, 45 nm in base width and 15 nm in tip width. X-ray diffraction and transmission electron microscopy showed that the wires were single-crystal zinc blende, although they contained a high density of rotational twins perpendicular to the <111> growth direction. Using triple-axis XRD, a rock curve width (FWHM) of only 42 arc-seconds was obtained. Room temperature photoluminescence spectra of the wires exhibited one peak centered at 912±10 nm. The prospects for using nanostructures to integrate III/V materials onto silicon will be discussed at the meeting.

See more of this Session: Nanowires A: Synthesis, Modeling and Applications
See more of this Group/Topical: Nanoscale Science and Engineering Forum

Investigation of Zeolite Mordenite as a Template for the Formation of Nanowires

Leonel Quinones, Chemical Engineering, University of Puerto Rico - Mayaguez Campus, Mayaguez, PR

Javier Huertas, Chemical Engineering, University of Puerto Rico - Mayaguez Campus, Mayaguez, PR

Maria Martinez-Inesta, Chemical Engineering, University of Puerto Rico - Mayaguez Campus, Mayaguez, PR

The use of templates has demonstrated to be a cost effective reliable method for the formation of uniform structures. Membranes of macropore (larger than 50 nm) and mesopore (between 2 and 50 nm) dimensions have been successfully used for the formation of metallic nanowires. However there is little evidence in the literature of studies using templates of micropore dimensions for the formation of nanowires. Here we will present work done using zeolites as templates for the formation of transition metal nanowires. Zeolites are crystalline aluminosilicate materials that possess 1D, 2D and 3D pore structures that range in sizes from 0.4 nm to 1.3 nm, thus its use as templates have the potential of yielding the smallest uniform nanostructures available through a bottom up method. Our group has studied this system using both computational and experimental methods. Computationally the effect of zeolite structure, zeolite composition, metal loading, and temperature has been studied on the formation of nanowires using Monte Carlo simulations and we have identified zeolites with optimal structures for the growth of nanowires. Experimentally we have used zeolite mordenite which is a zeolite with a unidimensional pore structure, as template to study the formation of Pt nanowires. Various reduction methods have been studied and the materials have been characterized using XRD, FTIR, and HRTEM. Thus far our group has been successful synthesizing polycrystalline Pt nanowires of 2nm in diameter, Pt nanoflowers, and single crystal Pt nanowires.

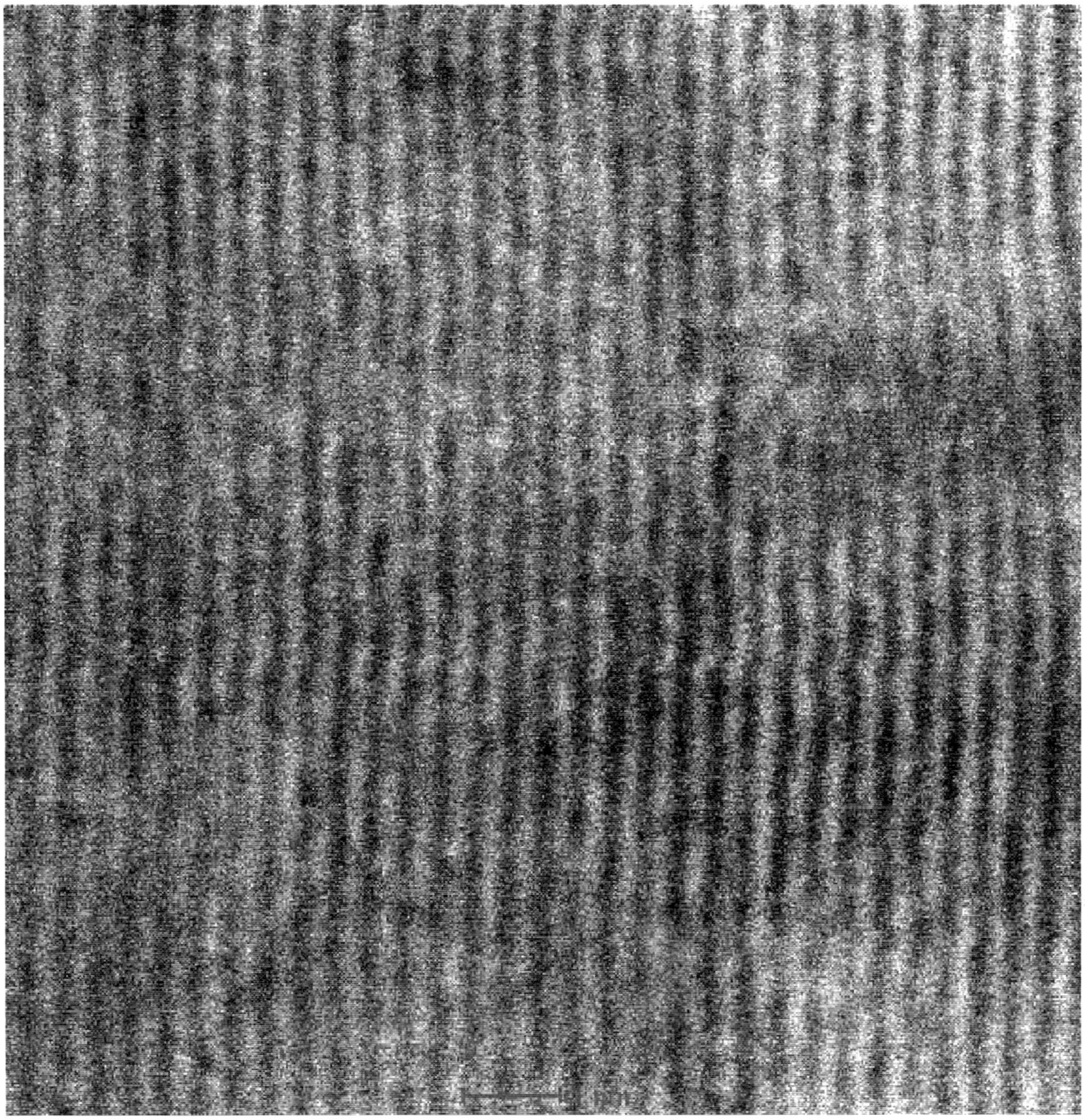

See more of this Session: Nanowires A: Synthesis, Modeling and Applications
See more of this Group/Topical: Nanoscale Science and Engineering Forum

Structural Characterization of the Rate-Dependent Energy Release Mechanism of Gold Nanowires Under Elongation

Christopher R. Iacovella, Chemical and Biomolecular Engineering, Vanderbilt University, Nashville, TN

Qing Pu, Brewer Science, Inc., Rolla, MO

Yongsheng Leng, Department of Mechanical and Aerospace Engineering, George Washington University, Washington, DC

Peter T. Cummings, Department of Chemical and Biomolecular Engineering, Vanderbilt University, Nashville, TN

Understanding the underlying mechanisms of the formation of break-junction structures in metal nanowires has important implications in the area of molecular- and nano-electronics. Gold nanowires have been demonstrated to form single-atom and helical chains when stretched. The temperature and rate of elongation strongly dictate the length of these chains, e.g. long chains are favored for low temperature systems with fast elongation rates [1]. In this work we use molecular dynamics simulations of gold nanowires that utilize the tight-binding second-moment approximation potential (TB-SMA) to explore the rate-dependent energy release mechanisms that govern chain formation. We employ shape matching techniques based on spherical harmonics [2] to quantify the structural changes in the nanowires. We utilize these techniques to quantitatively identify dislocations in the crystal structure and transitions to non-crystalline local structures, as related to the energy release mechanisms.

[1] Pu Q, Leng YS, Cummings PT, Rate-Dependent Energy Release Mechanism of Gold Nanowires under Elongation, JACS 130 (52), 2008

[2] Iacovella CR, Keys AS, Horsch MA, Glotzer SC, Icosahedral packing of polymer-tethered nanospheres and stabilization of the gyroid phase, PRE 75(4), 2007

See more of this Session: Nanowires A: Synthesis, Modeling and Applications
See more of this Group/Topical: Nanoscale Science and Engineering Forum

Evaporation-Induced Hierarchical Assemblies of Rigid Silicon Nanowires

Chih-Hung Sun, Chemical Engineering, University of Florida, Gainesville, FL

Xuefeng Liu, School of Chemical and Material Engineering, Jiannan University, Wuxi, China

Peng Jiang, Chemical Engineering, University of Florida, Gainesville, FL

Periodic arrays of silicon nanowires (SiNWs) over large areas are of great technological importance in developing novel electrical, optical, biosensing, and electromechanical devices. Traditional metal-catalyzed vapor-liquid-solid (VLS) growth of SiNWs requires high-temperature and high-vacuum operations. Additionally, the precise control over the diameter and location of SiNWs, the unwanted doping of gold in SiNWs, and the controlled epitaxial growth of SiNWs on conventional (100) silicon wafers, are challenging. This paper reports a facile and scalable bottom-up technology for fabricating periodic arrays of single-crystalline SiNWs over wafer-sized area. Spin-coated 2-D nonclose-packed silica colloidal crystals are utilized as etching mask to generate ordered polymer posts whose size can be much smaller than the templating silica spheres. Using the resulting polymer posts as deposition mask to create gold nanoholes, highly ordered SiNWs with tunable structural parameters can be easily fabricated by a room-temperature wet-etching technology. As the SiNWs are produced by direct wet-etching in single-crystalline silicon wafer, they are relatively free of volume defects and thus their bending strength approaches the predicted theoretical maximum. Most importantly, the nonclose-packing of the colloidal template and the close-to-ideal mechanical property enable the formation of unusual open-structured hierarchical assemblies of rigid SiNWs during solvent evaporation. These assemblies exhibit excellent broadband antireflective and dewetting properties, promising for developing self-cleaning antireflection coatings for crystalline silicon solar cells.

See more of this Session: Nanowires A: Synthesis, Modeling and Applications
See more of this Group/Topical: Nanoscale Science and Engineering Forum

Intrinsic Property Characterization with Length and Chiral Vector Enriched Single Wall Carbon Nanotubes

Jeffrey A. Fagan, Polymers Division, National Institute of Standards and Technology, Gaithersburg, MD

Ji Yeon Huh, Polymers Division, National Institute of Standards and Technology, Gaithersburg, MD

Angela R. H. Walker, Physics Lab, National Institute of Standards and Technology, Gaithersburg, MD

Erik K. Hobbie, Polymers Division, National Institute of Standards and Technology, Gaithersburg, MD

Solutions of single wall carbon nanotubes sorted by length and chiral vector display significantly improved properties compared to unsorted materials. With published techniques, use of sequential ultracentrifugation based separations has allowed for the production of fractions that can be used to isolate specific optical and physical properties of the nanotubes. These include the frequency and chirality dependent anisotropic polarizability, the contribution of end caps to Raman D band scattering, and magnetic susceptibility. Characterization of the intrinsic properties of the sorted solutions will be discussed.

See more of this Session: Carbon Nanotubes II: Characterization, Functionalization, and Applications
See more of this Group/Topical: Nanoscale Science and Engineering Forum

pH-Responsive Copolymer Functionalized Single Wall Carbon Nanotubes for Switchable Aqueous Dispersibility

Pu Zhang, Chemical & Biological Engineering, University of Missouri-Rolla, Rolla, MO

David Henthorn, Chemical & Biological Engineering, University of Missouri-Rolla, Rolla, MO

Carbon nanotubes have been extensively studied for over a decade because of their unique mechanical, electrical, optical and thermal properties. However, in order to make carbon nanotubes useful in biomedical or biotechnologies applications, the materials must be modified first to improve aqueous dispersibility. The ultimate fate of dispersible nanomaterials, however, is a subject of great environmental concern. In this work, we aim to create carbon nanotubes with switchable dispersibility through the use of grafted pH-responsive polymer chains.

In this work, we successfully used an ultraviolet light initiated "graft from" polymerization method to fabricate polymer functionalized single wall carbon nanotubes with pendant pH-responsive polymer chains. The copolymer formed from poly (methacrylic acid) (PMMA) and polyethylene glycol methyl ether methacrylate (PEGMA) monomers is welled established for its pH-responsive swelling/deswelling behavior of the copolymer was utilized to control the carbon nanotube aqueous dispersibility. The morphology of the carbon nanotubes was characterized by SEM before and after functionalization. Dynamic light scattering was employed to observe the swelling/collapsing behavior of the grafted chains. In addition, covalent bonding of the copolymer to the carbon nanotubes was studied with Raman, FT-IR, and XPS spectroscopic techniques.

See more of this Session: Carbon Nanotubes II: Characterization, Functionalization, and Applications
See more of this Group/Topical: Nanoscale Science and Engineering Forum

Potential of Mean Force Between Aqueous Single Walled Carbon Nanotubes in Surfactant Solutions

Naga Rajesh Tummala, School of Chemical Biological and Materials Engineering, The University of Oklahoma, Norman, OK

Alberto Striolo, School of Chemical Biological and Materials Engineering, The University of Oklahoma, Norman, OK

Molecular dynamic simulations were conducted to calculate the effective potential of mean force (PMF) between two (6,6) SWNTs in the presence of aqueous surfactants. We employed two sodium dodecyl sulfate (SDS) concentrations, with SDS adsorbed on the (6,6) SWNTs, to calculate the PMF as a function of nanotube-nanotube distance. For comparison, we also computed the PMF between two pristine (6,6) SWNTs in water, as well as between two (6,6) SWNTs covered by flavin mononucleotide (FMN) surfactants. In the absence of surfactants our results show, as expected, a strongly attractive SWNT-SWNT PMF at short nanotube-nanotube separations. The presence of SDS surfactants modulates the PMF profile by introducing mid-range nanotube-nanotube repulsions. Because FMN surfactants, the phosphorilated form of vitamin B2, contain an aromatic isoalloxazine moiety and a chiral phosphate group, they couple more tightly with the SWNTs and yield a much more pronounced repulsive force between the approaching SWNTs than that observed in the presence of SDS. Our results will help us identify the surfactant properties that allow us to manipulate nanotube-nanotube effective interactions. This is the key for designing nanotube-specific dispersing agents.

See more of this Session: Carbon Nanotubes II: Characterization, Functionalization, and Applications
See more of this Group/Topical: Nanoscale Science and Engineering Forum

Langmuir
Article
pubs.acs.org/Langmuir

Emulsions Stabilized by Carbon Nanotube–Silica Nanohybrids

Min Shen and Daniel E. Resasco*

School of Chemical, Biological and Materials Engineering and Carbon Nanotube Technology Center (CANTEC), University of Oklahoma, Norman, Oklahoma 73021

Received April 18, 2009. Revised Manuscript Received May 23, 2009

Emulsions have been stabilized by carbon nanotube–silica nanohybrids. The as-prepared nanotube–silica particles generate water-in-oil (w/o) emulsions, regardless of the water/oil volume ratio used. The emulsion volume fraction was much higher than that obtained with nanotubes only, and it was found to increase with water/oil ratio due to an increasing amount of water retained in the emulsion droplets. However, beyond a certain water/oil ratio, the emulsion fraction rapidly decreased. This point of collapse is a strong function of the amount of carbon nanotube–silica particles in the system. Oxidation of the nanohybrids by nitric acid can effectively modify the particle surfaces and change the resulting emulsion properties. The treatment in nitric acid increases the density of functional groups on the carbon nanotubes. Depending on the extent of functionalization, the effect of oxidation can vary from increasing the volume of emulsion stabilized to reversing the type of emulsion to oil-in-water (o/w). The emulsions are remarkably stable against coalescence and sedimentation and can be easily separated by filtration or centrifugation, which make them suitable for applications in interfacial catalytic processes in which the catalyst can be easily recovered after reaction.

1. Introduction

Emulsions stabilized by solid particles have been known for more than a century[1] and were named after S. Pickering, who discovered that coalescence of droplets is suppressed when solid particles are adsorbed at the oil–water interface.[2] It is widely accepted that this suppression in coalescence is a kinetic effect caused by a combination of the formation of a rigid interfacial film and the increase in viscosity of the continuous phase.[3–9] One of the attractive features of Pickering emulsions is the great easiness for breaking the emulsion and recovering the two phases when needed.[3,10]Applications of solid-stabilized emulsions in various fields have been described in several articles and patents. For example, emulsifier-free cosmetic preparations have been proposed, in which finely dispersed particles are adsorbed at the interface generating oil-in-water emulsions. Bragg[11] has described a method for petroleum recovery by injection of emulsions stabilized by undissolved solid particles, which are partially oleophilic. The resulting particle-stabilized emulsion can be used as a driving fluid for displacing hydrocarbons from the porous formations. Many other types of solid particles have been applied as emulsification agents for obtaining Pickering emulsions.[9] To segregate to the interface of two liquid phases instead of to one of the phases, particles need to be amphiphilic/amphiphobic, partly wetted by each of the two phases.[12] Depending on whether the particles are more hydrophilic or more hydrophobic, they tend to stabilize oil-in-water or water-in-oil emulsions, respectively.[9–13] Particle size,[14] particle concentration,[15] oil/water ratio,[16,17] pH,[18] salt concentration,[19] and solvent type[20] all have been found to be crucial in determining the resulting emulsion properties.[10]

In this contribution we describe the behavior of emulsions stabilized by carbon nanotube–silica hybrids. Single-walled carbon nanotubes (SWNT) have unique structural and electronic properties.[21] Their hydrophobic nature[22,23] is well recognized, but it is also known that they can be poorly suspended in organic solvents.[24] It has been shown that nanotubes can undergo self-assembly at the oil–water interfaces.[25–27] Specifically, water-in-toluene emulsions can be stabilized by pristine SWNTs[28] and single-stranded-DNA-modified SWNT[29] in which the emulsion droplet diameters decrease as a function of the SWNT concentration. Ziegler and co-workers[30] have used the amphiphobic nature of SWNTs to develop an effective purification method. They

*Corresponding Author. E-mail: resasco@ou.edu; phone: (405) 325-4370.

(1) Ramsden, W. *Proc. Royal. Soc* **1903**, *72*, 156.
(2) Pickering, S. U. *J. Chem. Soc.* **1907**, *91*, 2001.
(3) Binks, B. P.; Lumsdon, S. O. *Phys. Chem. Chem. Phys.* **1999**, *1*, 3007.
(4) Hunter, T. N.; Pugh, R. J.; Franks, G. V.; Jameson, G. J. *Adv. Colloid Interface Sci.* **2008**, *137*, 57.
(5) Arditty, S.; Whitby, C. P.; Binks, B. P.; Schmitt, V.; Leal-Calderon, F. *Eur. Phys. J. E* **2003**, *11*, 273.
(6) Aveyard, R.; Clint, J. H.; Horozov, T. S. *Phys. Chem. Chem. Phys.* **2005**, *5*, 2398.
(7) Kralchevsky, P. A.; Ivanov, I. B.; Ananthapadmanabhan, K. P.; Lips, A. *Langmuir* **2005**, *21*, 50.
(8) Binks, B. P. *Curr. Opin. Colloid Interface Sci.* **2002**, *7*, 21.
(9) Binks, B. P.; Rodrigues, J. A. *Angew. Chem., Int. Ed.* **2005**, *44*, 441.
(10) Tambe, D. E.; Sharma, M. M. *Adv. Colloid Interface Sci* **1994**, *52*, 1.
(11) Bragg, J. R., US Pat. 5,855,243 (1999)
(12) Leal-Calderon, F.; Schmitt, V. *Curr. Opin. Colloid Interface Sci.* **2008**, *13*, 217.
(13) Finkle, P.; Draper, H. D.; Hildebrand, J. H. *J. Am. Chem. Soc.* **1923**, *45*, 2780.
(14) Binks, B. P.; Lumsdon, S. O. *Langmuir* **2001**, *17*, 4540.
(15) Binks, B. P.; Philip, J.; Rodrigues, J. A. *Langmuir* **2005**, *21*, 3296.
(16) Binks, B. P.; Whitby, C. P. *Langmuir* **2004**, *20*, 1130.
(17) Binks, B. P.; Lumsdon, S. O. *Langmuir* **2000**, *16*, 2539.
(18) Midmore, B. R. *Colloid Surf., A* **1998**, *132*, 257.
(19) Whitby, C. P.; Fornasiero, D.; Ralston, J. *J. Colloid Interface Sci.* **2009**, *329*, 173.
(20) Binks, B. P.; Lumsdon, S. O. *Phys. Chem. Chem. Phys.* **2000**, *2*, 2959.
(21) Baughman, R. H.; Zakhidov, A. A.; de Heer, W. A. *Science* **2002**, *297*, 787.
(22) Li, X. M.; Reinhoudt, D.; Crego-Calama, M. *Chem. Soc. Rev* **2007**, *36*, 1350.
(23) Zhang, L; Resasco, D. *Langmuir* **2009**, *25*, 4792.
(24) Bahr, J. L.; Mickelson, E. T.; Bronikowski, M. J.; Smalley, R. E.; Tour, J. M. *Chem. Commun* **2001**, *2*, 193.
(25) Zhang, X. J.; Shen, Y. F.; Kuehner, D.; Wu, S. X.; Su, Z. M.; Ye, S.; Niu, L. *Chem. Commun* **2008**, *36*, 4273.
(26) Matsui, J.; Yamamoto, K.; Miyashita, T. *J. Mater. Chem* **2007**, *17*, 3806.
(27) Matsui, J.; Yamamoto, K.; Inokuma, N.; Orikasa, H.; Kyotani, T.; Miyashita, T. *Carbon* **2009**, *47*, 1444.
(28) Wang, H.; Hobbie, E. K. *Langmuir* **2003**, *19*, 3091.
(29) Hobbie, E. K.; Bauer, B. J.; Stephens, J.; Becker, M. L.; McGuiggan, P.; Hudson, S. D.; Wang, H. *Langmuir* **2005**, *21*, 10284.
(30) Wang, R. K.; Park, H. O.; Chen, W. C.; Silvera-Batista, C.; Reeves, R. D.; Butler, J. E.; Ziegler, K. J. *J. Am. Chem. Soc.* **2008**, *130*, 14721.

found that bundles are selectively removed from aqueous dispersions by trapping them at the water/toluene interface, making high-quality gum arabic-suspended SWNTs comparable to those prepared by ultracentrifugation.

Surface wettability and particle size are two important parameters for controlling emulsion properties,[8] but they are not easily controlled in pristine SWNTs. The SWNT−silica hybrids discussed in this contribution allows us to control these properties. The CoMoCAT synthesis process developed by our group several years ago is capable of consistently producing SWNTs of high quality.[31,32] Moreover, the process employs sub-micrometer silica particles as a catalyst support. Consequently, the as-produced SWNT/catalyst composite is an interesting combination of hydrophobic/hydrophilic species. We have recently used this nanocomposite as an electrically conductive polymer filler that greatly helps dispersing the nanotubes in polymers.[33] For example, we found that for a given nanotube load it is possible to greatly vary the conductivity of polymer composites by simply varying the type of silica employed. Because a significant portion of the cost of nanotube materials is the purification step, the SWNT−silica nanohybrids offer the added benefit that they do not need to be further purified, since essentially they are the as-produced product of the CoMoCAT process. Therefore, SWNT−silica appears as an interesting system to be used in the stabilization of emulsions.

As mentioned, the amphiphobic nature of pristine SWNTs[34] makes them accumulate preferentially at the water/oil interface instead of dispersing in any of the bulk phases.[28] At the same time, silica particles have been extensively studied for oil-in-water Pickering emulsions as hydrophilic emulsifiers.[3,15,16] Therefore, we anticipate that the SWNT−silica combination offers a unique structure that can offer great flexibility in controlling the surface wettability by adjusting the carbon/silica ratio, silica particle size, nanotube length, and surface functionalities, which in turn can be varied by varying reaction conditions and surface treatments. Thus, it is possible to modify the distributions of partitioning coefficients between oil and water.

In this contribution, we report the preparation of water-in-oil and oil-in-water emulsions with variable fractions of emulsion volume by using the carbon nanotube−silica nanohybrids as stabilizers. For a given water/oil ratio, the volume fraction of emulsion obtained with the nanotube−silica nanohybrids is much higher than the fractions obtained under the same conditions with either nanotubes or silica particles alone. The effects of varying the water/oil volume ratios, the concentration of added solids, and the functionalization of the nanotube surface on the type of emulsion formed, emulsion droplet size, and total emulsion volume, have been investigated. Potential applications of these solid emulsifiers as interfacial catalysts and for emulsion polymerization are proposed.

2. Experimental Section

2.1. Materials. Decalin (>98%), cyclohexane (>99%), and toluene (99.9%) were provided by Fluka, Fischer Chemical, and Avocado Research Chemical, LTD, respectively. Water was purified by circulating through two ion exchangers (Cole-Parmer Inc.). SWNTs were obtained via our own CoMoCAT synthesis method. A detailed description of the CoMoCAT method can be found in previous publications and patents.[29,32,35,36] In brief, the silica-supported Co−Mo catalyst is synthesized by impregnating appropriate amounts of cobalt nitrate and ammonium heptamolybdate over SiO_2; the silica used in this case was a HiSil 210 (primary particles $d < 50$ nm, BET area 135 m^2/g, from PPG). The total metal loading was 2 wt % at a Co/Mo molar ratio of 1:3. After reduction in H_2 at 500 °C, the catalyst is heated in He flow to 750 °C and then exposed to a flow of pure CO for 1 h. The as-produced material was passed through a microfluidizer to achieve an uniform SWNT−silica aggregate size of about 1 μm. The carbon weight percentage of the products was characterized by a total carbon analyzer and determined to be about 7%. The purified SWNT (SG 65) were provided by SouthWest Nanotechnologies Inc. (SWeNT).

2.2. Characterization Techniques. The SWNT−silica nanohybrids were characterized by Raman spectroscopy, electron microscopy (TEM), and X-ray photoelectron spectroscopy (XPS). The Raman spectra were acquired on a Jovin Yvon-Horiba Lab Ram equipped with a charge-coupled detector and a He−Ne laser (632 nm) as excitation source. The TEM images were obtained on a JEOL JEM-2000FX. The XPS spectra were recorded on a Physical Electronics PHI 5800 ESCA system, equipped with an Al Kα X-ray anode operated at 350 W and 15 kV. The base pressure of the main chamber was kept at about 1.0×10^{-8} torr.

2.3. Emulsion Preparation. Before preparation of the emulsions, the SWNT−silica nanohybrids were presonicated in water for 3 h (Fisher Scientific, 600 W, 20 kHz). It was found that this presonication step is important to achieve a high extent of emulsification and small droplet size. The aqueous solutions containing the sonicated SWNT−silica nanohybrids were mixed with different amounts of decalin in a second sonication step, for 15 min. During this step, a homogeneous emulsion is formed, which remains between the water (bottom) and the organic (top) phases. The interfaces between the three phases can be clearly distinguished given that essentially all of the SWNT−silica nanohybrids remain suspended in the emulsion fraction. The characteristics of the prepared emulsions were monitored on an Olympus Mplan microscope immediately after each preparation. The average sizes were estimated based on the images of several emulsion droplets. The volume fraction of the emulsion phase was measured after suspending the homogenized solution for 1 h.

3. Results and Discussion

3.1. Characterization of Dry SWNT−Silica Nanohybrids. Before being used in the stabilization of the emulsions, the SWNT−silica nanohybrids, ultrasonicated in water, were characterized by a combination of techniques. Figure 1a shows a TEM image of an SWNT−silica aggregate. It is seen that the sizes of the primary silica particles are rather uniform, around 50 nm in diameter. The SWNT bundles can be clearly seen, with a bundle diameter around 10 nm, which is typically observed in the product synthesized by the CoMoCAT method at 750 °C.[37] Also, it is well-known that this process produces SWNTs of average diameter of 0.8 nm, with the (6,5) chirality as the dominant species.[38] Figure 1b shows the Raman spectrum of the nanotubes present in the hybrid after the sonication step. Raman spectroscopy is typically employed to evaluate the purity of the SWNT product and to estimate the presence of defects based on two characteristic bands, an intense band below 1600 cm^{-1}, typical of sp^2 ordered carbon (G band), and a smaller band around

(31) Kitiyanan, B.; Alvarez, W. E.; Harwell, J. H.; Resasco, D. E. *Chem. Phys. Lett.* **2000**, *317*, 497.
(32) Alvarez, W. E.; Kitiyanan, B.; Borgna, A.; Resasco, D. E. *Carbon* **2001**, *39*, 547.
(33) McDaniel, N. D.; McDaniel, M. P.; Balzano, L.; Resasco, D. E. *J. Appl. Polym. Sci.* **2009**, *111*, 589.
(34) Li, H. J.; Wang, X. B.; Song, Y. L.; Liu, Y. Q.; Li, Q. S.; Jiang, L.; Zhu, D. B. *Angew. Chem., Int. Ed.* **2001**, *40*, 1743.
(35) Alvarez, W. E.; Pompeo, F.; Herrera, J. E.; Balzano, L.; Resasco, D. E. *Chem. Mater.* **2002**, *14*, 1853.
(36) Resasco, D. E.; Alvarez, W. E.; Pompeo, F.; Balzano, L.; Herrera, J. E.; Kitiyanan, B.; Borgna, A. *J. Nanopart. Res* **2002**, *4*, 131.
(37) Zhang, L.; Balzano, L.; Resasco, D. E. *J. Phys. Chem. B* **2005**, *109*, 14375.
(38) Bachilo, S. M.; Balzano, L.; Herrera, J. E.; Pompeo, F.; Resasco, D. E.; Weisman, R. B. *J. Am. Chem. Soc.* **2003**, *125*, 11186.

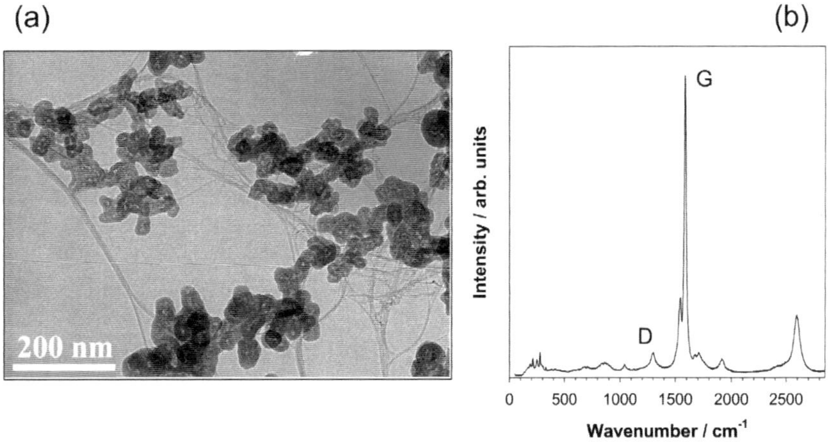

Figure 1. SWNT−silica nanohybrids (containing ~7 wt % SWNTs) used in the study after ultrasonication in water for 3 h (Fisher Scientific horn sonicator, 600 W, 20 kHz). (a) TEM image. (b) Raman spectrum.

Table 1. Surface Concentrations of Different Elements on the SWNT−silica Hybrids Ultrasonicated in Water for 3 h

element (sensitivity factors)	C 1s (0.314)	O 1s (0.733)	Si 2p (0.368)	Co 2p (3.529)	Mo 3d (3.544)
atomic concentration (%)	18.05	58.35	23.60	traces	traces

1300 cm^{-1}, typical of sp^3 carbon associated with defects (D band).[38] In this case, the G/D band ratio of the dry SWNT−silica nanohybrids is very high, indicating a low amount of carbon impurities and imperfection in the sonicated SWNT.[39] Table 1 summarizes the information obtained by XPS for the sonicated and dried SWNT−silica nanohybrids. The signals corresponding to Co and Mo, used as catalytic species in the synthesis of the SWNT, were practically negligible, indicating that these metals remain under the carbon species. From the relative intensity of the C(1s) signal compared to the other elements, the surface fraction of carbon for the sonicated nanohybrids can be quantified to be around 18.0%, which is significantly higher than the bulk concentration of 7 wt % C, as determined by TGA.

3.2. Water-in-oil Emulsions Stabilized by the SWNT−Silica Nanohybrids. The SWNT−silica nanohybrids characterized above were found to efficiently stabilize emulsions from oil/water mixtures. Figure 2a shows a diagram illustrating the appearance of an emulsion prepared by homogenizing sonicated nanohybrids in equal volumes of water and decalin to get a concentration of 2 g/L, and further sonicating for 15 min. The emulsion phase accounts for ~40% volume fraction of the total liquid. This fraction coexists between the two one-phase fractions, oil and water. The microscopic image taken for the liquid droplets acquired from the emulsion phase shows that the average emulsion droplet sizes appear to be around 3−20 μm, tens of times smaller than those prepared by only SWNT.[30] Figure 2b demonstrates that in this case the emulsion is of the w/o type, determined in the dye solubility test by tracing the colors of the emulsion droplets and the continuous phase after adding a green water-soluble dye (fluorescent Na salt) appearing inside the droplets, and a red oil-soluble dye (Sudan III), appearing in the continuous phase.

Figure 2c shows the well-known mechanism by which w/o emulsion droplets are stabilized by particles preferentially hydrophobic, as is the case of the SWNT−silica nanohybrids. The dominant hydrophobicity in the SWNT−silica nanohybrids make the particle surfaces wetted more by the oil phase, hence resulting in a high contact angle of the particles at the oil−water and the formation of the w/o structure. By contrast, as shown below, one can reverse the emulsion by increasing the hydrophilicity of the nanohybrids.

3.2.1. Effect of Varying the Oil/Water Ratio. We have found that the amount of emulsion stabilized by the nanohybrids and the size of the droplets depend on several factors that include the o/w ratio, concentration of SWNT−silica nanohybrids, sonication time, etc., and can be modified via surface modifications. For example, it is known that changing oil/water volume ratio can affect the emulsion stability and even invert the emulsion type.[14,17,40] However, as shown below, regardless of the water/oil ratio the emulsions stabilized by the as produced SWNT−silica nanohybrids are always w/o. This is in good accordance with the strong hydrophobicity previously reported for the pristine SWNT.[23]

Figure 3 summarizes the volume fraction of emulsion (left axis) and the droplet size (right axis) obtained at increasing water/decalin volume ratios for the two SWNT−silica particle concentrations (0.2 and 0.075 wt %). The variation in emulsion fraction with the water/oil ratio is remarkable. The emulsion fraction rapidly increases with the water/oil ratio up to a point at which it abruptly forms droplets. Different maximum values are obtained depending on hybrid particle concentration. For the higher particle concentration (0.2 wt %), more than 50% volume fraction of emulsion can be obtained at a water/decalin volume ratio of ~1.2/1. By contrast, at lower particle concentrations, not only the maximum emulsion fraction is significantly lower, but also it occurs at a lower water/oil ratio. That, is the maximum emulsion fraction is about 35%, and it happens at a water/oil ratio around 0.4/1.

The graph on the right panel shows that the emulsion droplet sizes also go through a maximum at intermediate water/decalin volume ratios, and this trend is particularly obvious for the 0.075 wt % case. The initial increase seen in both, emulsion volume and droplet size, with increasing amount of water phase in the mixture, is mostly due to a higher amount of water filling the

(39) Resasco, D. E.; Herrera, J. E. Structural Characterization of Single-walled Carbon Nanotubes. In *Encyclopedia of Nanoscience and Nanotechnology*; Nalwa, H. S., Ed.; American Scientific Publishers: Valencia, CA, 2003; Vol. *10*, p 1.
(40) Ashby, N. P.; Binks, B. P. *Phys. Chem. Chem. Phys.* **2000**, *2*, 5640.

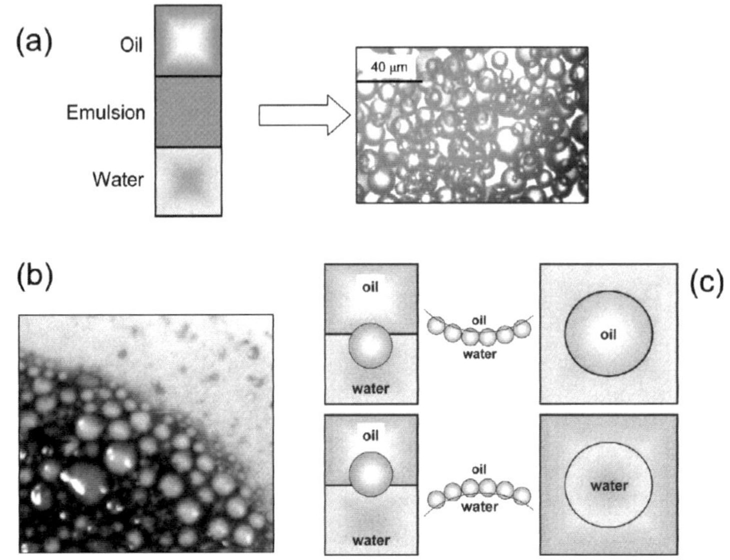

Figure 2. (a) Schematic description of the three fractions observed in a typical emulsion system together with a microscopy image of the emulsion fraction. (b) Color photograph of the emulsion prepared from SWNT–silica hybrids. Water-soluble dye (green color, fluorescent Na salt) and oil soluble dye (red color, Sudan III) were added to identify the water-in-oil type of the emulsion. (c) Schematic description of the stabilization mechanism of the oil-in-water and water-in-oil emulsions.

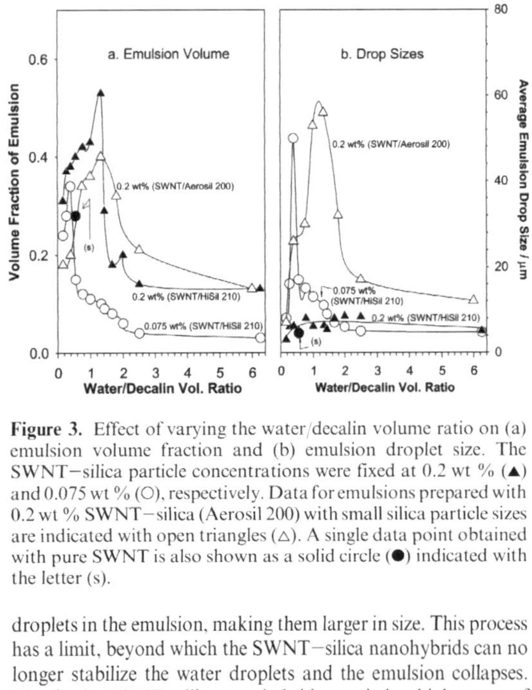

Figure 3. Effect of varying the water/decalin volume ratio on (a) emulsion volume fraction and (b) emulsion droplet size. The SWNT–silica particle concentrations were fixed at 0.2 wt % (▲) and 0.075 wt % (○), respectively. Data for emulsions prepared with 0.2 wt % SWNT–silica (Aerosil 200) with small silica particle sizes are indicated with open triangles (△). A single data point obtained with pure SWNT is also shown as a solid circle (●) indicated with the letter (s).

droplets in the emulsion, making them larger in size. This process has a limit, beyond which the SWNT–silica nanohybrids can no longer stabilize the water droplets and the emulsion collapses. That is, the SWNT–silica nanohybrids remain in a higher state of agglomeration within the decalin phase, as clearly observed by optical microscopy (data not shown). These aggregates may increase the viscosity[15] and negatively affect the emulsion formation.[16,41] It is worth mentioning that the trends reported here are quite general, and we have also observed similar behavior with a number of other organic solvents and conditions.

It has been previously shown that varying particle size of the emulsifiers can greatly affect the emulsion properties and even invert the emulsion type.[14] To illustrate this effect, we characterized the emulsions prepared by SWNT/silica hybrids synthesized by CoMoCAT using Aerosil 200 silica (primary particles $d \approx 12$ nm, BET area 200 m^2/g) as catalyst supports. The amount of SWNT in these hybrids (~7 wt %) was close to those using HiSil 210 silica as support, which allowed us to compare only the effect of silica particle size, at constant nanotube content. Overall, the results with the smaller particle silicas showed that the trends observed for the SWNT/silica (HiSil 210) hybrids are still valid for the smaller silica particle hybrids. Likewise, with these hybrids the type of the emulsions prepared was still w/o type, as observed with SWNT/silica (HiSil 210). However, a lower extent of emulsion volume fraction and much larger emulsion droplet sizes were found with smaller silica particles. This difference is in agreement with that observed by Binks and Lumsdon for polystyrene latex particles.[14] The difference in droplet size caused by changes in silica particle size might be due to a lower hydrophile-to-lipophile balance on the smaller particle size silica/SWNT hydbrids.[14]

Figure 3 also includes one data point obtained with pure SWNT used as emulsifiers instead of the SWNT–silica nanohybrids. Although in this comparison the total amount of solids is the same, the actual amount of nanotubes used when the emulsifier is the nanohybrid is 15 times lower than that required to obtain the same volume fraction of emulsion. That is, although pure nanotubes can also generate w/o emulsion, in agreement with the previous reports,[29,42] the effectiveness of the hybrids is remarkable.

3.2.2. Effect of Varying Nanohybrid Particle Concentration. In standard Pickering emulsions, particle concentration has

(41) Akartuna, I.; Studart, A. R.; Tervoort, E.; Gonzenbach, U. T.; Gauckler, L. J. *Langmuir* **2008**, *24*, 7161.

(42) Wang, R. K.; Reeves, R. D.; Ziegler, K. J. *J. Am. Chem. Soc.* **2007**, *129*, 15124.

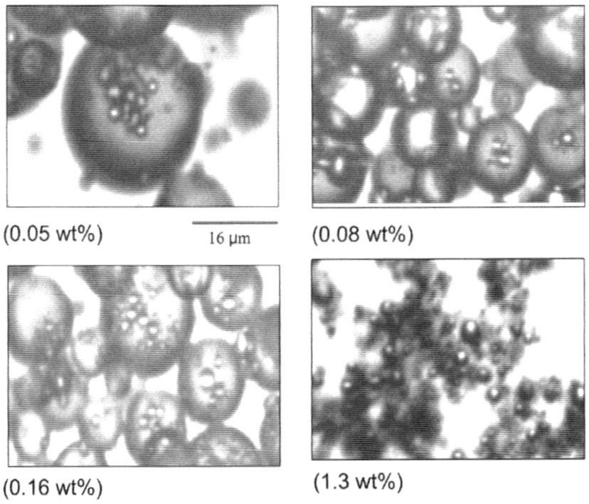

Figure 4. Microscopy images of the emulsions prepared at increasing SWNT−silica nanohybrid concentrations with a water/decalin volume ratio fixed at 0.65/1. The bar length represents 50 μm.

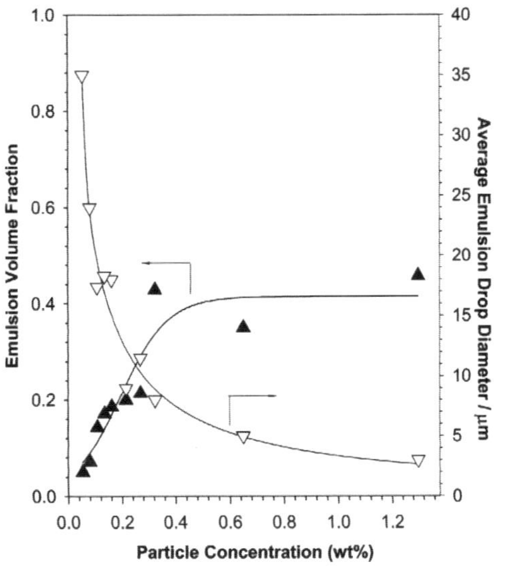

Figure 5. Emulsion volume fraction (solid symbols, left axis) and droplet average diameter (open symbols, right axis) as a function of SWNT−silica hybrid concentration.

been found to substantially affect both the emulsion stability and droplet size.[14,15] Figure 4 shows microscopy images taken for the emulsions prepared by the SWNT−silica nanohybrids at different particle concentrations, keeping the water/decalin volume ratio at 0.65. Quantification of the observed trends is done in Figure 5, which includes more samples than those illustrated in Figure 4. The droplet size is seen to decrease as a function of particle concentration, a trend previously observed for other emulsion systems.[15] At high particle concentrations a severe particle flocculation, similar to those observed at high water/decalin ratios, was observed; the high viscosity of the continuous phase may again play a role in stabilizing droplets of very small size. Interestingly, the emulsion droplets appear to cluster at all particle concentrations, perhaps as a result of nanotube−nanotube

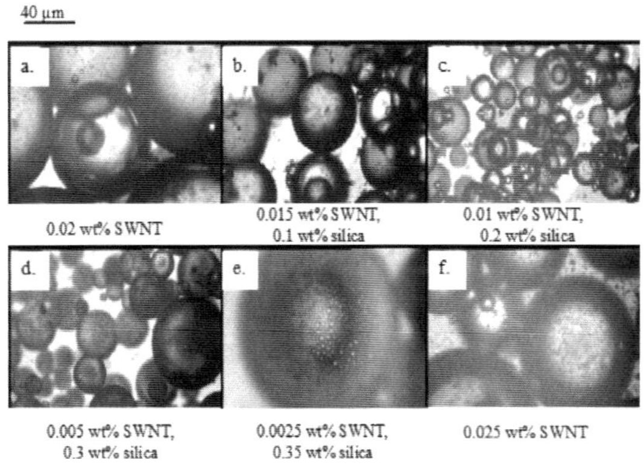

Figure 6. Microscopy images of emulsions prepared by physical mixtures with different amounts of purified SWNTs and silica particles. The water/decalin volume ratio was fixed at 0.4/1. The bar length represents 40 μm.

attraction. Similarly, these attractive forces may be responsible for the observed flocculation of the excess solid particles around the emulsions droplets with formation of a rather open network structure. The emulsion volume fraction and average droplet appear to follow opposite trends. For solid concentrations higher than 0.4 wt %, the addition of the nanohybrids does not increase the emulsion volume, and only result in increasing extent of the formation of the composite flocs as found in the oil phase (Figure 4), and a slight decrease in the average droplet diameter. However, for the values below 0.4 wt %, almost a linear increase as a function of particle concentration can be found, in accordance with the previous reports using other solid particles as emulsifiers.[15] Decreasing solid particle concentration not only leads to larger emulsion droplets but also to a greater heterogeneity of the droplet size, implying a less efficient emulsification under these conditions.

3.2.3. Formation of Multiemulsions. Interestingly, multiemulsion droplets (o/w/o) can be observed for particle concentration lower than 0.16 wt %, and more than half of the emulsion droplets were found to possess this structure for the lowest particle concentration (0.05 wt %). The inner o/w emulsion droplets are around 1 μm in diameter. For particle-stabilized emulsions, multiemulsion structures have been proposed to form in the presence of both hydrophobic and hydrophilic particles[43] or in the case of phase inversion.[15] In this case, it might be expected that some silica particles with low coverage of SWNT may also migrate to the decalin/water interface and form the o/w type of emulsion. The TEM images (Figure 1) and XPS data (Table 1) clearly show that some silica particles are barely covered by carbon nanotubes. It is interesting that this multiemulsion structure can only be observed when decalin is used as organic phase, but not when using toluene, hexane, or cyclohexane. This disparity may be due to the different emulsion stabilities obtained when using decalin as organic phase compared to other organic compounds. Decalin generated the highest amounts of emulsions with the smallest average droplet sizes, which might be related to the significantly higher surface tension of decalin (~32 mN/m) compared to that of the other hydrocarbons investigated (~18 to 25 mN/m).

The formation of multiemulsions is related to the different hydrophobic properties of the SWNT−silica particles. When we

(43) Midmore, B. R.; Herrington, T. M. *Prog. Colloid Polym. Sci.* **1999**, *112*, 115.

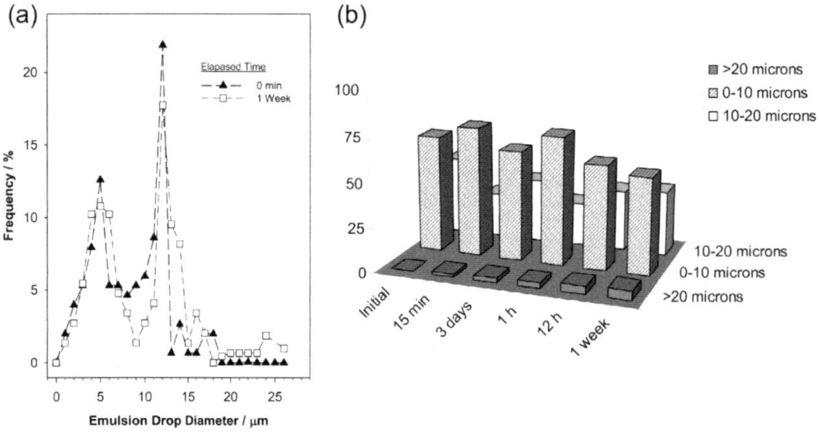

Figure 7. (a) Distribution of droplet diameters for a freshly prepared emulsion (▲) and after one week (□). The water/oil ratio was 0.65/1 and the SWNT−silica hybrids concentration was 0.15 wt %. (b) Bar graph showing the time evolution of the three major droplet diameter fractions for the same emulsion.

compared the emulsions stabilized by SWNT−silica nanohybrids to those stabilized by either purified SWNT or physical mixtures of SWNT and silica particles, interesting differences were observed. Figure 6 shows the microscopy images taken for the emulsion droplets formed after adding physical mixtures containing different amounts of silica (hydrophilic) and purified SWNT (hydrophobic) into decalin/water solutions. All the emulsions were prepared keeping the water/decalin volume ratio fixed at 0.4/1. As seen in the sequence in Figure 6, panels a to e, as the amount of the silica particles added increased and SWNT decreased, the emulsion droplet sizes initially decreased; but further addition of silica particles resulted in an opposite effect. At this point, the decrease in SWNT concentration becomes the dominant factor affecting the droplet diameters. Most importantly, under these conditions multiemulsion droplets can be observed (Figure 6, panels d and e), indicating that low concentration of hydrophobic particles may lead to large droplets and formation of these multiemulsion structures. Figure 6f shows one image taken for the same concentration of SWNT as that in Figure 6e, but without the addition of silica particles. As mentioned above, pure SWNTs produce much less amount of emulsion, the droplets are of smaller size, and no multiemulsion structure can be observed, confirming the role played by the silica particles.

3.2.4. Stability of Emulsions. We have observed that the emulsions prepared with the SWNT−silica nanohybrids exhibit high stability, even when heated as high as 70 °C or when placed in acidic media (i.e., pH < 1). On the other hand, these emulsions can be easily separated by mechanical means. We found that two clean liquid phases are readily obtained after centrifugation or after filtration. Moreover, the filtered nanohybrids can be reused to form a new emulsion.

3.2.5. Droplet Size Distribution. Figure 7 shows the droplet diameter distribution of the different emulsions prepared with a water/oil ratio of 0.65/1 and 0.15 wt % SWNT−silica nanohybrids. It is expected that coalescence, sedimentation, and Ostwald ripening may lead to changes in size distribution. In fact, small changes are observed as a function of time, but they are relatively small. Right after the preparation, the emulsion appears to be bimodal, exhibiting two main diameter distributions centered at around 5 and 12 μm; the elapsed time of even a week does not significantly affect the location of the center and width of the diameter distributions, evidencing the remarkable stability typical observed for solid-stabilized emulsions. It is worth noting that after three days a few emulsion droplets of diameters larger than 24 μm start being seen, indicative of a small extent of Ostwald ripening. This fraction seems to increase slowly with time.

3.3. Oil-in-water Emulsions Stabilized by the SWNT−Silica Nanohybrids. To gain more control on the type of emulsions that we can obtain, we attempted to modify the hydrophilicity of the SWNT−silica nanohybrids, based on the concept that the emulsions are o/w or w/o depending on the hydrophilic/hydrophobic balance. Accordingly, we attempted to enhance the hydrophilicity of the hydrophobic component (SWNT) and determine whether emulsion inversion can be achieved.

3.3.1. Nitric Acid Modification of SWNT−Silica Nanohybrids. Oxidization by nitric acid has proven to be a convenient way for grafting functional groups (−COOH) on wall defects and particularly at the ends of SWNT, a process that greatly changes the hydrophilicity of SWNTs.[44−46] Yi et al. have recently found that carbon nanotube capsules can be self-assembled in w/o emulsions when MWNTs are treated with nitric acid.[47] We have found that the same strategy is highly effective for modifying the emulsification properties of SWNT−silica nanohybrids. By controlling the extent of SWNT functionalization we have been able to manipulate the resulting emulsion properties. To reach varying degrees of functionalization, the SWNT−Silica nanohybrids were first bath sonicated in 8 M nitric acid at different temperatures (25−90 °C) and for different times (1−3 h), then thoroughly washed with water until the pH was neutral. The resulting solids were dried, then horn-ultrasonicated in water for 3 h, and homogenized in different amounts of decalin and water to prepare the emulsions. The extent of functionalization was monitored by following the absolute intensity of the G band, as well as the D/G band ratio in the Raman spectrum.

(44) Hamon, M. A.; Chen, J.; Hu, H.; Chen, Y.; Rao, A. M.; Eklund, P. C.; Haddon, R. C. *Adv. Mater.* **1999**, *11*, 834–840.
(45) Chen, J.; Hamon, M. A.; Hu, H.; Chen, Y.; Rao, A. M.; Eklund, P. C.; Haddon, R. C. *Science* **1998**, *282*, 95.
(46) Tchoul, M. N.; Ford, W. T.; Lolli, G.; Resasco, D. E.; Arepalli, S. *Chem. Mater.* **2007**, *19*, 5765.
(47) Yi, H. X.; Song, H. H.; Chen, X. H. *Langmuir* **2007**, *23*, 3199.

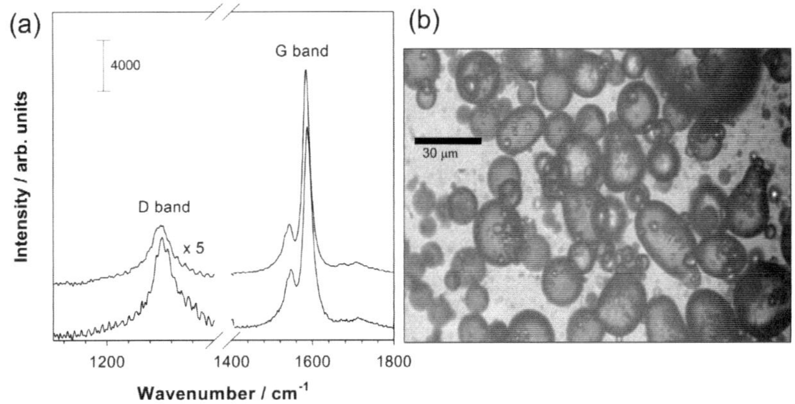

Figure 8. Characterization of the emulsions prepared by the SWNT–silica hybrids oxidized by bath sonication in 8 M HNO_3 for 3 h at 25 °C. (a) Comparison of the Raman spectra acquired for the untreated (upper spectrum) and oxidized (lower spectrum) SWNT–silica hybrids. (b) Microscopy image of the emulsion containing 0.2 wt % oxidized (25 °C) SWNT–silica hybrids (water/decalin volume ratio = 1.8/1).

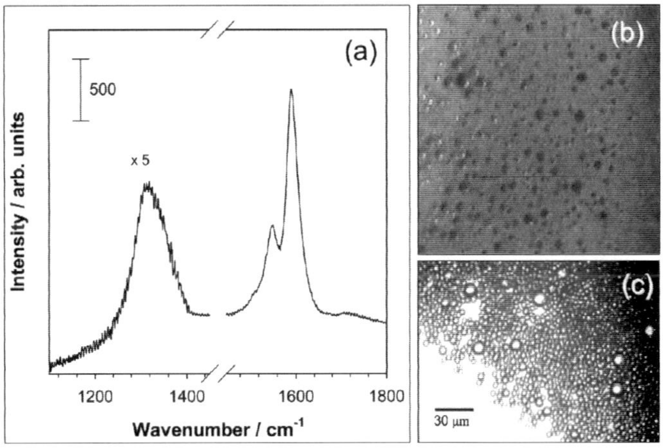

Figure 9. (a) Characterization of the emulsions prepared by the SWNT–silica hybrids heavily oxidized by heating in 8 M HNO_3 at 90 °C for 3 h. (b) Photographs of the emulsion using the dye test, that is, water-soluble green dye and oil-soluble red dye. (c) Microscopy image of the oil-in-water emulsion with heavily oxidized nanohybrid concentration of 0.2 wt % and water/decalin volume ratio of 1/1.

For example, Figure 8a shows the Raman spectrum acquired for the SWNT–silica nanohybrid sample oxidized in nitric acid at room temperature for 3 h. Clearly, compared with the untreated nanohybrids (Figure 1b), the mild oxidation leads to a higher intensity of the D band at ~1300 cm^{-1}, indicating the generation of oxidized defects, which increase the hydrophilicity of the SWNT.[44] Figure 8b shows one microscopic image taken for the emulsion prepared by 0.2 wt % RT oxidized SWNT–silica nanohybrids with a water/decalin volume ratio of 1.8/1. For this mildly oxidized sample, the emulsion type still was w/o. However, an interesting phenomenon starts to occur. Many of the emulsion droplets were not spherical in shape. This clearly indicates that surface modification of nanohybrids dramatically changes the particle hydrophobicity and modifies the interfacial tension of the emulsion droplets.[48]

3.3.2. Emulsion Inversion from w/o to o/w. Emulsion inversion in standard emulsions can be accomplished by simply varying the water/oil ratio. This is not the case on the emulsions stabilized by SWNT–silica particles. As the oxidation

(48) Thareja, P.; Velankar, S. S. *Colloid Polym. Sci.* **2008**, *286*, 1257.

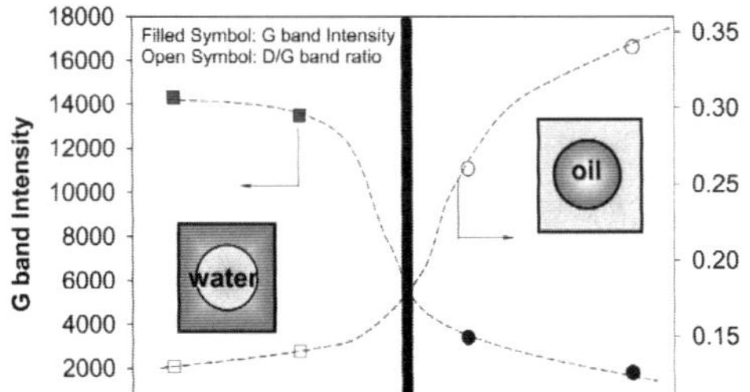

Figure 10. Characterization of SWNT/silica hybrids oxidized at different temperatures by Raman spectroscopy. G band intensity (solid symbols, left axis) and D/G band ratio (open symbols, right axis).

temperature increases beyond 60–70 °C, a more dramatic change in behavior occurs. As illustrated in Figure 9, the now heavily

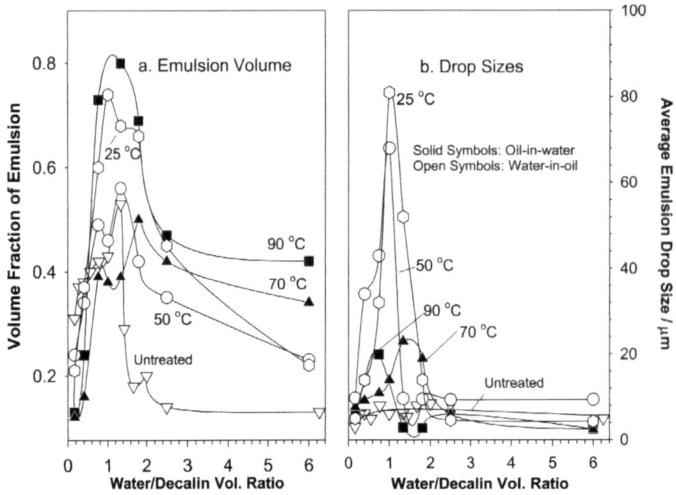

Figure 11. Effect of the oxidizing pretreatment on emulsion characteristics. Emulsion volume fraction (left panel) and emulsion drop diameter (right panel) as a function of water/decalin volume ratio for the SWNT–silica hybrids oxidized at different temperatures with hybrid concentration as 0.2 wt %.

oxidized nanohybrids are able to stabilize o/w emulsions, rather than w/o. The photo taken for the emulsion prepared by nanohybrids oxidized at 90 °C (Figure 9a) clearly shows that the dispersed phase is the oil (red oil-soluble dye added), whereas the continuous phase is water (green water-soluble dye added). In addition, the microscopic images taken for this emulsion (Figure 9b) also exhibit interesting differences compared to those taken for the w/o emulsion stabilized by untreated nanohybrids. In this case, the emulsion droplet sizes are remarkably small and uniform, and the droplets appear to be more closely packed than in the w/o emulsions.

This inversion parallels the greater extent of functionalization obtained at higher oxidation temperatures. Figure 10 shows that when the oxidation temperature exceeds 60–70 °C, the intensity of the G band greatly decreases, the D/G ratio increases, and the emulsion phase inversion occurs. Clearly, the higher extent of oxidation enhances the hydrophilic character of the nanohybrids, so the curvature of the interface changes, forcing the oil phase to the inner part of the droplet.

Figure 11 shows the emulsion volume fraction (left axis) and emulsion droplet sizes (right axis) of emulsions prepared from the oxidized and untreated nanohybrids. Similar trends were observed for all these emulsions regardless of the emulsion types, in terms of maxima in both emulsion volume and droplet sizes reached at intermediate water/decalin volume ratios. However, the emulsions stabilized by the oxidized nanohybrids can retain a larger amount of emulsion at higher water/decalin volume ratio. For example, at a water/decalin volume ratio of 1.6/1, all of the oxidized SWNT–silica nanohybrids can still produce >42% volume fraction of emulsion, whereas the untreated nanohybrids only makes about 13% volume fraction of emulsion. As a result, a much lower extent of particle aggregation is found when using these oxidized nanohybrids. In addition, the average diameters of the emulsion droplets stabilized by the oxidized nanohybrids are strongly dependent on the oxidation temperature used in the nitric acid treatment. The nanohybrids oxidized at 25 °C produce emulsion droplets with an average diameter of 80 μm for a water/decalin volume ratio of 1/1, much larger than those produced by untreated nanohybrids. However, the emulsion droplet diameters decrease as a function of the oxidation temperature. In particular, a remarkable decrease is seen after the point at which the emulsion phase inversion happens. For example, the sample treated at both 70 and 90 °C stabilize the droplets with diameter of ~2.7 μm when water/decalin volume ratio higher than 1.4, even smaller than the values obtained for the untreated nanohybrids. This indicates that the initial w/o type of emulsions might pass through a regime of instability and then invert to stable o/w emulsions as the surface hydrophilicities increase after oxidation treatment at higher temperatures.

It is important to point out that these wide variations in emulsion characteristics are highly reproducible and controllable by relatively simple pretreatments, which make the SWNT–silica nanohybrids as an attractive solid emulsifier. The emulsion volume fractions and droplet sizes can be easily manipulated by adjusting the particle concentrations and water/oil volume ratios. Compared with previously reported emulsions employing purified SWNT,[49] the ones reported in this contribution not only avoid the complex purification of SWNT but also result in much higher volume fractions of emulsions. Hence, this method appears promising in making a large quantity of stable emulsions given that these types of SWNT–silica nanohybrids are already being produced on commercial scales.

3.3.3. Potential Applications of Emulsions Stabilized by SWNT–Silica Nanohybrids. These solid-stabilized emulsions are particularly suitable for creating catalysts that operate at the water/oil interface in biphasic systems, found in important processes such as bio-oil refining or fine chemical synthesis. After carrying out the interfacial reaction, the SWNT–silica nanohybrids can be easily separated by filtration or centrifugation. Without the solids, which can be reused in subsequent reactions, the emulsion is broken and the two liquid phases are readily separated. An interesting application that we are currently investigating is using either fraction of the nanohybrid (i.e., nanotube side or the silica side) as a support for heterogeneous catalyst on which reactions can selectively take place at either the

(49) Lau, K. K. S.; Bico, J.; Teo, K. B. K.; Chhowalla, M.; Amaratunga, G. A. J.; Milne, W. I.; Mckinley, G. H.; Gleason, K. K. *Nano. Lett.* **2003**, *3*, 1701.

aqueous or oil phase. As shown here, we can modify the type of emulsion from w/o to o/w, so different types of liquid biphasic reactions can be carried out. We envision depositing different catalysts on either side of the interface and selectively convert functional groups of molecules that are water-soluble or oil-soluble at the same time in a selective manner. For example, this "phase selectivity" might allow us to work with complex mixtures and selectively hydrogenate water-soluble molecules in the aqueous phase in the presence of oil-soluble molecules that we do not want to hydrogenate. Working with emulsions, one can have the added benefit of catalytically converting at the interface molecules that are soluble in one phase so they become insoluble products in that phase and migrate to the other phase for easy separation.

Another important application of these emulsions is the preparation of nanotube-based composites via emulsion polymerization. In-situ emulsion polymerization has been reported for Pickering emulsions stabilized by Janus particles.[50] In previous studies, we have used surfactant-stabilized emulsions to generate a dispersed medium for polymerization, after which the resulting composite had a good dispersion of the nanotubes.[51] Using the solid emulsifiers reported in this contribution, particularly those stabilizing o/w emulsions, we plan to expand the studies to produce SWNT–silica-polymer composites, which already have shown promise to enhance electrical and mechanical properties.[33]

4. Conclusions

We have shown that particle-stabilized w/o and o/w emulsions can be prepared by SWNT–silica nanohybrids. Water/oil volume ratio, particle concentration, and nitric acid oxidation pretreatment greatly affect the emulsion type, emulsion volume fractions, and average droplet diameters, in a reproducible and controllable way. The resulting emulsions exhibit good resistance to coalescence and sedimentation. This novel way to prepare emulsions by supported SWNT shows great advantage in achieving high emulsion volume compared with using only SWNT, and might open new opportunities for utilizing these materials as supports for heterogeneous catalysis to be used in biphasic systems. These catalyst/emulsifiers would be easily removed from the reaction system, breaking the emulsion, and obtaining the two liquid phases containing products. Also, the solid-stabilized emulsions of the o/w type can be used to conduct in situ polymerization with nanotubes kept in a highly dispersed state at the oil/water interface.

Acknowledgment. The project was supported by the Department of Energy—Basic Energy Sciences (Grants DEFG03-02ER15345 and DE-FG02-06ER64239), the Oklahoma Center for Advancement of Science and Technology (OCAST), and the Advanced Energy Consortium (AEC).

(50) Walther, A; Hoffmann, M; Mueller, A. H. E. A. *Angew. Chem., Int. Ed.* **2008**, *47*, 711.

(51) Barraza, H.; Pompeo, F.; O'Rear, E.; Resasco, D. E. *Nano. Lett.* **2002**, *2*, 79.

Aligned Carbon Nanotube Films by Lyophilization

Matthew J. Kayatin, Department of Chemical Engineering, Auburn University, Auburn, AL

Virginia A. Davis, Department of Chemical Engineering, Auburn University, Auburn, AL

We have developed a novel method for assembling oxidized single-walled carbon nanotubes (SWNTs) into anisotropic films using lyophilization. The initial and final SWNT surface chemistry was characterized using attenuated total reflectance Fourier transform infrared spectroscopy and X-ray photoelectron spectroscopy. The effect of oxidation treatment time was studied under scanning electron microscopy and Raman spectroscopy revealing the evolution into an ordered structure arising from an increased degree of oxidation and increased mobility by length shortening of the tubes. In addition, the use of HNO_3 and H_2SO_4 for nanotube oxidation is particularly attractive since unpurified carbon nanotubes can potentially be purified and assembled in one process. The resultant carboxylic acid functional groups provide a plethora of possibilities for further functionalization routes. Finally, lyophilization studies were also performed from surfactant stabilized dispersions using both ionic and steric surfactants. Potential film applications include layered nanocomposites, specifically those where anisotropic property enhancement is desired.

See more of this Session: Carbon Nanotubes II: Characterization, Functionalization, and Applications
See more of this Group/Topical: Nanoscale Science and Engineering Forum

Optimizing the Synthesis of Carbon Nanotube Fibers Utilizing Biocompatiable Components

John M. Landers, Department of Chemical and Biochemical Engineering, Rutgers University, Piscataway, NJ

Alexander V. Neimark, Department of Chemical and Biochemical Engineering, Rutgers University, Piscataway, NJ

The eloquent properties of carbon nanotubes (CNT) with their high conductivity, high aspect ratio and high tensile strength make them an optimal material with broad applications. Of these, the production of fibers to serve as artifical actuators can be achieved by wet spinning a dispersion of CNT into a viscous polymer bath. Upon injecting the dispersion into the bath, the polymer will coagulate around the liquid dispersion thus fortifying it as an usable material. However,in order to serve as an actuator a sensitive balance is required among the fibers inherent strength, conductivity and biocompatability. Of these three, the last dictates that the fiber produced is not subject to an attack by the body upon insertion. In order to circumvent this issue, various biomolecules are utilized as both the dispersant agent to separate the CNT bundles and as the coagulant agent in the production of these CNT fibers.

See more of this Session: Carbon Nanotubes II: Characterization, Functionalization, and Applications
See more of this Group/Topical: Nanoscale Science and Engineering Forum

Effects of Nanoparticle Properties On Cellullar Uptake and Magnetic Fluid Hyperthermia

Madeline Torres-Lugo, Chemical Engineering, University of Puerto Rico, Mayaguez Campus, Mayaguez, PR

Hector Rodriguez-Luccioni, Chemical Engineering, University of Puerto Rico, Mayaguez Campus, Mayaguez, PR

Adriana Herrera, Chemical Engineering, University of Puerto Rico, Mayaguez Campus, Mayaguez, PR

Magda Latorre-Estevez, Chemical Engineering, University of Puerto Rico, Mayaguez Campus, Mayaguez, PR

Janet Mendez, Chemical Engineering, University of Puerto Rico, Mayaguez Campus, Mayaguez, PR

Carlos Rinaldi, Chemical Engineering, University of Puerto Rico, Mayaguez Campus, Mayaguez, PR

Suspended ferrite magnetic nanoparticles are known to dissipate energy under an oscillating magnetic field. Such energy dissipation could be employed to locally raise temperature inside a tumor to 41-45¢XC (hyperthermia), promoting cell death. This novel treatment is known as magnetic fluid hyperthermia (MFH). Even though clinical work has been conducted in relation to this phenomenon, little is known regarding the mechanism of cell death and the main factors involved. This work focused on the examination of cell death mechanisms when an AC magnetic field is applied. The temperature range was maintained between 41"aC and 45"aC at a frequency of 237 kHz and a magnetic field of 2.8 kA/m. Two cell models (Caco-2 and MCF-7) were employed with distinct known properties. Magnetite nanoparticles were synthesized by coprecipitation or by thermodecompositon. Particles were functionalized with different polymers, including commercial carboxymethyldextran, with 23-COOH or with 5COOH, and poly (ethylene glycol). Cytotoxicity studies performed during one week of exposure indicated that coprecipitated particles with commercial carboxymethyldextran demonstrated cytotoxic effects at concentrations higher than 0.9 mg magnetite/mL, while cell viability was not affected in the concentration range of 0.05-1.5 mg magnetite/mL with nanoparticles synthesized by thermodecomposition. Nanoparticle uptake was performed utilizing carboxymethyl dextran with various degrees of substitution. The purpose of these experiments was to elucidate the effects of charge on particle internalization utilizing the same polymer chain. Results indicated that particle uptake was in the range of 0.5-0.6 pg magnetite/cell with magnetite-CMDX (~23 COOH groups/chain). Magnetite-CMDX5COOH demonstrated no internalization in both cell models, suggesting that the degree of substitution showed an effect on particle uptake. On the other hand Magnetite-PEG nanoparticles demonstrated the highest uptake by MCF-7 cells of 0.8-1.0 pg/cell. These results suggest that nanoparticle uptake was affected by nanoparticles functionalization and cell type. Magnetite-CMDX nanoparticles were selected to perform MFH experiments. Hot water hyperthermia was performed and compared with magnetic fluid hyperthermia. Magnetic fluid hyperthermia exposure times (i.e application of magnetic field) were 60 min and 120 min. Cell viability was measured immediately after treatment, 24 or 48 hours after treatment. Results related to hot water hyperthermia indicated that cell viability effects were observed 48 hours after treatment, while the effects on cell viability of magnetic fluid hyperthermia were detected 24 hours after treatment. Results indicated a significant reduction in cell viability for a magnetic field exposure of 120 min and a resting time of 48 hours when compared to hot water hyperthermia. Hot water hyperthermia decreased cell viability to 30-45% while magnetic fluid hyperthermia decreased cell viability almost completely, 2-5%. MCF-7 cells appeared to be more sensitive to MFH than Caco-2 cells. The process of cell death appears to be mainly apoptotic. The results presented here suggest that apoptosis was likely induced during MFH and it has a higher effect on cell viability when compared to hot water hyperthermia. These results also suggest that MFH might be causing additional effects on cell viability when compared to hot water hyperthermia.

See more of this Session: Magnetic Nanoparticles in Biotechnology and Biomedicine II

See more of this Group/Topical: Nanoscale Science and Engineering Forum

A Study of Magnetic Nanoparticles for Hyperthermia and Their Behavior in Biological Systems and Thermally-Responsive Hydrogels for Drug Delivery

Mary Kathryn Sewell, Chemical and Biological Engineering, University of Alabama, Tuscaloosa, AL

Christopher S. Brazel, Department of Chemical and Biological Engineering, The University of Alabama, Tuscaloosa, AL

A combination cancer therapy using an environmentally-responsive treatment platform has the potential to improve the efficacy of cancer treatment, while decreasing negative side effects currently associated with such treatments. The proposed platform consists of a thermally-responsive hydrogel embedded with magnetic nanoparticles (MNPs) and an anti-cancer therapeutic agent. The novel design allows for concurrent hyperthermia and chemotherapy treatments, as application of an external magnetic field causes the MNPs to heat, triggering the thermal responsive of the gel. Thermally-responsive poly(N-isopropylacrlamide-co-acrylamide) hydrogels were synthesized and embedded with magnetic nanoparticles. Drug release studies show that increasing the concentration of FePt nanoparticles in a gel has negligible effect on the diffusion of a model drug. $CoFe_2O_4$ magnetic nanoparticles were synthesized in house and tagged using the red fluorescent dye rhodamine. These MNPs were then introduced to two biological systems: HeLa cancer cells and *C. elegans* worms. The localization and toxicity of the MNPs were studied. At concentrations below 5 wt% in media, both HeLa cells and the worms, the rhodamine-tagged $CoFe_2O_4$ showed negligible toxicity. Images obtained from fluorescence microscopy show the nanoparticles internalized in the cytoplasm of HeLa cells and the GI tract of the *C. elegans* worms.

See more of this Session: Magnetic Nanoparticles in Biotechnology and Biomedicine II
See more of this Group/Topical: Nanoscale Science and Engineering Forum

Magnetic Nanoparticle-Triggered Liposomal Drug Delivery for Cancer Imaging and Treatment

Jae-Ho Lee, Center for Cancer Research Nanobiology Program, National Cancer Institute, Frederick, MD

Robert Blumenthal, Center for Cancer Research Nanobiology Program, National Cancer Institute, Frederick, MD

We will describe drug release from liposomes formulated with superparamagnetic iron oxide nanoparticles (SPION) and drugs for potential applications in magnetic resonance imaging (MRI) and hyperthermia-mediated drug release. SPION are a class of MRI contrast agents and depending on the crystal size of the iron oxide they can also be used as hyperthermia agents via excitation by an alternative magnetic field (AMF). Liposomes can encapsulate drugs in either the bilayer or inner aqueous phase. Therefore, combination of liposomes and SPION can be used to meet the requirements for cancer treatment. In this study, we use calcein and doxorubicin (DOX) as the model drug agents. Doxorubicin loading protocols using the pH gradient strategy and methods for SPION incorporation in aqueous phase of liposomes (magnetoliposomes) have been reported in literature. However, attempts to co-encapsulate doxorubicin and SPION in the liposomes have been limited due to interactions of negative charged SPION with the positively charged DOX resulting in aggregation. Our studies show that the encapsulation of DOX and SPION results in complexation of the two agents followed by precipitation in solution. In addition, the current pH gradient method for maximizing doxorubicin encapsulation causes release of iron from SPION. We have investigated new strategies to prevent the precipitation of the two agents and maximize drug encapsulation, by modifying SPION surface electrostatically and hydrophobically. We screened out several SPION for hyperthermia agent using thermo-sensitive liposome formulation and tested nanoparticle-triggered liposomal drug delivery in a gel system. We will also discuss our approaches for nanoparticle-triggered liposomal drug delivery and how the surface properties of SPION affect the encapsulation of SPION and drugs.

See more of this Session: Magnetic Nanoparticles in Biotechnology and Biomedicine II
See more of this Group/Topical: Nanoscale Science and Engineering Forum

PEG-Based Magnetic Hydrogel Nanocomposites for Combined Chemotherapy and Hyperthermia Treatment of Cancer

Samantha A. Meenach, Chemical & Materials Engineering, University of Kentucky, Lexington, KY

Chinedu G. Otu, Department of Biology, University of Kentucky, Lexington, KY

Christopher R. Barton, Chemical & Materials Engineering, University of Kentucky, Lexington, KY

J. Zach Hilt, Chemical & Materials Engineering, University of Kentucky, Lexington, KY

Kimberly W. Anderson, Chemical & Materials Engineering, University of Kentucky, Lexington, KY

Hyperthermia, the heating of tissue to 41 to 45°C, has been shown to improve the efficacy of cancer therapy when used in conjunction with irradiation and/or chemotherapy. Current limitations include but are not limited to non-homogeneous distribution of temperature over the cancer site, low specificity, patient discomfort, and lack of localized delivery of heat. These shortcomings can be overcome through the utilization of polymeric biomaterials which can be delivered at tumor sites and remotely heated from outside the body while simultaneously delivering therapeutic drugs. Hydrogel nanocomposites which are composed of biocompatible crosslinked polymeric material and nanoparticulates capable of heating upon exposure to an electromagnetic field are a potential platform of polymeric biomaterials for this application. In addition to the remote heating, the network structure of the hydrogel allows for the tailored delivery of therapeutic agents. In this work, hydrogel nanocomposites have been developed that can be implanted or injected within or near a tumor site and deliver both heat and a chemotherapeutic agent (e.g., paclitaxel). In previous studies, the efficacy of paclitaxel has been proven to increase when combined with elevated temperatures. Here, the nanocomposites studied involve a stealth, poly(ethylene glycol) and iron oxide nanoparticle-based system. The iron oxide nanoparticles were physically entrapped in the hydrogel matrix which was comprised of poly(ethylene glycol) methyl methacrylate macromer and poly(ethylene glycol) dimethacrylate as the crosslinker.

Thermal analysis showed the capability of the hydrogels to be heated in an alternating magnetic field to various target surface temperatures depending on the strength of the field and composition of the hydrogel matrix. Hydrogels with lower swelling ratios proved to heat better than those with higher ratios due to the greater polymeric content in the gels which corresponds to higher iron oxide loading in the gels. The heating of the hydrogel nanocomposites was controlled by the strength of the magnetic field and hyperthermia-specific temperature ranges were reached for all hydrogel systems at field strengths ranging from 12.7 to 17.4 kA/m. To examine the cytocompatibility of the hydrogels, NIH 3T3 murine fibroblasts were exposed to the hydrogel systems, and the viability of the cells was not affected by leachants from the hydrogels. The delivery of a model drug, rifampicin, from the hydrogel nanocomposites was investigated and was shown to be controlled via the hydrogel matrix. This drug was chosen due to its similarity in hydrophobicity and molecular weight to paclitaxel. The delivery of paclitaxel from the hydrogels is also being investigated. An initial proof-of-concept demonstration was completed where M059K glioblastoma multiforme cells were heated to thermoablative temperatures (above 55°C) with the hydrogel nanocomposites exposed to the AMF and cell death was demonstrated. Studies with both M059K glioblastoma and MDA MB 231 breast adenocarcinoma cells where the cells are exposed to both hyperthermia-specific temperatures and paclitaxel (i.e., delivered directly in the cell media) to study if the efficacy of paclitaxel in inducing cell death with and without moderate heat are underway. In summary, the ability of these systems to heat and their capability to deliver drugs to specific sites make them viable systems for treating of deep-seated tumors or reoccurring cancer.

See more of this Session: Magnetic Nanoparticles in Biotechnology and Biomedicine II

See more of this Group/Topical: Nanoscale Science and Engineering Forum

Radiofrequency Actuation of Iron Oxide-Hydrogel Nanocomposites: Experimental Analysis and Modeling

Nitin S. Satarkar, Chemical & Materials Engineering, University of Kentucky, Lexington, KY

Samantha A. Meenach, Chemical & Materials Engineering, University of Kentucky, Lexington, KY

Christopher R. Barton, Chemical & Materials Engineering, University of Kentucky, Lexington, KY

Kimberly W. Anderson, Chemical & Materials Engineering, University of Kentucky, Lexington, KY

J. Zach Hilt, Chemical & Materials Engineering, University of Kentucky, Lexington, KY

Hydrogels and hydrogel nanocomposites are attractive materials for a variety of biomedical/pharmaceutical applications. In recent studies, magnetic hydrogel composites of N-isopropylacrylamide (NIPAAm) hydrogels have been demonstrated as pulsatile drug delivery systems with radiofrequency(RF) at 293 kHz as a trigger.1 In another study, magnetic nanocomposites were demonstrated as actuators in microfluidic device, with ability to control the flow by application of RF.2 Some of the other applications being pursued with these systems is use of RF induced heating for hyperthermia cancer treatment. All these applications rely heavily on design of the nanocomposite systems. In order to achieve desired response from hydrogel nanocomposite system, it is very important to understand the heating induced by application of RF and resultant change in the hydrogel properties.

Goal of this study was to analyze and predict nanocomposite response based on hydrogel composition, nanoparticle loading, RF strength, and nanocomposite dimensions. Magnetic nanocomposites of poly (ethylene glycol) hydrogels were synthesized with iron oxide (Fe_3O_4) nanoparticles. Different nanocomposite systems were obtained by variation of nanoparticle loadings and hydrogel composition. 293 kHz RF was applied to nanocomposite discs and surface temperatures were analyzed using infrared (IR) thermography. The strength of RF was varied to further analyze the heating effect.

A heat transfer model was developed to account heat generation due to RF as well as heat loss to surroundings. By fitting the temperature data, a correlation was obtained for dependence of heat generation on nanoparticle loadings, and RF strength. It was demonstrated that the model could successfully predict the resultant temperatures by using a hydrogel system with different swelling properties. The heating analysis was further extended for potential hyperthermia cancer treatment applications. The bioheat equation was used to account the heat transport in vivo, and resultant nanocomposite temperatures were predicted.

In the case of NIPAAm nanocomposites, RF heating is accompanied by collapse, which forces the fluids out of hydrogel network. Magnetic NIPAAm nanocomposites of different dimensions (thickness) were synthesized and subjected to RF heating. The resultant collapse and recovery process was quantified by weight measurements at different time points. The heat transfer model was modified to account change in the mass as hydrogel collapsed. Furthermore, a mass transfer model was developed for transport of liquid as the nanocomposite collapsed and recovered.

1. Satarkar, N. S.; Hilt, J. Z. J. Control. Rel. 2008, 130, 246-251.

2. Satarkar, N. S.; Zhang, W.; Eitel, R. E.; Hilt, J. Z. Lab Chip 2009, In Press.

See more of this Session: Magnetic Nanoparticles in Biotechnology and Biomedicine II
See more of this Group/Topical: Nanoscale Science and Engineering Forum

Effect of AEM Probe Configuration On Heating of Magnetic Nano Particles for the Application of Cancer Hyperthermia

Krishna Kanth Sanapala, Chemical Engineering, University of Louisville, Louisville, KY

Kapila Hewaparakrama, Chemical Engineering, University of Louisville, Louisville, KY

Mustafizur Rahman, Chemical Engineering, University of Louisville, Louisville, KY

Kyung A. Kang, Chemical Engineering, University of Louisville, Louisville, KY

Magnetic nanoparticles, when subjected to alternating electromagnetic (AEM) field, generate heat mainly due to Neel and Brown relaxation. In magnetic particle mediated hyperthermia, the particles accumulated in the tumor are subjected to a non-invasively applied AEM field. For the low heat hyperthermia to be highly effective the temperature of the magnetic particle containing tumor is maintained in the range of 42~45 ^{o}C. Iron oxide (Fe_3O_4) nano particles have attracted considerable interest in biomedical application purely because of their biocompatibility. The most important issue in this therapy is applying an appropriate amount of energy to kill tumors of various sizes and at various depths, and to avoid the normal tissue damage, in a non-invasive and user-friendly way. The heat energy generated by the magnetic nano particles is dependent on (1) AEM unit operating parameters (power, time of AEM field application etc.), (2) configuration of the probe, and (3) size of the particles.

In this study, the effectiveness of three configurations of the probe (solenoid, pancake and sandwich) on the heating nano particles was analyzed. The study correlates the heating with the configuration of the probe and concentration of iron oxide particles and size and depth of the tumor. Our study results indicated that the solenoid type probe, which is most commonly used, generates maximum AEM field in the core but was not user-friendly. The probe can also be a pancake type or sandwich configuration with adjusted distance. Pancake probe was better for treating surface tumors but the field doesn't penetrate deep. Sandwich probe(s) provide controlled intensity of magnetic field that can treat deep seated tumors.

See more of this Session: Magnetic Nanoparticles in Biotechnology and Biomedicine II
See more of this Group/Topical: Nanoscale Science and Engineering Forum

Dissipative Particle Dynamics Simulation On the Effect of Polymeric Coatings in Magnetic Fluid Hyperthermia

MIchael Tomasini, Biomedical Engineering, Rutgers University, Piscataway, NJ

M. Silvina Tomassone, Rutgers University, Piscataway, NJ

Expanding on previous work, we attempt to study Magnetic Fluid Hyperthermia (MFH) using molecular simulations. MFH involves magnetic nanoparticles (MNPs) functionalized with a biocompatible coating for the treatment of cancerous tumors. When the MNPs are placed in an oscillating magnetic field, the particles begin to rotate resulting in local mechanical stresses and local heating. While it is generally assumed that the mechanism by which MFH results in cell death is through hyperthermia, it is plausible that the mechanical forces generated by the rotating magnetic nanoparticles could induce rupture of the cell membrane, enhancing the effect of MFH. In this study, we aim to use Dissipative Particle Dynamics (DPD) simulations to elucidate the effect of a MNP polymeric coating, PEO-PEE (poly ethylene oxide and poly ethyl ethylene) block copolymer, on a model lipid bilayer (DPPC) under uniform shear.

The DPD simulation technique is a coarse-grained method in which several heavy atoms are lumped together to form an effective pseudo-particle or bead. The interaction between the beads includes only soft interactions enabling large length and long time scales. This allows for the study of MFH using systems on the order of tens of nanometers and times in the microsecond regime, the order of time for the magnetic field to switch directions.

Using a coarse graining of 3 water molecules per DPD bead (1 DPD bead per polymer monomer and 13 beads per DPPC lipid) we examine the forces and energies necessary to rupture a lipid bilayer under a shear stress. To quantify the rupture and rupture energies, we measured the surface tension the bilayer can withstand just prior to rupture, the energy necessary for rupture, and measure the diffusion of the water molecules in the plane perpendicular to the bilayer surface. We also study systematically the polymer length, rigidity, and composition to determine the optimal polymer properties for enhancing MFH through cell membrane rupture.

See more of this Session: Magnetic Nanoparticles in Biotechnology and Biomedicine II
See more of this Group/Topical: Nanoscale Science and Engineering Forum

FORC Diagrams as a Diagnostic Tool for Evaluating Nanoparticle Coating

Alexandra Teleki, Department of Mechanical and Process Engineering, Swiss Federal Institute of Technology Zurich, Zurich, Switzerland

Ann M. Hirt, Department of Earth Sciences, ETH Zurich, Zurich, Switzerland

Sotiris E. Pratsinis, Department of Mechanical and Process Engineering, Swiss Federal Institute of Technology Zurich, Zurich, Switzerland

The use of magnetic nanoparticles has become important in a wide range of biomedical and polymer nanoparticle applications. Particles are frequently coated to improve thermal stability and functionality but also clustering and therefore retention of superparamagnetic (SP) behavior. The effectiveness of coating is often evaluated using electron microscopy and magnetic hysteresis loops. First-order reversal curves (FORC) allow for a more rigorous evaluation of effective particle size and interaction. Predominantly γ-Fe_2O_3 particles with low amounts of α-Fe_2O_3 were produced by scalable flame technology and the resulting aerosol was in situ coated with SiO_2 by oxidation of hexamethyldisiloxane (HMDSO) vapor.[1,2] The individual Fe_2O_3 grains were encapsulated by thin SiO_2 films and the coating thickness was controlled by the HMDSO concentration in the reactor. The mean particle size obtained from TEM images is 22.4 nm for uncoated and slightly higher for coated particles, although particle size can vary from about 10 to 80 nm. Particles were also produced by co-oxidizing Fe and Si precursors in the flame reactor. Co-oxidation reduces the Fe_2O_3 grain size to about 8 nm and results in particles segregated in crystalline Fe_2O_3 and amorphous SiO_2 domains. FORC analysis shows that co-oxidized particles with 7- 46 wt% SiO_2 display predominantly SP behavior due to the small Fe_2O_3 grain size. The uncoated Fe_2O_3 and 7 - 23 wt% SiO_2-coated particles show a broader coercivity spectrum and therefore a larger spectrum of effective particle size; there is also a broader range of interaction fields. The FORC diagrams for samples with higher wt% SiO_2 coating and thus higher coating thickness suggest SP behavior with a narrow grain size spectrum and low degree of particle interaction. FORC diagrams are a powerful tool for assessing the effectiveness of coating treatment in the production of non-interacting SP nanoparticles.

[1] Teleki, A., M. Suter, P. R. Kidambi, O. Ergeneman, F. Krumeich, B. J. Nelson, and S. E. Pratsinis, Hermetically-coated superparamagnetic Fe_2O_3 particles with SiO_2 nanofilms, *Chem. Mat.* **in press**, (2009).

[2] Teleki, A., M. C. Heine, F. Krumeich, M. K. Akhtar, and S. E. Pratsinis, In situ coating of flame-made TiO_2 particles with nanothin SiO_2 films, *Langmuir* **24**, 12553 (2008).

See more of this Session: Magnetic Nanoparticles in Biotechnology and Biomedicine II
See more of this Group/Topical: Nanoscale Science and Engineering Forum

Shape-Controlled Tin Nanoparticle Synthesis and Its Application as Nanosoldering Materials

Karunaharan Rajathurai, Chemical Engineering, University of Massachusetts Lowell, Lowell, MA

Qingzhou Cui, Chemical Engineering, University of Massachusetts Lowell, Lowell, MA

Zhiyong Gu, Chemical Engineering, University of Massachusetts Lowell, Lowell, MA

Shape-controlled synthesis of metal nanoparticles has opened many new possibilities to design ideal nano-building blocks for future nanofabrication and nanomanufacturing. These shaped controlled nanoparticles have been proposed for a wide variety of scientific, engineering, and technical applications. Surfactant-assisted synthesis and growth is one of the mostly used methods to obtain shape-controlled nanoparticles. Nanosolders, as a concept, have been proposed and many types of nanoparticles from silver, gold, copper, etc have been fabricated to prepare solder paste. However, the study on constructing ideal nanosolders from low melting-point soldering materials including tin (Sn), indium (In), and corresponding alloys is still limited. In the present study, we report a simple strategy to prepare tin nanostructures with various sizes and shapes by controlled growth in the presence of an anionic surfactant - sodium dodecyl sulfate (SDS). We demonstrate that by adjusting SDS concentration, stirring speed, and growth time, different shapes of tin nanoparticles can be obtained.

The tin nanoparticles are characterized by field emission scanning electron microscopy (FE-SEM), transmission electron microscopy (TEM), and differential scanning calorimetry (DSC). Thermal reflow properties for the nanosolder particles were also studied. It was found that an inert environment of nitrogen (N2) was very important for the nanoparticle melting and reflow. Those tin nanoparticles show great promise as soldering material, which could enable the newly emerged nanodevice fabrication through forming robust interconnects by soldering. Constructing nanoink from these tin nanoparticles is currently underway in our laboratory.

See more of this Session: Nanoelectronic Materials I
See more of this Group/Topical: Nanoscale Science and Engineering Forum

InP/InAs Core-Shell Nanopillars on InP (111)B

Vanessa Evoen[1], Li Gao[1], Marta Pozuelo[2], Sarah Chowdhury[1], Jun Tatebeyashi[3], Baolai Liang[3], Suneel Kodambaka[2], Diana L. Huffaker[3], Robert F. Hicks[,1]*

[1]*Department of Chemical and Biomolecular Engineering, University of California, Los Angeles*

[2]*Department of Materials Science and Engineering, University of California, Los Angeles*

[3]*Department of Electrical Engineering, University of California, Los Angeles*

ABSTRACT

In this study, we demonstrate the growth of InP/InAs/InP core-shell nanopillars by metalorganic vapor-phase epitaxy (MOVPE) using indium droplets to catalyze deposition. At 400°C and a V/III ratio of 86, hexagonal pillars were grown with smooth side walls and average widths and heights of 45 and 70 nm, respectively. Transmission electron microscopy revealed that the pillars were single crystal wurtzite and were bounded by $\{1\bar{1}00\}$ sidewalls. Indium arsenide quantum wells were deposited on the nanopillars at 420°C and a V/III ratio of 120, and then capped with a thin layer of InP. The thickness of the quantum well was ~5 nm. Photoluminescence spectra at 77 K yielded a single intense band at 1750 nm (0.7 eV) with a full width at half maximum of 350 nm.

* Corresponding author. E-mail: rhicks@ucla.edu

Introduction

Compound semiconductor core-shell nanostructures represent an important class of one-dimensional systems that are of great interest for high performance electronic and photonic devices [1-4]. A capping layer protects the quantum well from harmful surface states. The energy band offset at the core-shell interface produces internal fields that lead to the accumulation of high-mobility carriers [5]. For example, the energy band line-up at the interface of Ge/Si core-shell NWs [2-4] produces a high concentration of hole carriers. This has enabled fundamental studies of quantum transport [2,3] and the realization of the best performing p-channel, nanopillar field effect transistors to date [4].

Heterojunctions formed in bulk crystalline materials with different lattice constants are subject to misfit dislocations that impair device performance. In 1D nanoscale structures, lattice mismatch can be more easily accommodated due ti the much smaller distance over which the strain field develops [6]. Hence, nanostructures have the potential for forming mismatched junctions without threading dislocations [7].

The InP/InAs/InP heterostructure is attractive for high mobility transistors because the conduction band offset of ~0.52 eV provides a good confinement potential for electrons, and because the type I quantum well structure also traps holes that may be generated thermally or as a result of impact ionization in the channel [8-9]. Both InP and InAs have a wurtzite crystal structure in the nanopillar form, although the bulk materials crystallize to a zinc-blende structure [10]. Despite a lattice mismatch of 3.1% between InAs and InP, epitaxial growth of the InAs quantum well can occur along the InP core in the lateral direction [11-14].

In this paper, we present the controlled growth of InP/InAs core shell nanopillars via metalorganic vapor-phase epitaxy (MOVPE). The InP/InAs core-shell nanopillars were prepared by the self-catalyzed growth of InP nanopillars from indium droplets, followed by deposition of the InAs quantum well and deposition of the InP cap layer. Transmission electron microscopy studies revealed single-crystal InP cores with epitaxial InAs quantum wells 5 to 10 nm in thickness. Self-catalyzed InP/InAs nanopillars have been fabricated by selective area metalorganic vapor-phase epitaxy on patterned In P (111)B substrates [15-16]. The catalyst free fabrication of self-assembled InP/InAs nanopillars presented in this work offers a new strategy for fabricating nanoscale transistors.

Experimental Methods

The experiments were carried out in a Veeco D125 MOVPE reactor using trimethylindium

(TMIn), tertiarybutylphosphine (TBP) and tertiarybutylarsine (TBA). The nanostructures were grown on InP (111)B +/- 0.5^0 substrates doped with Fe. The samples were placed in the reactor and annealed at 550 °C in 1.34×10^{-3} mol/min of TBP for 5 min. Then the temperature was lowered to between 390 and 400 °C, and indium droplets were deposited by feeding 5.06×10^{-5} mol/min of TMIn for 0.2 min. Then both Tmln and TBP were introduced into the reactor to grow the InP nanopillar. The flow rate of TBP was varied from 4.46×10^{-4} to 1.34×10^{-3} at a constant TMIn flow of 9.45 μmol/min. The corresponding V/III ratio ranged from 30 to 90. After 9 min, the growth of the InP core was terminated by switching off the TMIn and TBP supply. The reactor was purged with H_2 for 10 s and then with TBA for 3 min while the temperature was ramped up to 420^{0}C. Following the interrupt, InAs quantum wells were deposited on the InP nanopillars by varying the flow of TBA from 7.59 x 10^{-5} mol/min to 9.82 x 10^{-3} mol/min at a constant TMIn flow of 2.09 x 10^{-5} mol/min. The corresponding V/III ratio of the InAs shell ranged from 1 to 120. The InAs growth time was varied between 1 to 30 s to optimize PL intensity. After depositing the InAs shell, the reactor was purged with H_2 for another 10 s and with TBP for 1 min. Then an InP cap layer was deposited by introducing 9.45 μmol/min of TMIn and 1.34 x 10^{-3} mol/min of TBP into the reactor for 1 min. The V/III ratio of the InP capping layer was 90. The sample was cooled down and removed from the reactor for characterization. The total pressure in the reactor was maintained at 60 Torr.

The surfaces were imaged using a Hitachi S4700 field emission scanning electron

microscope (SEM). The data were processed with the aid of ImageJ. Multiple SEM images were analyzed to obtain the average diameter, width, height, and areal density. High resolution TEM images were obtained on a FEI Titan 300 kV scanning transmission electron microscope. In some cases, the wires were exfoliated from the substrate and placed on a TEM grid. To image the quantum well, the sample was prepared by a lift-off technique. First, Pt was deposited onto the nanostructures by ion beam. Then a Pt electron beam was used to slice a thin vertical cross-section in an FEI Nova 600 focused ion beam (FIB)-SEM system. The PL measurements were made at 77K using a 659 nm line of a diode laser. Double axis XRD measurements were made on a BEDE D[1] diffractometer equipped with a Maxflux focusing graded X-ray mirror and a two bounce Si 220 channel-cut collimator crystal, using Cu $K_{\alpha 1}$ radiation.

Results and Discussion

Images of the indium droplets were obtained prior to growth. The droplet diameters were 25 nm and their densities were 10^7 per cm^2 for deposition at 370 °C. A side view SEM image of the nanopillars before and after InAs quantum well deposition is shown in Figure 1. The nanopillars were grown at 390 °C and a V/III ratio of 30 with average widths and heights of 40 nm and 300 nm, respectively. The InAs shell was deposited at 420 °C and a V/III ratio of 129 for 10 s, followed by the growth of an InP capping layer for 1 min. The average width and height is 45 nm and 300 nm, respectively. The InP pillars are cylindrical with some tapering and are surrounded by 150 nm wide hexagonal skirts at the bottom. After coreshell growth, the skirts are thicker as seen in Figure 1b.

(a) (b)

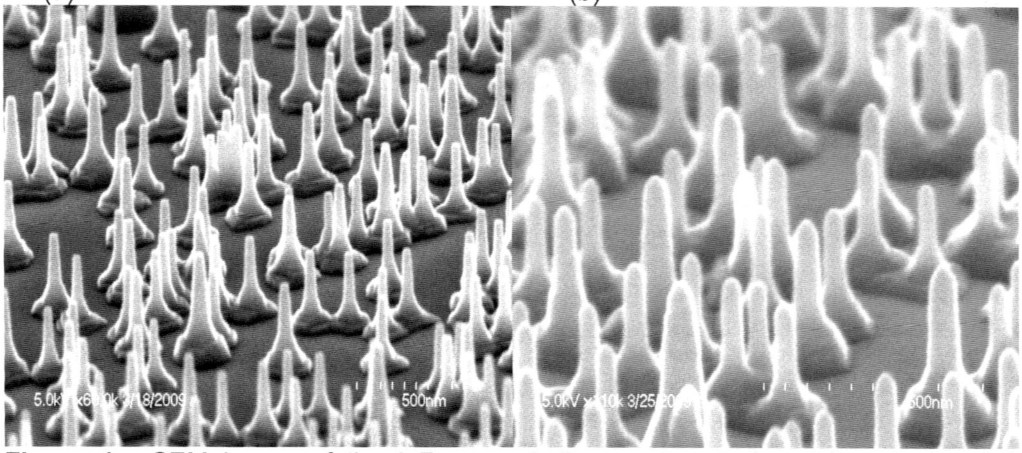

Figure 1. SEM image of the InP core shell nanowires before (a) and after InAs quantum well and InP capping layer deposition (b).

Indium phosphide nanopillars with hexagonal cross-sections were grown at 400 °C and a V/III ratio of 90. Then an InAs shell was deposited at 420 °C and a V/III ratio of 120. Figure 2a and 2b are top and side view SEM images respectively, showing a large-scale overview of the core-shell nanopillars. These structures have average widths and heights of 200 and 250 nm, respectively. These structures have hexagonal skirts that are 275nm wide. The pillars are shorter and thicker than the previously shown pillars because at higher growth temperature, the growth mechanism favors vapor phase epitaxy along the side walls [17].

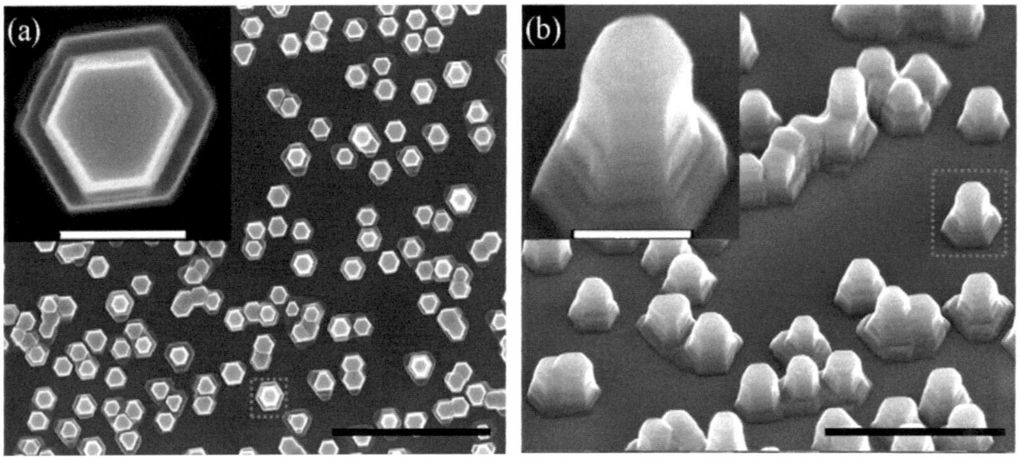

Figure 2. (a) Top- view SEM. The scale bar is 2 mm. (b) 52° tilted-view SEM image showing vertically aligned pillars. The scale bar is 1 mm. The insets in both images correspond to higher resolution images of the pillars highlighted by red squares. The scale bar is 200 nm in the insets.

A cross-sectional transmission electron microscopy image of a hexagonal InP/InAs core-shell nanopillar is shown in Figure 3a. The arrows indicate the interface between the core and the shell structures determined by the image contrast and the difference in lattice spacing as revealed by the FFT analysis (data not shown). The InP core in Figure 3a has a diameter of 70 nm, while the shell has a thickness of 5 nm. There is no distinct appearance of an InP cap layer on the InAs shell. The TEM image reveals that there is a distinct $InP_{1-x}As_x$ layer on the core. Selected area electron diffraction (SAED) obtained from the pillar in (a) is shown in Figure 3b. This reveals that the Zone axis = [0001] and the crystal structure is wurtzite. Arrows highlight the diffraction spots corresponding to InP and $InP_{1-x}As_x$ structures.

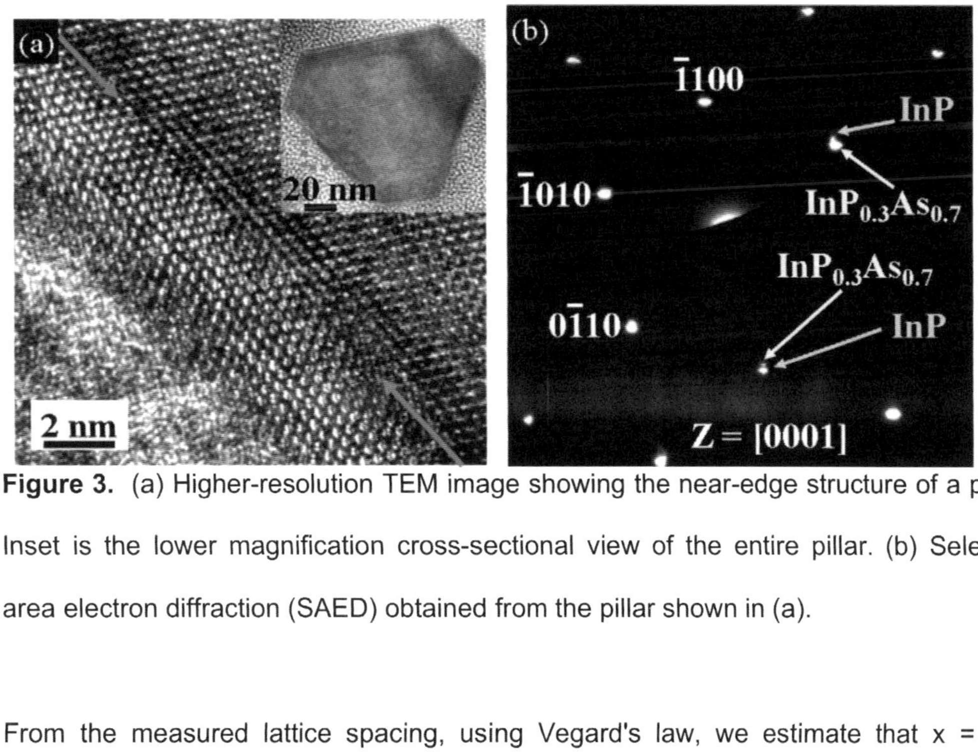

Figure 3. (a) Higher-resolution TEM image showing the near-edge structure of a pillar. Inset is the lower magnification cross-sectional view of the entire pillar. (b) Selected area electron diffraction (SAED) obtained from the pillar shown in (a).

From the measured lattice spacing, using Vegard's law, we estimate that x = 0.7. Energy dispersive X-ray (EDX) spectra acquired along the line of a pillar is shown in Figure 4. Scanning transmission electron microscopy (STEM) image of the pillar is shown inset. Note the increase in As peak intensities towards the edges of the pillar. From the line profiles, we confirm that the InP core is ~70 nm in diameter. Bright field scanning transmission electron microscopy (STEM) images as well as energy-dispersive X-ray (EDX) mapping confirm that the core/shell contrast observed in Figure 3a arises from the well-defined spatial distribution of As and P elements in the core and shell, respectively, of the NW structures.

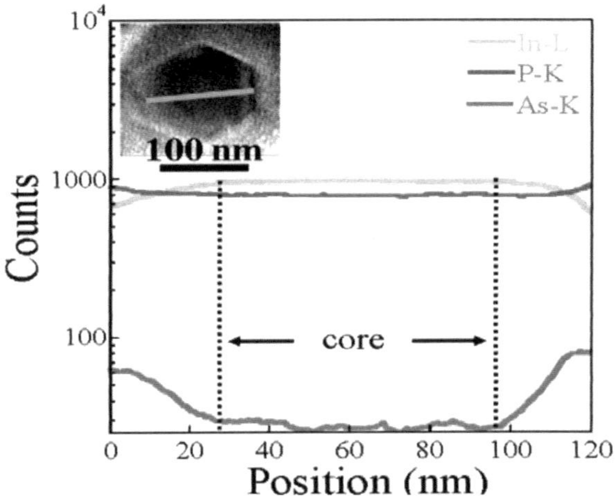

Figure 4. Energy disperse X-ray (EDX) spectra acquired along the line from a pillar. Scanning transmission electron microscopy (STEM) image of the pillar is shown as inset.

The optical properties of the InP/InAs/InP core-shell structures grown on the InP nanopillars were characterized by photoluminescence. These results are presented in Figure 5. Curve 2 represents the sample with the strongest PL emission. A single intense band is recorded at 0.7 eV (1750 nm) with a full-width-at-half-maximum of 350 meV. These results indicate that the group V elements intermixed with each other forming an $InAs_xP_{1-x}$, as confirmed by the cross-sectional TEM image and the EDX spectra from the pillars.

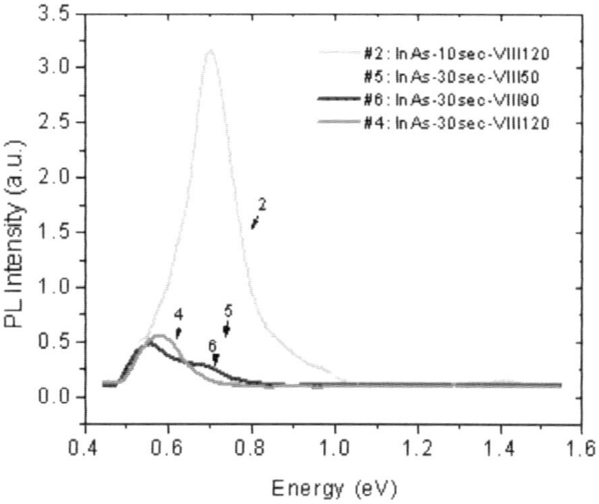

Figure 5. 77K photoluminescence spectra of InP/InAs core-shell structures grown under different conditions.

Conclusion

We have demonstrated the catalyst-free growth of single-crystal InP/InAs core/shell nanopillar heterostructures exhibiting electron mobilities more than 10 times that of silicon. Transmission electron microscopy studies revealed single-crystal InP cores with epitaxial InAs shells 5 nm in thickness, and energy-dispersive X-ray spectroscopy analysis further confirmed the composition of the designed heterostructure. The growth of high electron mobility InP/InAs core-shell NWs can enable fundamental studies of quantum transport and be incorporated as a platform for high-speed, low-power nanoelectronics. Future work would involve the integration of III-V core-shell nanowires into next generation integrated silicon circuits.

References

1. Lauhon, L. J.; Gudiksen, M. S.; Wang, D.; Lieber, C. M. *Nature* **2002**, 420, 57.
2. Lu, W.; Xiang, J.; Timko, B. P.; Wu, Y.; Lieber, C. M. *Proc. Natl. Acad. Sci. U.S.A.* **2005**, 102, 10046.
3. Xiang, J.; Vidan, A.; Tinkham, M.; Westervelt, R. M.; Lieber, C. M. *Nat. Nanotechnol.* **2006**, 1, 208.
4. Xiang, J.; Lu, W.; Hu, Y. J.; Wu, Y.; Yan, H.; Lieber, C. M. *Nature* **2006**, 441, 489.
5. Tambe, M.J.; Lim, S.K.; Smith, M.J.; Allard L.F.; Gradecak S. *Appl. Phys. Lett.*, **2008**, 93, 151917.
6. Mohan P.; Motohisa J.; Fukui T. *Appl. Phys. Lett.*, **2006**, 88 133105.
7. Mathews, J. W.; Blakeslee A.E.; *J. Cryst. Growth.,* **1974**. 27, 118.
8. Xiang, J.; Xiong, Q.; Nam, S.; Qian F.; Li. Y; Lieber, C. M. *Nanolett.* **2007**, 7, 3214.
9. Sze, S. M. *Physics of Semiconductor Devices*, 2nd ed.; John Wiley & Sons: New York, 1981.
10. Pal B.; Goto K.; Ikezawa M.; Masumoto Y.; Mohan P.; Motohisha K.; Fukui T.; *Appl. Phys. Lett.*, **2008,** 93, 073105.
11. Fukui, T.; Mohan, P.; Motohisa, J. in *Proceedings of the 19th International Conference on Indium Phosphide and Related Materials*, Matsue, Japan, **2007**, 384.
12. Matthews, J. W.; Blakesleee, A. E. *J. Cryst. Growth* **1974**, 27, 118.
13. Kobayashi, Y.; Kobayashi, N. *Jpn. J. Appl. Phys.* **1992**, 31, 3988.
14. Hopkinson, M.; David, J. P. R.; Claxton, P. A.; Kightley, P. *Appl. Phys. Lett.* **1992**, 60, 841.
15. Mohan, P.; Motohisa, J.; Fukui, T.; Appl. *Phys. Lett.* **2006,** 88, 133105.

16. Mohan, P.; Motohisa, J.; Fukui, T.; *Nanotechnology*. **2005**, 16, 2903.

17. Woo. R.L.; Gao L.; Kodambaka S.; Nanoletters. 2009, 9, 2223.

Atomic-Scale Modeling of the Mechanical Behavior of Ultra-Low-Dielectric-Constant Mesoporous Amorphous Silica Films

M. Rauf Gungor, Department of Chemical Engineering, University of Massachusetts, Amherst, MA

James J. Watkins, Department of Polymer Science and Engineering, University of Massachusetts, Amherst, MA

Dimitrios Maroudas, Department of Chemical Engineering, University of Massachusetts, Amherst, MA

High-performance microelectronic devices require ultra-low-dielectric-constant (ULK) dielectric materials in order to reduce the capacitive coupling between the interconnect lines due to increased density and larger-scale integration of transistors in small areas (on the order of 10^9 cm^{-2}). All candidates for ULK dielectric materials must also satisfy certain integration requirements, such as a minimum mechanical strength, threshold of thermal and electrical breakdown, and suitability for lithographic processing during chip manufacturing. Mechanical strength is particularly important in order to retain the structural integrity of the microelectronic device under the thermomechanical loading conditions characteristic of semiconductor manufacturing processes, chip packaging, and device service. Among materials that are being developed for ULK applications, porous amorphous silicas are particularly appealing due to their compatibility with current semiconductor manufacturing technologies.

In this presentation, we report results of molecular-dynamics (MD) simulations aiming at a fundamental understanding of the physical mechanisms that control the mechanical behavior of mesoporous amorphous silica film structures with pore diameters of a few nanometers, as well as prediction of the response of such structures to various mechanical loading conditions. The MD simulations employ a realistic classical potential that includes two-body and three-body interatomic interactions. The normal-density amorphous silica structures are prepared through MD starting from a crystalline beta-cristobalite solid structure and following a thermal processing sequence that includes melting, rapid quenching, and a thermal annealing schedule. We have generated "regular" and "disordered" mesoporous structures through the introduction of various geometric arrangements of spherical or cylindrical pores by removal of atoms from the normal-density amorphous silica matrix and subsequent thermal annealing at the temperature of interest to ensure proper structural relaxation.

We present a systematic analysis of the mechanical response of regular mesoporous amorphous silica structures with spherical pores under applied strains within the elastic limit near room temperature; the analysis is based on isostrain MD simulations using large-size computational supercells. The elastic moduli of the mesoporous structures are computed and their structural stability under tensile and compressive straining is analyzed as a function of density and pore diameter. In addition, we calculate the mechanical properties of these mesoporous structures for various pore geometries and geometric arrangements, focusing on spherical and cylindrical pores arranged in simple cubic and hexagonal lattices, respectively, as well as disordered worm-like arrangements of cylindrical pores. Furthermore, we analyze the effects of high-temperature thermal treatment of the mesoporous amorphous structures on their structural stability and mechanical strength in correlation with the underlying microstructural evolution.

See more of this Session: Nanoelectronic Materials I
See more of this Group/Topical: Nanoscale Science and Engineering Forum

Molecular-Dynamics Simulation Study of the Mechanical Behavior Under Biaxial Strain of Pre-Strained FCC Metallic Ultrathin Films

Kedarnath Kolluri, Department of Chemical Engineering, University of Massachusetts, Amherst, MA

M. Rauf Gungor, Department of Chemical Engineering, University of Massachusetts, Amherst, MA

Dimitrios Maroudas, Department of Chemical Engineering, University of Massachusetts, Amherst, MA

Nanometer-scale-thick metallic films are used increasingly in modern technologies, from microelectronics to various areas of nanofabrication. The mechanical behavior of metallic small-volume structures is very different from that of bulk metals; for example, recent experimental studies on nanopillars of face-centered cubic (fcc) metals have shown that the strength of such fcc small-volume structures increases with decreasing characteristic lengths and that the dislocation density always decreases during the application of the strain. Improvement of the reliability, functionality, and performance of nano-scale devices requires a fundamental understanding of the physical mechanisms that govern the thin-film response to mechanical loading in order to establish links between the films' structural evolution and their mechanical behavior.

In this presentation, we report results of large-scale molecular-dynamics (MD) simulations for the dynamic deformation (i.e., at constant strain rate and temperature) under biaxial tensile strain of nanometer-scale-thick films of various fcc metals. The MD simulations employ an embedded-atom-method parameterization for the description of the interatomic interactions. The initial dislocation density of the thin films is on the order of 1017 m-2. Our results indicate that films of metals with moderate-to-high propensity for formation of stacking faults (such as Ni and Cu) exhibit an extended easy-glide regime followed by a sharp increase in the material stress, whereas those with low propensity for stacking-fault formation (such as Al) exhibit a monotonic increase in the stress during dynamic loading.

The mechanical response of the Ni and Cu films to applied biaxial tensile strain exhibits three stages of deformation. In the first stage, the response is nearly elastic, with a sharp increase in the stress; only a few dislocations are annihilated in this stage. During the second stage, gliding dislocations interact with stacking faults. These interactions lead to dislocation dissociation and aid in dislocation annihilation. Almost always, the obstacle stacking faults are unzipped during these interactions. We identify three classes of dislocation-stacking fault interactions, where the stacking faults act as barriers to dislocation glide and as sources for dislocation cross-slip. As the dislocation density decreases, the thin-film strength increases and the material's mechanical behavior is found to resemble that of a perfectly elastic film. In the third stage, continued application of strain leads to increase in the thin-film stress, which leads eventually to nucleation of new dislocations in the thin film. Dislocation nucleation and depletion in the thin film continues in cycles until thin-film failure. In Al thin films, however, we find that the stress increases monotonically until thin-film failure and that the dislocation depletion rate is much lower than that in the Ni and Cu thin films. The analysis of the defect interaction mechanisms in Al thin films reveals that in metals with low propensity for stacking-fault formation dislocation annihilation is due to collinear interactions between dislocations. Furthermore, the plastic strain rate in Cu and Ni thin films is far greater than that in Al thin films, causing significant stress dissipation and an extended easy-glide regime in the Cu and Ni thin films.

See more of this Session: Nanoelectronic Materials I
See more of this Group/Topical: Nanoscale Science and Engineering Forum

Towards Ordered Layer-by-Layer Growth of Organic Semiconductors: Calculating the Ehlich-Schwöebel Barrier for Step Edge Descent

Paulette Clancy, Cornell University, Ithaca, NY

Joseph E. Goose, Chemical & Biomolecular Engineering, Cornell University, Ithaca, NY

Semiconducting organic thin films are under intense scrutiny as materials for large area, mechanically flexible electronic displays, for "smart tags" (RFID), for OLEDs (organic light emitting diodes) and for solar cells. The most promising candidate materials are π-conjugated molecules with high field effect mobility if they can be grown in an ordered 2D layer-by-layer thin film. It remains a challenge to determine effective nucleation and growth strategies to grow such films.

Our most recent computer simulation work has studied one phenomenon that affects the ability to grow smooth films, namely, the tendency of a molecule diffusing over the surface of a thin film to encounter a significant uphill energy barrier (the so-called Ehrich-Schwöebel barrier) near the edge of an island (the step edge). If this energy barrier is high, the molecule will tend to stay on the higher terrace rather than dropping down to the incomplete layer below, and thereby promote unwanted 3D growth.

We use a combination of energy minimization techniques and Molecular Dynamics simulations with semi-empirical models (MM3-π) whose accuracy has been tested successfully against a variety of Density Functional Theory models. We will show results for the way that a group of organic semiconductor molecules (the acenes, DIP, rubrene, and C60) fall over a step edge and determine the factors that affect the energy barrier that they encounter.

We unravel the mystery why results in a 2008 Science paper show a very high barrier for step-edge descent by sexiphenyl molecules, highlighting the importance of molecular rotational degrees of freedom and the unexpected influence that the tendency of a molecule to twist and bend can have on the approach to the step edge. Finally, we offer a way to predict the Schwöebel barrier without doing any computer simulations.

See more of this Session: Nanoelectronic Materials I
See more of this Group/Topical: Nanoscale Science and Engineering Forum

A Nanochannel/Microwell Array Device for Controlled Gene Delivery to Individual Cells by Localized Electroporation

Pouyan E. Boukany, Nanoscale Science and Engineering Center for Affordable Nanoengineering of Polymeric Biomedical Devices, The Ohio State University, Columbus,, OH

Orin L. Hemminger, Chemical and Biomolecular Engineering, Ohio State University, Columbus, OH

Xin Hu, NSEC center for Affordable Nanoengineering of Polymer Biomedical Devices (CANPBD), The Ohio State University, Columbus, OH

Zhengzheng Fei, Department of Chemical and Biomolecular Engineering, The Ohio State University, Columbus, OH

Nan-rong Chiou, Department of Chemical and Biomolecular Engineering, The Ohio State University, Columbus, OH

Ly James Lee, The Ohio State University, Columbus, OH

Most gene and macromolecule drug delivery methods, such as viral and non-viral nanoparticles, electroporation, gene gun, etc., are stochastic processes where 10^5-10^6 cells are randomly transfected by a very large number of genes/drugs. It is impossible to control the injected dose, particularly for multiple gene/drug delivery. In this study, we introduce a novel, simple and low-cost method capable of fabricating large and well-defined arrays of polymeric planar nanochannels, together with microscale features that allow cells to be loaded for electroporation. This nanochannel/microwell array system can manipulate either a small (one to several) or a large (>100,000) number of cells and inject multiple genes/drugs through nanochannels into individual cells with minimum damage and high efficiency by localized electroporation. The dosage can be controlled by changing the length of nanochannel and the diameter of microwells. Another advantage of this design is the uniform electric field distribution leading to much more uniform gene transfection when compared to bulk electroporation where the shading effect always exists due to the random cell distributions.

See more of this Session: Nanotechnology for Biotechnology and Pharmaceuticals II
See more of this Group/Topical: Nanoscale Science and Engineering Forum

Tracking Biodistribution of NIR-Labeled Polymeric Micelles of Varying Shape and Surface Charge

David A. Christian, Chemical & Biomolecular Engineering, University of Pennsylvania, Philadelphia, PA

Olga Garbuzenko, Ernest Mario School of Pharmacy, Department of Pharmaceutics, Rutgers, The State University of New Jersey, Piscataway, NJ

Tamara Minko, Ernest Mario School of Pharmacy, Department of Pharmaceutics, Rutgers, The State University of New Jersey, Piscataway, NJ

Dennis E. Discher, Chemical & Biomolecular Engineering, University of Pennsylvania, Philadelphia, PA

Polymeric micelles synthesized from the self-assembly of amphiphilic block copolymers in solution have received increasing attention for applications in drug delivery. While the effect of size of these micelles on biodistribution and circulation kinetics has been well characterized, questions of the effects of micelle shape and surface charge remain largely unanswered. In this study, the shape of micellar carriers can be tuned from spheres to short and long cylinders (filomicelles) to bilayer vesicles (polymersomes) with simple changes in block copolymer architecture. In addition, the surface charge of polymersomes can be tuned to mimic the charge of red blood cells without chemical modification of the block copolymers. To efficiently track the biodistribution of these polymeric micelles in vivo, a near infrared fluorophore (NIRF) was non-covalently incorporated without affecting micelle morphology or surface charge. Preliminary results for the effect of surface charge on polymersomes indicate a distinct shift in the carrier biodistribution where anionic polymersomes are directed almost exclusively to the liver rather than the spleen and liver for neutral polymersomes. The tracking of NIRF-labeled filomicelles in healthy mice shows that the majority of carriers - independent of length - remained in circulation after 24 hours. In tumor bearing mice, NIRF imaging suggests that filomicelles convect into a solid tumor and some fragments can penetrate into the tumor stroma, confirming that filomicelles can effectively deliver encapsulated anti-cancer therapeutics to tumors.

See more of this Session: Nanotechnology for Biotechnology and Pharmaceuticals II
See more of this Group/Topical: Nanoscale Science and Engineering Forum

Incorporation of Various Forms of Dexamethasone Acetate Into Biodegradable Nanoparticles with Encapsulated Magnetite

Brian Grady, University of Oklahoma, Norman, OK

Ken Dormer, Physiology, University of Oklahoma-Health Sciences, Oklahoma City, OK

Rick Kopke, Hough Ear Institute, Oklahoma City, OK

Youdan Wang, Hough Ear Institute, Oklahoma City, OK

Steroids have been reported to be effective for the treatment of sensorineural hearing loss. Dexamethasone (DEX) is one of the corticosteroids which have shown hair cell protection in previous studies. The majority of current treatments using corticosteroids involve middle ear direct injection which has limited inner ear distribution and it is hard to maintain drug levels over an extended period of time. Clearance from the middle ear is another problem. Poly(lactic-co-glycolic acid) polymeric nanoparticles encapsulating dexamethasone might represent a way to deliver corticosteroids to inner ear and to release therapeutics locally in a controlled and sustained manner. In this study, we tested three forms of dexamethasone: the basic form, the phosphate form and the acetate form for efficiency of encapsulation as well as release profile. With the water-soluble phosphate form, a water-oil-water double emulsion process was used, while with the two other forms, an oil-in-water procedure was used. The most hydrophobic of these forms, the dexamethasone acetate, yielded the highest encapsulation efficiency as well as minimal burst release followed by a slow, steady release over many days. The incorporation of oleic acid coated superparamagnetic magnetite nanoparticles (size ~10 nm) did not change either the amount encapsulated or release profile. The magnetite is well dispersed and the PLGA particle size/shape and surface charge does not change dramatically with magnetite encapsulation.

See more of this Session: Nanotechnology for Biotechnology and Pharmaceuticals II
See more of this Group/Topical: Nanoscale Science and Engineering Forum

Enhanced Drug Loading in Polymeric Nanoparticles Via Micellization in Supercritical Fluids

Zachary Tyrrell, Chemical & Petroleum Engineering, University of Wyoming, Laramie, WY

Youqing Shen, University of Wyoming, Laramie, WY

Maciej Radosz, University of Wyoming, Laramie, WY

Block copolymer micelles used for drug delivery are normally formed and processed in aqueous solutions. A major challenge in this area is how to increase the concentration of hydrophobic drugs in the nanoparticle core. Other issues are the lengthy dialysis procedure and the inefficient freeze dry procedure. We address these issues by first processing the polymeric nanoparticles in highly compressible non-aqueous solvents, such as subcritical and supercritical polar fluids, followed by dispersion in an aqueous medium. The resulting nanoparticle solution is obtained in a fraction of the time with enhanced drug loading as compared to conventional preparation methods.

See more of this Session: Nanotechnology for Biotechnology and Pharmaceuticals II
See more of this Group/Topical: Nanoscale Science and Engineering Forum

Study of Different Aging Mechanisms in Perfluorocarbon Nanoemulsions

Andrew Akbar Shalá, Chemical Engineering, Universidad De Los Andes, Bogotá, Colombia

David Eduardo Torrado, Chemical Engineering, Universidad De Los Andes, Bogotá, Colombia

Oscar Alberto Alvarez, Chemical Engineering, Universidad De Los Andes, Bogotá, Colombia

Johanna María Galindo, Chemical Engineering, Universidad De Los Andes, Bogotá, Colombia

Camila Irene Castro, Chemical Engineering, Universidad De Los Andes, Bogotá, Colombia

Juan Carlos Briceño, Mechanical Engineering, Universidad De Los Andes, Bogotá, Colombia

Perfluorocarbon (PFC)-in-water nanoemulsions may serve as synthetic oxygen carriers for different clinical procedures. To prepare these nanoemulsions a biocompatible surfactant as lecithin can be used, forming PFC droplets and liposomes with different particle size. However, the simultaneous presence of liposomes and PFC droplets on global nanoemulsion stability has not been widely studied.

Preliminary observations from nanoemulsion behavior have made possible to identify a sedimentation phenomena as the first instability mechanics, producing two different phases: supernatant and sediment, without nanoemulsion breakage. The aim of this work is to study the evolution of each phase and their influence on global nanoemulsion stability during the time.

Nanoemulsions are prepared with perfluorooctyl bromide 99% (PFOB, Exfluor Research Corp, USA) as dispersed phase and soybean lecithin (Epikuron 170®, DEGUSSA, Germany) as surfactant. Analysis of supernatant and sediment is done following the evolution of: i) particle size distribution, through dynamic light scattering and near infrared (NIR) spectroscopy, ii) rheological behavior, by viscosity changes, and (iii) pH measurements. As reference system a PFC-free dispersion is also prepared and characterized through time.

The results of particle size analysis show a multimodal distribution in both phases. In this distribution, each population and its evolution could be associated to a specific kind of particle by a comparison of supernatant and sediment with the reference system. This behavior is supported by the analysis of NIR spectra, which show difference between droplets and liposomes.

The difference in the evolution of high shear rate viscosity value between supernatant and sediment, allows characterizing macroscale behavior. Finally, there is a pH decrease over time suggesting lecithin oxidation.

In conclusion, in the time supernatant and sediment shall rich in liposomes and droplets respectively, with different aging mechanisms.

See more of this Session: Nanotechnology for Biotechnology and Pharmaceuticals II
See more of this Group/Topical: Nanoscale Science and Engineering Forum

Using Nanotechnology to Improve the Water Solubility and Bioavailability of Health Care Products-Soy Isoflavones

Jane Jinjie Xu, Department of Chemical and Biomolecular Engineering, Hong Kong University of Science and Technology, Kowloon, Hong Kong and Kathy Qian Luo, School of Chemical and Biomedical Engineering, Division of Bioengineering, Nanyang Technological University, Singapore

Abstract (ID# 161939)

Low water solubility can lead to poor bioavailability of many hydrophobic pharmaceuticals in a biological system. One way to solve this problem is to reduce their particle sizes to nanometer scales so that the compounds can be dissolved in water by dispersion. Our group has successfully increased the water solubility and bioavailability of water insoluble compounds, soy isoflavones, using supercritical CO_2 as an antisolvent to precipitate out isoflavones in the form of nanoparticles. Isoflavones are known as phytoestrogens since their chemical structures are similar to that of the female hormone, estrogen. Isoflavones can thus be used as a replacement for estrogen to treat women with menopausal symptoms, since prolonged intake of synthetic estrogen can increase the risk of developing breast cancer. Soy isoflavones can exist in nature in twelve different kinds of chemical entities; and genistein has the strongest estrogenic activity. In this study, we evaluated the impact of various process parameters of supercritical CO_2-based antisolvent precipitation method on particle formation. Among the three parameters tested in this study-pressure of precipitation, concentration of drug solution, and feed rate ratio of CO_2/drug solution, we found that the pressure is the most important parameter which can greatly affect the particle size. Using the optimized experimental conditions of supercritical CO_2, the size of genistein particle has been successfully reduced from its original size of more than 50 μm to around 220 nm as determined by scanning electron microscopy. Finally, the 24-hour pharmacokinetic study performed in rats showed that the plasma concentration of genistein has increased by 75% after nanomization demonstrating the possibility of improving the bioavailability of water insoluble compounds using the supercritical antisolvent (SAS) process. The generic technique developed in this study can be used to generate more efficacious pharmaceuticals and health care products, including water-insoluble compounds isolated from Chinese herbal medicines.

Introduction

Isoflavones are a flavonoid subgroup, in which the ring B is attached to the C-3 position of ring C (Fig.1). They are considerably widespread compounds in many plants, especially soybeans. To date, 12 main isoflavones have been characterized in soybeans and soybean products including genistein, daidzein, glycitein, and their respective malonyl, acetyl, and glucosyl forms [1, 2]. Isoflavones are regarded to be phytoestrogen because they have the ability to produce estrogenic effect due to their structural similarity to the human hormone estradiol (17β-estradiol) [3]. Nowadays, isoflavones are increasingly used instead of traditional hormone replacement therapy for estrogen deficient women in menopause or postmenopause [4], since they may decrease the risk of estrogen-related cancers, which are one of the main health risks related to the hormone replacement therapy [5, 6].

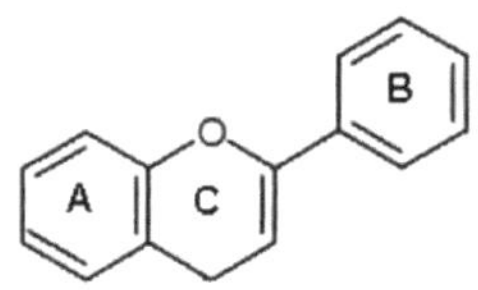

Fig. 1. The basic structure of flavonoid.

Genistein has many biological activities that have made it a subject in over 3600 published studies (listed in Biological Abstracts) in the last 10 years [7]. Most of these studies have focused on its phytoestrogen activity [8, 9] and other beneficial effects include antioxidant [10] and anticancer activities [11-13]. Although genistein has shown multiple health benefits, its therapeutic application can be limited by its low water solubility and low bioavailability. In order to improve its water solubility, some research groups have tried to use different types of carriers such as cyclodextrins [14-16] and PEG [17] to complex with genistein. There are some limitations to these methods, such as the residual organic solvent and the instability of drugs and carriers.

Recently, micro- and nano-particle formation processes based on the use of supercritical fluids as co-solvents or anti-solvents for poorly water soluble active pharmaceutical ingredients have been introduced as a viable means of controlling particle formation to improve physicochemical properties in solid state. Interest in supercritical fluids and their potential usages for process improvements has significantly increased in the past decade because of its mild critical conditions (T_C=31.1°C, P_C=7.38 MPa), non-toxicity, non-flammability and low cost. The use of supercritical CO_2 for micro-particle preparation has been widely reported as co-solvent using rapid expansion of supercritical solution (RESS) [18-20] and a variety of antisolvent processes such as gas antisolvent [21, 22], supercritical antisolvent (SAS) [23, 24], aerosol solvent extraction systems [25] and solution enhanced dispersion by supercritical fluids [26]. Even though the particles with controlled size could be produced by RESS, it has some disadvantages: high ratios of gas/substance owing to the low solubility of the substance, high pressures and sometimes high temperatures; difficult separation of small particles from large volumes of expanded gas and supercritical solutions established discontinuously, and the requirement of large-volume pressurized equipment. In comparison to RESS, the milder operating condition of gas anti-solvent process makes it more attractive for scale-up applications.

Supercritical antisolvent precipitation process is a semicontinuous precipitation technique in which a supercritical fluid (SCF) and a liquid solution are continuously delivered to a high-pressure precipitator which can form the precipitation of the dissolved solid by using the anti-solvent effect of the SCF. The supersaturation of the solid is rapidly reached by the volumetric expansion of the liquid solution with a time scale of approximately 10^{-5}s [27]. To date, despite the fact that many works have been published on the generation of particles by SAS, the mechanisms of the process have not been fully uncovered. Indeed, these mechanisms are quite complex because the precipitation is the result of the interplay of fluid dynamics (jet breakup), mass transfer, nucleation kinetics, and thermodynamics (high-pressure ternary phase equilibrium). Moreover, particle morphology can be affected by many process parameters such as temperature, pressure, solution concentration, and solution to antisolvent flow rate ratio. The process arrangement and the apparatus can also heavily influence the product properties.

Our group aim to improve the bioavailability of genistein by using precipitation with a compressed fluid antisolvent (PCA) process. In this study, the effects of process variables on particle size and distribution of genistein during particle formation were investigated. The *in vitro* dissolution rate and *in vivo* bioavailability were also compared between nano-structured and unprocessed genistein. The generic technique developed in this study can be used to generate more efficacious pharmaceuticals and health care products, including water-insoluble compounds isolated from Chinese herbal medicines.

Materials and methods

Materials

Genistein with a minimum purity of 98% was obtained from Allway Minerals and Science Technology Co., Ltd. Sulfatase (S9626) extracted from *Helix pomatia* was a product of Sigma Chemical Company and was used to hydrolyze the glucuronide and sulfate conjugates of genistein. Tween-80 was also a product from Sigma. The carbon dioxide used in the solubility study was supplied by Hong Kong Specialty Gases with a purity of 99.99%. Ethanol was obtained from Merck with the purity not less than 99.9%. Acetone with a minimum purity of 99.5% was obtained from Scharlau Chemie, S.A. Ethyl acetate with a purity of 99% was from Merck. Materials required for scanning electron microscope (SEM) sample preparation are 24 mm aluminum specimen mount stubs used for sample mounting purposes, double-sided carbon 12 mm adhesive tabs and silica wafers for sample attachment.

Apparatuses and Procedures

The experimental set-up was schematically shown in Fig. 2. The CO_2 supply was connected to a cooler that keeps the CO_2 in the liquid phase to be pumped to the heater at experimental operating conditions. A flow meter was also installed to constantly monitor the flow rate of CO_2 through the system. Supercritical CO_2 was first pumped into the top of the high-pressure vessel by a high-pressure pump. Once the pressure and temperature were equilibrated, the genistein dissolved liquid solution was then co-introduced into the particle precipitation vessel through a coaxial nozzle. The coaxial nozzle consisted of a 1/16 inch inner tube with an outlet of 0.1 mm and a 1/8 inch outer tube. This tube arrangement was used to enhance the mixing effect of the two fluids to produce smaller solution droplets. The residual solvent was drained out of the particle precipitation vessel by the backpressure regulator (BPR) and sent to the cyclone separator, where the remaining solvents were recollected while the vapor was then vented out of the system. Temperature control was monitored by a computerized system governing the whole PCA system, including the outer and inner vessel temperatures. At washing step, an additional SC-CO_2 continued to flow into the precipitation vessel for further 30 min to wash out the residual content of solvent dissolved in the supercritical antisolvent. The precipitation vessel was slowly depressurized down to the atmospheric pressure and finally the particles were collected from a metal filter located at the bottom of the vessel with the pore size of 5 μm.

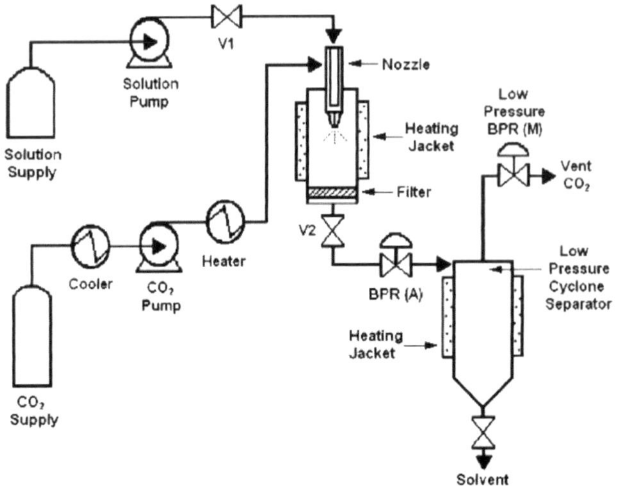

Fig. 2. Schematic diagram of the PCA equipment [28].

Scanning electron microscopy (SEM)

The morphology of particles was examined by scanning electron microscopy (SEM; JSM-6300 F, Jeol Ltd., Japan). A gold sputter coater was used in sample preparation to allow the sample to be conductive for viewing.

Dissolution study

The dissolution rate was preformed according to the USP XXVIII paddle method. The stirring speed used was 50 rpm, and the temperature was maintained at 37 ± 0.1 °C. Each test was carried out in 900 ml of distilled water containing 0.1 % Tween 80. For each run, 10 mg of the genistein particles were placed in the dissolution medium. Then, 2 ml of aliquot samples was withdrawn at predetermined time intervals, which was 5, 10, 15, 20, 30, 45, 60, 75, 90, 120 min and filtered by using a 0.2 μm nylon syringe filter. The released genistein concentration was determined by HPLC.

Bioavailability study in rats

Female SD rats (8-10 weeks old) weighing between 200 and 250 g were used to perform bioavailability study. The rats were housed in a cage and maintained on a 12 h light/dark at room temperature. After 24 hours fasting, gelatin minicapsules (Size 9, Torpac, Fairfield, NJ, USA) filled with unprocessed and PCA-processed genistein particles equivalent to 20 mg/kg of genistein, were administered orally to rats using minicapsule dosing syringe. After administration of the capsules, the rats were immediately given 1 mL of distilled water. The blood samples were obtained from the tail vein of the rats at 0, 0.5, 1, 2, 4, 6, 8, 12, 24 h. All the samples were first stored at 4°C and centrifuged at 13,200 rpm, 4°C for 10 min. Plasma was transferred to individual centrifuge tube and stored at -80°C until analyzed.

Sample preparation

Plasma samples (200 μL) were incubated with 200 units of the sulfatase at 37°C, 1000 rpm on a mixshaker for 3 h. Then 600 μL of ethyl acetate were added to samples for extraction. Extraction was conducted by vortexing vigorously for 30 min and centrifuged at 4000 rpm for 20 min. The supernatant was evaporated to dryness under a gentle stream of nitrogen at 40°C. The residue was reconstituted with 100 μL of ethanol for HPLC analysis.

HPLC analysis

Genistein in all samples were analyzed by using a reversed-phase C18 column (Phenomenex, 5 μm, 250×4.6 mm i.d.) on the Agilent 1100 series liquid chromatograph (including the pumping system, vacuum degasser, autosampler, and UV-DAD detector). The sample injection volume was 20 μL. The mobile phase was water with 0.1% acetic acid (A) and methanol (B). A linear gradient elution was applied from 20 to 80% B starting from 0 to 30 min, at a flow rate of 1.0 mL/min. The temperature of the column was maintained at 40°C, and the detection wavelength was set at 262 nm, where absorbance peak areas were quantified.

Results

In the PCA process, genistein was precipitated from acetone at various experimental conditions. The experimental conditions are summarized in Table 1. All tests reported in this part of the work were performed at 40°C. In this study, we evaluated the impact of various process parameters of supercritical CO_2-based antisolvent precipitation method on particle formation, such as pressure, genistein concentration and feed ratio between CO_2 and drug solution.

Table 1. Summary of the conditions used in the PCA experiments

Run	Pressure (bar)	Concentration (mg/mL)	Feed ratio, *R* (CO_2/liquid)	Temperature (°C)
1	100	4	60 (30/0.5)	40
2	70	4	60 (30/0.5)	40
3	100	2	60 (30/0.5)	40
4	75	2	30 (30/1)	40
5	75	2	60 (30/0.5)	40
6	70	2	60 (30/0.5)	40
7	100	1	60 (30/0.5)	40
8	70	1	60 (30/0.5)	40
9	120	0.4	60 (30/0.5)	40
10	100	0.4	60 (30/0.5)	40
11	70	0.4	60 (30/0.5)	40

1. Effect of pressure on particle size formation

Several experiments were performed to evaluate the effect of pressure ranging from 70 to 120 bar on particle size formation. In these experiments, all the operation conditions were identical except pressure. These conditions are: Temp=40°C, the concentration of genistein in acetone=0.4 mg/mL, and the ratio between CO_2 flow rate and liquid flow rate R=60. The results in Figure 3, shows that the size of genistein particle reduced significantly when the pressure of the system decreased from 120 bar to 70 bar. This observation can be explained as the supercritical antisolvent (SAS) precipitation process involves atomization of the solution of the solid solute to be precipitated out. As the atomized droplet comes in the contact with the antisolvent CO_2, the latter diffuses into the solution and the solvent evaporates from the droplet surface. This two-way mass transfer process requires the attainment of extremely high, very rapid and uniform at supersaturation condition. Since the decreased pressure caused a sharp rise in the supersaturation, leading to a reduction in the liquid solvent power which resulted in the formation of small and uniform particles.

Fig. 3. SEM images of original genistein (a) and genistein precipitated by the PCA process under conditions listed in Table 1 Run 9 (b), Run 10 (c), and Run11 (d).

2. Effect of concentration on particle size formation

A gradient of genistein concentration in acetone from 4 mg/mL to 0.4 mg/mL (40°C, R=60) was used in this study. It was found that at higher pressure such as 100 bar, reducing drug concentration could increase the mean diameter of the particles (Fig.4: a-d). In contrast, at lower pressure such as 70 bar, reducing genistein concentration could decrease the particle size (Fig. 4: e-h). These opposite effects of drug concentration on particle size formation under different pressure conditions can be explained in terms of nucleation and growth processes. At lower pressure, solute can be quickly precipitated during the expansion process, then grow into larger particles in the case of high drug concentration. On the other hand, the higher pressure can increase the mass-transfer rate of the solvent (acetone) out of the droplets, and increase the supersaturation of the solute (genistein) in the case of high drug concentration, which make genistein into smaller particles.

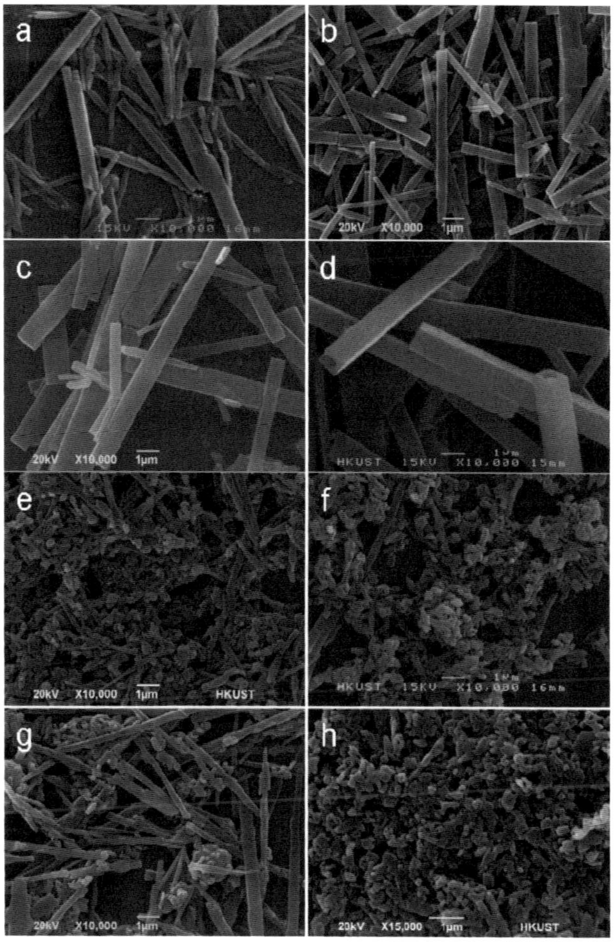

Fig. 4. SEM images of genistein precipitated by the PCA process: at 100 bar, Run 1 (a), Run 3 (b), Run 7 (c), Run 10 (d); and at 70 bar, Run 2 (e), Run 6 (f), Run 8 (g), Run 11 (h).

3. Effect of the feed rate ratio of CO_2/drug solution on particle size formation

In this experiment, two feed ratios between CO_2 and genistein solution at *R* values of 30 and 60 were used. The pictures in Fig. 5 show that increasing feed ratio from 30 to 60 slightly reduced the particle size. This may be because the raised feed ratio between CO_2 and genistein solution resulted in higher supersaturation which yields smaller particles.

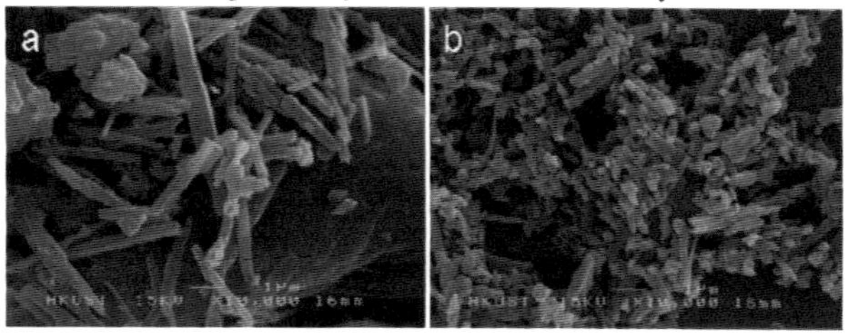

Fig. 5. SEM images of genistein precipitated by the PCA process: Run 4 (a), Run 5 (b).

4. Dissolution rate study

In theory, a decrease in particle size can increase its surface areas, thus improve its solubility. To test whether this theory holds true for genistein, we have compared the solubility of genistein before and after nanomization. The results in Figure 6 show that the solubility of genistein has increased significantly after nanomization. For example, at 60 min, the percentage of dissolved genistein has increased from 36% to 60 %. Even at the end of 2 hours experiment, the solubility of processed genistein was about 56%, while the solubility of raw genistein was below 40%. This huge increase in the solubility of genistein shows that indeed we can improve the water solubility of genistein by reducing its particle size.

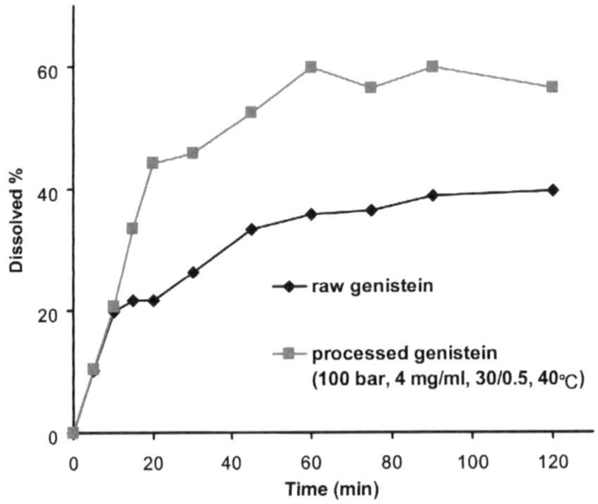

Fig.6. Dissolution rate comparison of original genistein particles and PCA processed particles from Run 1.

5. Pharmacokinetic study of nano-structured genistein on rats

Finally, we wanted to test whether nanomization can improve the bioavailability of genistein by comparing its plasma concentration before and after PCA process. To do this, three different genistein produces (1) original without processing, (2) PCA processed genistein at high pressure and concentration: 100 bar, 4 mg/mL, 30/0.5, 40°C, and (3) PCA processed genistein at low pressure and concentration: 70 bar, 0.4 mg/mL, 30/0.5, 40°C were put into different gelatin minicapsules (size 9, Torpac, Fairfield, NJ, USA). These minicapsules were orally administered to rats at the dosage of 20 mg/kg. At different time points during the course of 24 hours pharmacokinetic study, plasma samples were removed from the tail vein of the rats and plasma concentration of genistein was measured by HPLC. The graphs in Figure 7 show that the genistein concentrations in Run 1 using high pressure and high genistein concentration are significantly increased during the first 12 hours. In comparison, the genistein generated in Run 11 under low pressure and low genistein concentration increased the genistein plasma concentration after 6 hours of the experiment, but more importantly, this product continued to increase genistein concentration and maintained its high level until about 18 hours. After comparing the areas under the curve between the original genistein (blue graph) and that of Run 11 (yellow graph), we found that the plasma concentration of genistein has increased about 75% (Fig. 7).

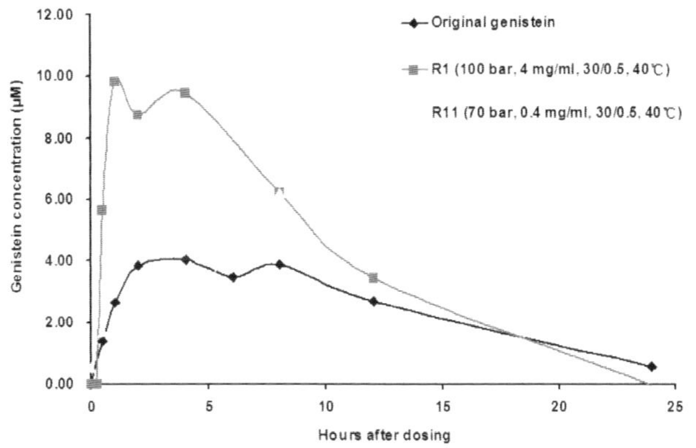

Fig.7. Plasma concentration of genistein in rats after oral administration of PCA processed nanoparticles or unprocessed drug at a dose equivalent to 20 mg of genistein/kg of body weight.

Conclusion

Generally speaking absorption of orally administrated drugs occurs in two steps: the first is to dissolve the drug in the gastrointestinal fluid and the second is for the dissolved drugs to cross the gastrointestinal membrane. The absorption of genistein should be controlled by the first step: dissolving in the aqueous environment, since genistein is hardly soluble in water. Although it has been demonstrated by several investigators that reduction of drug's particle size could enhance their dissolution and absorption, no one has reported its application on genistein. In this study, genistein nanoparticles with approximately 200 nm in mean particle sizes were successfully prepared from acetone using PCA process. After examine various parameters that could affect particle size formation, it was found that the size of genistein particle was mainly determined by two factors: operating pressure and drug concentration. The dissolution rate and the plasma concentration of genistein have been significantly increased after nanomization demonstrating the possibility of improving the bioavailability of water insoluble compounds using PCA process. It is estimated that almost 41% of the new chemical entities are poorly soluble and most do not progress to preclinical evaluation because of formulation challenges [29]. The technique developed in this study can be used to generate more efficacious pharmaceuticals and health care products.

Reference

[1] Apers, S.; Naessens, T.; Van Den Steen, K.; Cuyckens, F.; Claeys, M.; Pieters, L.; Vlietinck, A. Fast high-performance liquid chromatography method for quality control of soy extracts. *J Chromatogr A* **1038:**107-112; 2004.

[2] Rostagno, M. A.; Palma, M.; Barroso, C. G. Pressurized liquid extraction of isoflavones from soybeans. *Analytica Chimica Acta* **522:**169-177; 2004.

[3] Beck, V.; Rohr, U.; Jungbauer, A. Phytoestrogens derived from red clover: an alternative to estrogen replacement therapy? *J Steroid Biochem Mol Biol* **94:**499-518; 2005.

[4] Glazier, M. G.; Bowman, M. A. A review of the evidence for the use of phytoestrogens as a replacement for traditional estrogen replacement therapy. *Archives of Internal Medicine* **161:**1161-1172; 2001.
[5] Ettinger, B.; Grady, D.; Tosteson, A. N.; Pressman, A.; Macer, J. L. Effect of the Women's Health Initiative on women's decisions to discontinue postmenopausal hormone therapy. *Obstet Gynecol* **102:**1225-1232; 2003.
[6] Beral, V. Breast cancer and hormone-replacement therapy in the Million Women Study. *Lancet* **362:**419-427; 2003.
[7] Dixon, R. A.; Ferreira, D. Genistein. *Phytochemistry* **60:**205-211; 2002.
[8] Branham, W. S.; Dial, S. L.; Moland, C. L.; Hass, B. S.; Blair, R. M.; Fang, H.; Shi, L.; Tong, W.; Perkins, R. G.; Sheehan, D. M. Phytoestrogens and mycoestrogens bind to the rat uterine estrogen receptor. *J Nutr* **132:**658-664; 2002.
[9] Kinjo, J.; Tsuchihashi, R.; Morito, K.; Hirose, T.; Aomori, T.; Nagao, T.; Okabe, H.; Nohara, T.; Masamune, Y. Interactions of phytoestrogens with estrogen receptors α and β (III). Estrogenic activities of soy isoflavone aglycones and their metabolites isolated from human urine. *Biological and Pharmaceutical Bulletin* **27:**185-188; 2004.
[10]Wei, H.; Bowen, R.; Cai, Q.; Barnes, S.; Wang, Y. Antioxidant and antipromotional effects of the soybean isoflavone genistein. *Proc Soc Exp Biol Med* **208:**124-130; 1995.
[11]Li, Z.; Li, J.; Mo, B.; Hu, C.; Liu, H.; Qi, H.; Wang, X.; Xu, J. Genistein induces G2/M cell cycle arrest via stable activation of ERK1/2 pathway in MDA-MB-231 breast cancer cells. *Cell Biol Toxicol* **24:**401-409; 2008.
[12]Majid, S.; Dar, A. A.; Ahmad, A. E.; Hirata, H.; Kawakami, K.; Shahryari, V.; Saini, S.; Tanaka, Y.; Dahiya, A. V.; Khatri, G.; Dahiya, R. BTG3 tumor suppressor gene promoter demethylation, histone modification and cell cycle arrest by genistein in renal cancer. *Carcinogenesis* **30:**662-670; 2009.
[13]Wang, J.; Eltoum, I. E.; Lamartiniere, C. A. Genistein chemoprevention of prostate cancer in TRAMP mice. *J Carcinog* **6:**3; 2007.
[14]Daruhazi, A. E.; Szente, L.; Balogh, B.; Matyus, P.; Beni, S.; Takacs, M.; Gergely, A.; Horvath, P.; Szoke, E.; Lemberkovics, E. Utility of cyclodextrins in the formulation of genistein part 1. Preparation and physicochemical properties of genistein complexes with native cyclodextrins. *J Pharm Biomed Anal* **48:**636-640; 2008.
[15]Stancanelli, R.; Mazzaglia, A.; Tommasini, S.; Calabro, M. L.; Villari, V.; Guardo, M.; Ficarra, P.; Ficarra, R. The enhancement of isoflavones water solubility by complexation with modified cyclodextrins: a spectroscopic investigation with implications in the pharmaceutical analysis. *J Pharm Biomed Anal* **44:**980-984; 2007.
[16]Lee, S. H.; Kim, Y. H.; Yu, H. J.; Cho, N. S.; Kim, T. H.; Kim, D. C.; Chung, C. B.; Hwang, Y. I.; Kim, K. H. Enhanced bioavailability of soy isoflavones by complexation with beta-cyclodextrin in rats. *Biosci Biotechnol Biochem* **71:**2927-2933; 2007.
[17]Motlekar, N.; Khan, M. A.; Youan, B. B. C. Preparation and characterization of genistein containing poly(ethylene glycol) microparticles. *Journal of Applied Polymer Science* **101:**2070-2078; 2006.
[18]Matson, D. W.; Fulton, J. L.; Petersen, R. C.; Smith, R. D. Rapid expansion of supercritical fluid solutions: solute formation of powders, thin films, and fibers. *Industrial and Engineering Chemistry Research* **26:**2298-2306; 1987.
[19]Mohamed, R. S.; Debenedetti, P. G.; Prud'homme, R. K. Effects of process conditions on

crystals obtained from supercritical mixtures. *AIChE Journal* **35:**325-328; 1989.
[20]Charoenchaitrakool, M.; Dehghani, F.; Foster, N. R.; Chan, H. K. Micronization by rapid expansion of supercritical solutions to enhance the dissolution rates of poorly water-soluble pharmaceuticals. *Industrial and Engineering Chemistry Research* **39:**4794-4802; 2000.
[21]Gallagher, P. M.; Krukonis, V.; Botsaris, G. D. Gas Anti-Solvent (GAS) Recrystallization. Application to particle design. *AIChE Symposium Series*; 1991: 96-103.
[22]Yeo, S. D.; Lim, G. B.; Debendetti, P. G.; Bernstein, H. Formation of microparticulate protein powder using a supercritical fluid antisolvent. *Biotechnol Bioeng* **41:**341-346; 1993.
[23]Kim, M. S.; Lee, S.; Park, J. S.; Woo, J. S.; Hwang, S. J. Micronization of cilostazol using supercritical antisolvent (SAS) process: Effect of process parameters. *Powder Technology* **177:**64-70; 2007.
[24]Kim, M. S.; Jin, S. J.; Kim, J. S.; Park, H. J.; Song, H. S.; Neubert, R. H.; Hwang, S. J. Preparation, characterization and in vivo evaluation of amorphous atorvastatin calcium nanoparticles using supercritical antisolvent (SAS) process. *Eur J Pharm Biopharm* **69:**454-465; 2008.
[25]Rogers, T. L.; Johnston, K. P.; Williams, R. O., 3rd. Solution-based particle formation of pharmaceutical powders by supercritical or compressed fluid CO2 and cryogenic spray-freezing technologies. *Drug Dev Ind Pharm* **27:**1003-1015; 2001.
[26]Ghaderi, R.; Artursson, P.; Carlfors, J. Preparation of biodegradable microparticles using solution-enhanced dispersion by supercritical fluids (SEDS). *Pharm Res* **16:**676-681; 1999.
[27]Mawson, S.; Kanakia, S.; Johnston, K. P. Coaxial nozzle for control of particle morphology in precipitation with a compressed fluid antisolvent. *Journal of Applied Polymer Science* **64:**2105-2118; 1997.
[28]Lin, C.; Ng, K. M.; Wibowo, C. Producing nanoparticles using precipitation with compressed antisolvent. *Industrial and Engineering Chemistry Research* **46:**3580-3589; 2007.
[29] Lipinski, C. A. Avoiding investment in doomed drugs. *Curr. Drug Discov.*:17-19; 2001.

Bulk Production of Metal Oxide Nanowires Using A Novel Microwave Plasma Reactor

Vivekanand Kumar, Department of Chemical Engineering, University of Louisville, Louisville, KY

Jeong H. Kim, Chemical Engineering, University of Louisville, Louisville, KY

Mahendra K. Sunkara, Department of Chemical Engineering, University of Louisville, Louisville, KY

Nanowires are beginning to be used in electronics, optoelectronics, electrochemical cells, nanoelectromechanical devices, catalysis, and several other fields. Synthesis of bulk quantities of nanowires with controlled composition, crystallinity and morphology is important to continued development of nanowire technology. For many applications, nanowire quantities of several grams or more are needed. Bulk quantities of nanowires have been difficult to produce using the methods known so far.

A novel method is used to produce bulk quantities of high purity metal oxide nanowires using a custom made high throughput microwave plasma reactor. Low cost, very fast, large scale production of different metal oxide nanowires can be accomplished using this technique. Also, there is no need of any catalyst, substrate or template to produce the nanowires. So far we have produced bulk quantities of tin oxide, zinc oxide, copper-zinc oxide and titanium oxide nanowires. The underlying synthesis concept is the direct reaction between a metal or metal precursor and oxygen containing plasmas resulting in the production of nanowires directly in the gas phase. Initial results show theoretical nanowire production capacity to the tune of kilograms per day. Pure nanowires are obtained by a dispersion in 1-methoxy 2-proponal followed by a horn sonication and gravity sedimentation.

See more of this Session: Nanowires B: Nanowires and Their Composites for Sensing, Energy Conversion and Energy Storage
See more of this Group/Topical: Nanoscale Science and Engineering Forum

Nanowire Assembly and Interconnect Formation Using Nano-Solders

Fan Gao, Chemical Engineering, University of Massachusetts Lowell, Lowell, MA

Subhadeep Mukherjee, Chemical Engineering, University of Massachusetts Lowell, Lowell, MA

Xiaopeng Li, Chemical Engineering, University of Massachusetts Lowell, Lowell, MA

Zhiyong Gu, Chemical Engineering, University of Massachusetts Lowell, Lowell, MA

Nanowires are receiving increasing interest as essential components in future nanoelectronics. To date the synthesis and assembly techniques of nanowires have been well developed. In this research, we show our recent results in using nano-solders to assemble and join nanowires for nanoelectronics applications. Tin-based and indium-based lead-free nano-solders have been synthesized directly onto multisegmented metal nanowires or conducting polymer nanowires by template assisted electrodeposition method. After synthesis, nanowires are assembled in a so called "liquid solder reflow" process, in which the nano-solder nanowires are assembled in a liquid medium and solder joints are formed between nanowires. 2D or 3D ordered nanowire arrays are formed by either electric field assisted alignment (dielectrophoretic assembly) or magnetic field assisted alignment. Nano-solders, when combined with assembling techniques, are used to efficiently connect nanowires and nanowire electronic components with low contact resistance, which are very well suited for sensor as well as nanoelectronics applications.

See more of this Session: Nanowires B: Nanowires and Their Composites for Sensing, Energy Conversion and Energy Storage
See more of this Group/Topical: Nanoscale Science and Engineering Forum

Model Nanowires: Mono- Versus Polycrystalline Nanostructure

Randy L. Vander Wal, Energy and Mineral Engineering, Pennsylvania State University, University Park, PA

Michael Kulis, NCSER c/o NASA-Glenn Research Center, Cleveland, OH

Gordon M. Berger, NCSER c/o NASA-Glenn Research Center, Cleveland, OH

Gary W. Hunter, The NASA-Glenn Research Center, Cleveland, OH

Jennifer C. Xu, The NASA-Glenn Research Center, Cleveland, OH

The superiority of nanomaterials for sensor applications is clear. High surface area and controlled structure are the hallmarks. Each aspect is particularly relevant to sensors. Surface area is critical to reactive gas adsorption and translates into high sensitivity. Relative to micron sized grains or powders or layers or films, nanoparticles offer a 10 to 100-fold increase in each parameter. Controlled structure provides the reactive sites for adsorption and their effect upon the overall conductance. Nanomaterials often possess enhanced reactivity due both to a strained surface lattice and/or increased prevalence of step, edge, corner and terrace sites. Such activity further enhances sensitivity and potentially lowers operational temperature. Operation at lower temperatures saves power, extends operating lifetime and maintains reproducibility by preventing sintering-induced grain growth. Finally, lower operational temperature, combined with structure control can advantageously yield selectivity.

Two competing approaches have been developed for synthesizing 1-dimensional forms of metal oxide semiconductors (MOSs): electrospinning and thermal-evaporative-condensation (TEC) synthesis. Each method produces 1-d sensor elements that can be incorporated into next generation sensors. Each synthesis method and product has attendant advantages and limitations. Apart from device fabrication and manufacturing issues, these two methods produce elements that differ primarily in their crystallinity. TEC synthesis can produce single crystalline nanowires. Electrospinning produces polycrystalline elements of the (as-spun) sol-gel fiber upon calcination. Though each form is a nanoscale material, an additional parameter, crystallinity, will have profound consequences upon the viability of each 1-d element for reactive gas sensing and devices incorporating them.

With a diameter approaching twice the nominal charge depletion layer thickness (~ 10's of nanometers, dependent upon temperature), the full conducting channel of a nanorod can be affected by adsorption of oxidizing gases, analogous to the electrical conductivity modulation in a field effect transistor. Although somewhat thicker, the polycrystallinity and cylindrical structure of a nanofiber potentially permits deeper penetration of the depletion layer into the structure. Essentially the same narrowing of the "effective" conduction channel occurs. Carrier depletion (or replenishment) throughout the "bulk" nanostructure will expand the sensor dynamic range by the virtue of oxidizing (reducing) adsorbates leading to full charge depletion (or replenishment) with corresponding near-zero or high conductivity, respectively. With the entire bulk of the material responsive to surface adsorbed species, sensitivity gains may be achieved relative to 2-d films where often only the outer surface layer is affected while the underlying portion remains unchanged (maintaining constant conductance).

Though it would appear that the nanorod is the limit of the summation describing the nanofiber case, in practice the nanorod diameter is generally larger than twice the depletion layer thickness. Moreover the uniform crystallographic surface planes may not support a high surface concentration of defect sites or chemisorbed oxygen species. Alternatively in the polycrystalline nanofiber not all junctions may be accessible to adsorbates. Such spots would correspond to a "short". Moreover there could be a considerable variation in potential boundaries between grains, given the random orientation of single crystal nanoparticles with respect to each other. Therein experimental tests of the sensing performance of each material were conducted. The validity of such a comparison is facilitated by the one-dimensional morphology of each differently nanostructured metal oxide.

In order to further enhance the sensitivity of these morphologies to gases, catalyst nanoparticles were deposited upon the surface of the sensing elements. The utility of catalysts upon active oxide supports is well known in catalysis but has yet received little attention in gas sensors. Most approaches using catalysts of either noble metals or other transition metal doped the material. With a homogeneous distribution, most of the catalyst (whether atomically distributed or as nanophase particles) is inaccessible to the near-surface region, the region where active sensing occurs. Secondly, an atomic dispersion will alter lattice and grain structure, but not specifically create catalytic sites upon the sensor surface. The latter is the desirable structure from a catalysis viewpoint and is exploited for the 1-d sensing elements reported here. We note that this approach is frequently used in catalysis where the noble metal nanoparticles and/or the interfacial region between the particle and oxide support greatly accelerates the reaction compared to the bare oxide surface.

Generally this particle coating will permit exposure of the underlying metal oxide support and most importantly establishes an interface between the particle and support oxide. These interfacial junctions will be self-polarized (forming a Schottky junction) by virtue of charge transfer, driven by the difference between the metal work function and oxide electron affinity. This interface is expected to be highly reactive for crystalline metal nanoparticles. Clearly this reactivity is critically dependent upon the nature of the oxide support material (composition, its crystallinity, chemisorbed oxygen sites and their binding energy, etc.) and availability of lattice oxygen from the semiconductor oxide surrounding the metal nanoparticle. Nevertheless such junctions are expected to be particularly reactive given the interfacial polarization. Additionally the catalyst nanoparticles will alter the free carrier charge density and band energy levels.

To summarize, the central goal of this investigation was to operationally test the difference between these two nanostructured materials forms, with and without catalysts. In the fabrication of the prototype devices we gained practical knowledge of each synthesis method and realized caveats for integration of each element type towards commercial device manufacture. Advantages and limitations of each method towards practical sensor fabrication will also be presented.

See more of this Session: Nanowires B: Nanowires and Their Composites for Sensing, Energy Conversion and Energy Storage
See more of this Group/Topical: Nanoscale Science and Engineering Forum

Electrochemical Scanning Tunneling Microscopy of Dye Sensitized Solar Cells

Sudhira Pasupuleti, Chemical Engineering, University of Louisville, Louisville, KY

Gerold.A Willing, Chemical Engineeering, University of Louisville, Louisville, KY

Photoelectrochemical cells provide a promising future for providing energy from the sun. Though bulk studies on PEC systems can suggest possible alternative materials and system designs, the fundamental knowledge of charge transport through a PEC required for the engineering and optimization of the cell calls for more intricate research in terms of spatial resolution to further explore the discrepancies and inefficiencies within the system. In one such attempt to provide detailed information on the functioning of a PEC system in terms of its electronic properties and its corresponding photocurrent generation, we have applied electrochemical scanning tunneling microscopy to the investigation of a dye sensitized solar cell. This system provides a measurement of photocurrent generation along the surface of the nanocrystalline metal oxide working electrode at nanoscale resolution through which electron/hole generation on the surface can be calculated. In this talk, we will compare the working nature of different materials such as nanowires, colloidal thin films and hybrid structures to understand the metal oxide/dye/electrolyte interface reactivity, improve the charge transfer of the electrons along the surface and provide better photon to electron conversion efficiency.

See more of this Session: Nanowires B: Nanowires and Their Composites for Sensing, Energy Conversion and Energy Storage
See more of this Group/Topical: Nanoscale Science and Engineering Forum

Reactive Bi-Metallic Aluminium/Nickel Nanowires as Nano-Heaters and Their Applications

Qingzhou Cui, Chemical Engineering, University of Massachusetts Lowell, Lowell, MA

Yingying Sha, Chemical Engineering, University of Massachusetts Lowell, Lowell, MA

Julie Chen, Mechanical Engineering, University of Massachusetts Lowell, Lowell, MA

Zhiyong Gu, Chemical Engineering, University of Massachusetts Lowell, Lowell, MA

Nanoscale heating sources ("nano-heaters") based on exothermic reaction between reactive materials, such as aluminum (Al) and nickel (Ni), have received great attention in recent years. In this presentation, we will show that Al-Ni bi-metallic nanowires can be fabricated by a combination of electrochemical deposition and physical evaporation methods. The heterogeneous nanostructures have been characterized by a variety of tools including scanning electron microscopy (SEM), transmission electron microscopy (TEM), and energy dispersive X-ray analysis (EDX). Ignition tests by heating and laser were performed. Potential applications of the nanostructures as nano-heaters are studied in this work. We incorporate the bimetallic Al/Ni nanowire nano-heaters onto multi-segmented nanowires and utilize the heat generated from the nano-heaters to bond or solder other nanowire structures.

See more of this Session: Nanowires B: Nanowires and Their Composites for Sensing, Energy Conversion and Energy Storage
See more of this Group/Topical: Nanoscale Science and Engineering Forum

ELECTROLESS ETCHED COMPOSITE SILCION NANOWIRE ANODE FOR LITHIUM ION BATTERY

W. Xu and J. C. Flake
Louisiana State University, Baton Rouge, LA, 70808

Introduction

Silicon is a promising candidate for use as an anode material in secondary lithium ion cells[1]. The fully lithiated $Li_{22}Si_5$ phase has the highest theoretical capacity (4200 $mA \cdot hg^{-1}$) among any known material[2]; however, the large volume expansion (100-300 %) associated with lithium insertion causes pulverization and degrades electronic and ionic conduction. These properties typically result in silicon anodes with high initial capacities and high levels of capacity loss[2]. In efforts to mitigate the pulverization and fade associated with silicon anodes, a variety of silicon structures have been considered[2]. Porous silicon, silicon pillars, and thin films have been studied as anode material for their excellent accommodation for volume change of silicon during charge/discharge cycles[3-5]. Recently, the nanowire structure attracts considerable attention as a promising active material for composite anode for its 1-D structure to stretch in axial direction and sustain volume change without mechanical failure[6]. Studies with Vapor Liquid Solid (VLS) grown silicon nanowire arrays on substrate show near theoretical capacities[6-9]. Although these silicon anodes mentioned above shows good cyclicability and reasonable reversible capacity after hundreds of cycles, they require complicated and high-cost fabrication procedures, limiting their use in commercial secondary lithium ion cells.

Silicon composite anodes in which silicon micro or nano-powders are combined with carbon have also been considered in several reports[10-15]. Silicon micro or nanoparticles may be dispersed homogeneously into inactive soft and ductile matrices to accommodate large volume changes on cycling; The composite anode enjoys great advantage in fabrication for its relatively easy and low-cost procedures, however, the performance of composite anodes are strongly dependent on particle size, particle electrical properties, shape, arrangement, active/inactive material ratios, preparation methods, binders, and additive species[2,15]. Silicon nanowire fabricated via electroless etching as proposed by Qiu et al.[9] is one possible structure for silicon composite anode for its excellent accommodation of volume change and low-cost facile fabrication process. Not like silicon particles, the nanowires in composite anode will expend and form network contact with other nanowires so as enhance the mechanical integrity of the anode, and large agglomeration can be avoided. Another interesting application for silicon nanowire is that the nanowires on substrate can be easily functionalized with organic groups by electrochemical method[16-17], so as to further enhance cell performance by facile functionalization of silicon surface[18]. In this work, we report a new silicon nanowire composite anode for lithium rechargeable battery so as to combine both the advantages of composite anode and nanowire structure. The anode contains 15% electrolessly fabricated silicon nanowire and graphite as active material. The electrochemical performance of the anode has been studied, and the morphology of the anode surface has been characterized by SEM as well.

Experiment

Silicon nanowire arrays were fabricated via electroless etching of boron-doped (100) silicon wafers in a 5.0 M hydrofluoric acid with 0.02 M silver nitrate at 50 °C for 30 minutes. Metallic silver was removed after etching followed by HF cleaning, de-ionized water rinsing and air drying. The fabricared silicon nanowires were separated from the parent wafer by sonication. Nanowires were centrifuged and separated from the transfer solvent and combined with graphite to create a 15:85 (w/w) active materials. Active components were combined with carbon black and polyvinylidene fluoride (PVdF) in n-methyl-2-pyrrolidone (NMP, Aldrich) and ball milled for 24 hours followed by deposition onto nickel foil using doctor blade technique. Anodes were thermally treated at 120 °C for 12 hours in argon to yield a dry anode composition of 60 % active material, 20 % carbon black and 20 % PVdF. Electrochemical experiments were performed in a typical three-electrode cell with excess electrolyte, a lithium foil counter and lithium rod reference electrode. The electrolyte is 1M lithium hexafluorophosphate ($LiPF_6$) in ethylene carbonate and dimethyl carbonate (EC/DMC, Aldrich) solution at a 1:1 volumetric ratio. Electrochemical experiments were performed in an argon filled glove box system (<1 ppm H_2O and O_2) using a Princeton Applied Research PARSTAT 2272 potentiostat. Charge and discharge cycles were performed galvanostatically at 0.1 $mA \cdot cm^{-2}$ (~C/10, normalized by current collector area) in the potential range from 0.01 V to 1.5 V versus Li/Li^+. Measured capacity data were normalized by active material weight. Samples were transferred to SEM chambers in sealed, argon-filled containers to prevent oxidation and imaged within 4 hours or transfer.

Results and Discussions

As shown in Figure 1, silicon nanowires may be separated from their parent substrate by sonication and concentrated via centrifugation. The nanowires are crystalline and retain the doping and orientation (<100> along the major axis) from the parent substrate.[3] Nanowires created in this process typically have diameters ranging from 100 to 200 nm and are 10 to 20 μm in length depending on etching time.[3, 4]

Figure 1. SEM image of silicon nanowire on substrate fabricated via electroless etching.

Figure 2 shows an SEM image of a composite anode surface created with 15 % silicon nanowires and graphite. Silicon nanowires are randomly distributed in the graphite matrix at this concentration.

Figure 3 shows an SEM image of the composite anode after 15 charge/discharge cycles showing the formation of a Solid Electrolyte Interphase (SEI) layer including intact nanowires at the surface.

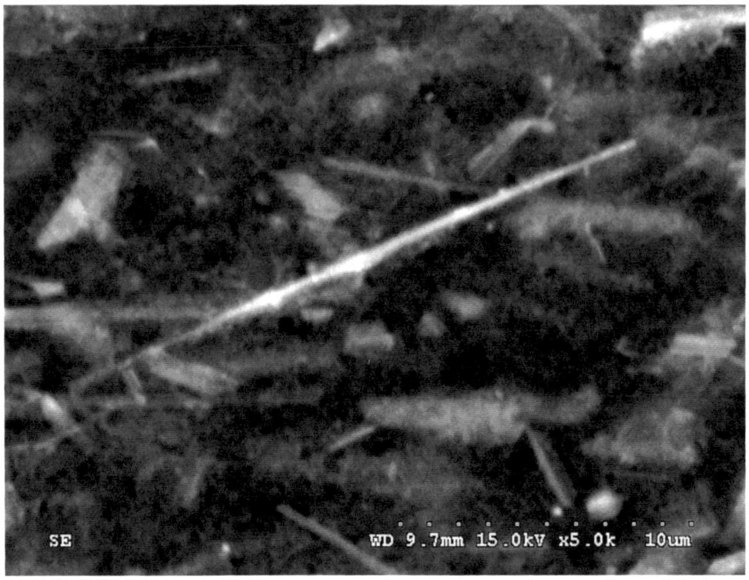

Figure 2. SEM image of silicon nanowire in composite anode as fabricated.

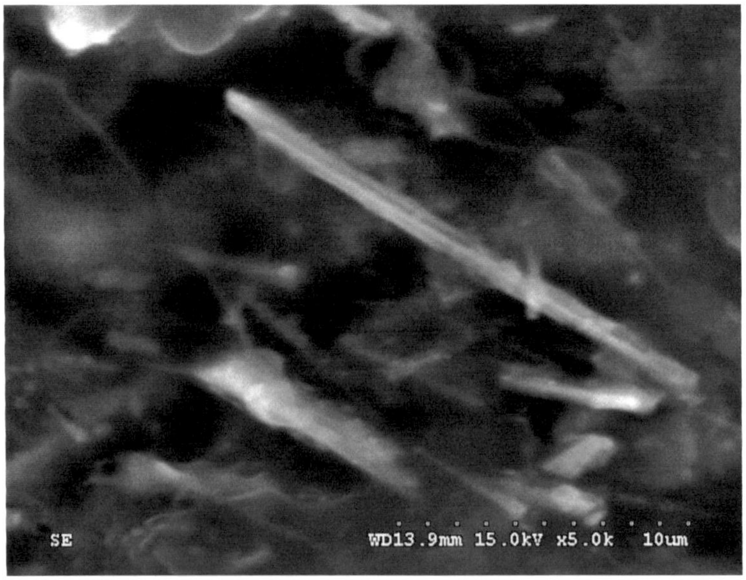

Figure 3. SEM image of silicon nanowire in composite anode after cell cycles.

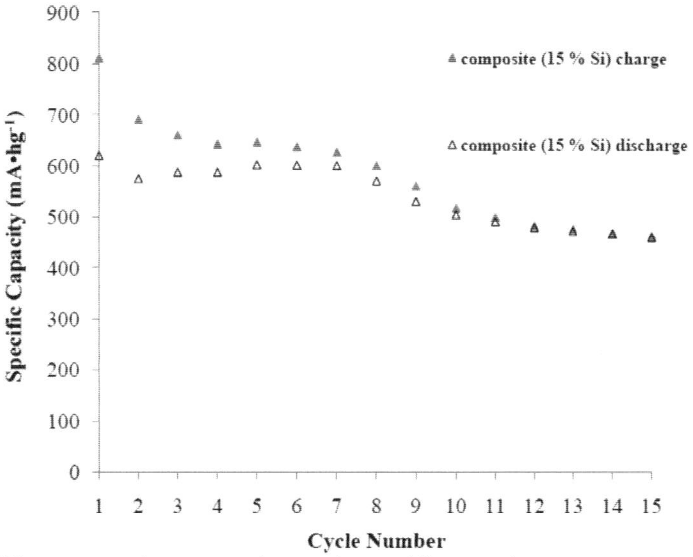

Figure 4. Silicon nanowire composite anode specific capacity versus cell cycle number.

As shown in Figure 4, the initial charge capacity is 811 $mA \cdot hg^{-1}$ and represents 95 % of predicted capacity of the composite anode (853 $mA \cdot hg^{-1}$) based on silicon concentration and theoretical capacity for $Li_{15}Si_4$ phase formed below 50 mV.[5] The initial composite anode capacity is approximately 2.7 times greater than the equivalent graphite anode, and the reversible capacity is approximately 1.7 times greater than graphite. Couloumbic efficiency for the first cycle is approximately 73.9 %. Reversibility improves as the cell cycles, resulting in a reversible capacity of approximately 512 $mA \cdot hg^{-1}$ after 10 cycles with a capacity fade of approximately 1.4 % per cycle. The capacity of the equivalent graphite anode improves slightly with cycling and a reversible capacity of 307 $mA \cdot hg^{-1}$ is seen after 10 cycles. This reversible capacity is equivalent to 82.5 % of theoretical graphite capacity at 372 $mA \cdot hg^{-1}$.

The capacity fade mechanism is still under extensive study for silicon composite anode. Besides replacing PVdF with CMC binders to improve anode binding, surface chemistry of silicon anode might be one of the most important factors in determining reversible capacity for prolonged cell cycles. Preliminary studies including functionalization of silicon nanowire surfaces with organic groups to form a passivation layer has been conducted to improve anode performance. Further works will be carried out to understand loss mechanisms and methods to mitigate their effects.

Conclusion

A novel silicon nanowire composite anode (15% Silicon, 85% graphite w/w) is demonstrated with initial capacity 2.7 times greater than graphite with a reversible anode capacity of 512 $mA \cdot hg^{-1}$ for 10 cycles. SEM analysis shows the silicon nanowire structure grows in diameter and remains intact after charge/discharge cycles, showing the anode structure can sustain large volume change during cell cycling. In comparison with uncoated silicon, the electrolessly-etched silicon nanowires are considered excellent candidates for use in composite anodes due to their; unique geometry, natural hydride termination, controllable doping, and a facile fabrication process.

References

1. Huggins, R. A., "Lithium alloy negative electrodes," *Journal of Power Sources* **1999,** 81-82, 13-19.
2. Kasavajjula, U.; Wang, C.; Appleby, A. J., "Nano- and bulk-silicon-based insertion anodes for lithium-ion secondary cells," *Journal of Power Sources* **2007,** 163, (2), 1003-1039.
3. Wanli, X.; Vadim, P.; John, C. F., "Nickel Monosilicide Contact Formation in Electrolessly Etched Silicon Nanowires Deposited onto Nickel Electrodes," *Journal of The Electrochemical Society* **2009,** 156, (7), H544-H547.
4. Peng, K. Q.; Yan, Y. J.; Gao, S. P.; Zhu, J., "Synthesis of large-area silicon nanowire arrays via self-assembling nanoelectrochemistry," *Advanced Materials* **2002,** 14, (16), 1164-1167.
5. Obrovac, M. N.; Christensen, L., "Structural Changes in Silicon Anodes during Lithium Insertion/Extraction," *Electrochemical and Solid-State Letters* **2004,** 7, (5), A93-A96.
6. Cui, L. F.; Ruffo, R.; Chan, C. K.; Peng, H. L.; Cui, Y., "Crystalline-Amorphous Core-Shell Silicon Nanowires for High Capacity and High Current Battery Electrodes," *Nano Letters* **2009,** 9, (1), 491-495.
7. Dimov, N.; Kugino, S.; Yoshio, M., "Carbon-coated silicon as anode material for lithium ion batteries: advantages and limitations," *Electrochimica Acta* **2003,** 48, (11), 1579-1587.
8. Dimov, N.; Xia, Y.; Yoshio, M., "Practical silicon-based composite anodes for lithium-ion batteries: Fundamental and technological features," *Journal of Power Sources* **2007,** 171, (2), 886-893.
9. Dimov, N.; Kugino, S.; Yoshio, M., "Mixed silicon-graphite composites as anode material for lithium ion batteries: Influence of preparation conditions on the properties of the material," *Journal of Power Sources* **2004,** 136, (1), 108-114.
10. Yoshio, M.; Tsumura, T.; Dimov, N., "Electrochemical behaviors of silicon based anode material," *Journal of Power Sources* **2005,** 146, (1-2), 10-14.
11. Chan, C. K.; Hailin, P.; Gao, L.; McIlwrath, K.; Xiao Feng, Z.; Huggins, R. A.; Yi, C., "High-performance lithium battery anodes using silicon nanowires," *Nature Nanotechnology* **2008,** 3, (1), 31-35.
12. Lestriez, B.; Bahri, S.; Sandu, I.; Rou, L.; "Guyomard, D., On the binding mechanism of CMC in Si negative electrodes for Li-ion batteries," *Electrochemistry Communications* **2007,** 9, (12), 2801-2806.
13. Chan, C. K.; Ruffo, R.; Hong, S. S.; Cui, Y., "Surface chemistry and morphology of the solid electrolyte interphase on silicon nanowire lithium-ion battery anodes," *Journal of Power Sources* **2009,** 189, (2), 1132-1140.
14. Kong, F.; Kostecki, R.; Nadeau, G.; Song, X.; Zaghib, K.; Kinoshita, K.; McLarnon, F., "In situ studies of SEI formation," *Journal of Power Sources* **2001,** 97-98, 58-66.
15. John C. Flake, Melissa M. Rieger, Gerard M. Schmid, and Paul A. Kohla, "Electrochemical Etching of Silicon in Nonaqueous Electrolytes Containing Hydrogen Fluoride or Fluoroborate," *Journal of The Electrochemical Society,* **146** (5) 1960-1965 **1999.**
16. Choi, N. S.; Yew, K. H.; Kim, H.; Kim, S. S.; Choi, W. U., "Surface layer formed on silicon thin-film electrode in lithium bis(oxalato) borate-based electrolyte," *Journal of Power Sources* **2007,** 172, (1), 404-409.
17. Song, S. W.; Baek, S. W., "Silane-Derived SEI Stabilization on Thin-Film Electrodes of Nanocrystalline Si for Lithium Batteries," *Electrochemical and Solid State Letters* **2009,** 12, (2), A23-A27.
18. Ngunjiri, J. V., Vengunta S. S.; Flake, J, "Cathodic Electrografting of Alkyl Nanopatterns

on(100) Silicon," *J. Electrochem. Society* **156** (7), H516-H521, 2009.

Palladium Nanofibers for Electrooxidation of Alcohols in Alkaline Medium

Liang Su, Chemical, Materials and Biomolecular Engineering, University of Connecticut, Storrs, CT

Wenzhao Jia, Chemical, Materials and Biomolecular Engineering, University of Connecticut, Storrs, CT

Ashley Schempf, Chemical, Materials and Biomolecular Engineering, University of Connecticut, Storrs, CT

Yu Ding, Chemical, Materials and Biomolecular Engineering, University of Connecticut, Storrs, CT

Yu Lei, Chemical, Materials and Biomolecular Engineering, University of Connecticut, Storrs, CT

Palladium (Pd) nanofibers with high surface area had been successfully prepared. The as-prepared membranes were characterized by field-emission scanning electron microscopy (FESEM), energy-dispersive X-ray spectrometry (EDX) and X-ray diffraction (XRD). Its application for ethanol oxidation was studied extensively. The results show that the Pd nanofibers greatly promote the oxidation of ethanol in alkaline medium for fuel cell application.

See more of this Session: Nanowires B: Nanowires and Their Composites for Sensing, Energy Conversion and Energy Storage
See more of this Group/Topical: Nanoscale Science and Engineering Forum

One-Dimensional Alignment of Monodispersed Silica Nanospheres with Block Copolymer

Tatsuya Okubo, Department of Chemical System Engineering, The University of Tokyo, Tokyo 113-8656, Japan

Masashi Fukao, Department of Chemical System Engineering, The University of Tokyo, Tokyo 113-8656, Japan

Ayae Sugawara, Department of Chemical System Engineering, The University of Tokyo, Tokyo 113-8656, Japan

Atsushi Shimojima, Department of Chemical System Engineering, The University of Tokyo, Tokyo 113-8656, Japan

Wei Fan, Department of Chemical Engineering and Materials Science, University of Minnesota, Minneapolis, MN

Michael Tsapatsis, Department of Chemical Engineering and Materials Science, University of Minnesota, Minneapolis, MN

Monodispersed silica nanospheres can be synthesized by hydrolysis and polycondensation of tetraethoxysilane (TEOS) in the presence of basic amino acids as catalyst. To fully realize their potential as new materials, these nanoparticles need to be fabricated into ordered arrays or predesigned structures. We have developed a very simple method of fabricating one-dimensionally aligned silica nanospheres in size of 15 nm by mixing with poly(ethyleneoxide) - poly(propylene oxide- poly(ethylene oxide)- type triblock copolymer. The average number of connected particles and the connecting pattern can be easily tuned by controlling the experimental conditions such as the pH, the temperature and the heating time. The details of the experiments as well as the formation mechanism will be reported in the presentation.

See more of this Session: Templated Assembly of Inorganic Nanomaterials I
See more of this Group/Topical: Nanoscale Science and Engineering Forum

Pulse-Field Gradient NMR Study of TAA^+-Silica Association in Zeolite Clear-Solution Syntheses

Xiang Li, Chemical Engineering, Texas A&M University, College Station, TX

Daniel F. Shantz, Department of Chemical Engineering, Texas A&M University, College Station, TX

Silicalite-1 syntheses from optically transparent dilute TPAOH / TEOS / H_2O mixtures have been widely studied and are of fundamental importance as a model system for understanding zeolite nucleation. These mixtures have been studied by many labs using methods including small-angle scattering, electron microscopy, calorimetry and various spectroscopies.[1-6]. Recently, Fedeyko et al. reported formation of silica nanoparticles in base solutions of tetraalkylammonium (TAA) cations above a critical aggregation concentration of silica originally dissolved in solution.[3] An important, and as yet unresolved question is how the interaction between the silica nanoparticles and the organocation results in zeolite nucleation and crystallization.

We will summarize our efforts in determining the forces between organocation-silica nanoparticles using ^1H Pulse-Field gradient NMR and Two-Dimensional Diffusion-Ordered NMR spectroscopy(DOSY). From measurements of the organocation self-diffusion coefficient one can determine the fraction of the organocations bound to the silica nanoparticle. From this information one can then determine thermodynamic properties for these mixtures as they relate to cation - nanoparticle interactions. We have found the diffusion coefficient of TPA^+ decreased significantly in the presence of silica nanoparticles and it decreased inversely with the silica/TPA^+ ratio. Hence, we could assume a two state model consisting of nanoparticle-bound TPA^+ in equilibrium with freely diffusing TPA^+ in bulk solution. Since the chemical exchange between the bound and free state of TPA^+ is fast compared to the relaxation time scale, the observed diffusivity of TPA^+ is a population weighted average of the two sites. From this analysis information about the adsorption strength and binding ratio of TPA^+ versus silica nanoparticles at adsorption saturation can be extracted. Further variation of the association strength of silica nanoparticles and TPA^+ with electrolyte concentration and organocations with different density charge (N^+/C) can provide more detailed illustration of the organic-inorganic interaction at the beginning of silicalite-1 crystallization. The NMR results for a series of TAA cations will be discussed in the context of existing literature, and the implications of the findings for zeolite growth will be discussed.

Reference

[1] Davis,.T.M., Drews, T.O., Rmanan, H., et.al., *nature materials*, **2006**, 5, 400-408

[2] Kumar, S., Wang, Z., Penn, R.L., *J. Am. Chem. Soc.* ,**2008**, 130, 17284-17286

[3] Fedeyko, J., Rimer, J., Lobo, R. et.al., *J. Phys. Chem. B*, **2004**, 108, 12271-12275

[4] Kinrade, S.D., Knight, C. T. G., Pole, D.L., *Inorg. Chem.* **1998**, 37, 4272-4277

[5] Cheng. C.H., Shanzt, D.F., *J.Phys.Chem.B*, **2005**, 109, 7266-7274

[6] Schoeman, B. J., *Microporous Mesoporous Mater.* **1998**, 22, 9-22

See more of this Session: Templated Assembly of Inorganic Nanomaterials I
See more of this Group/Topical: Nanoscale Science and Engineering Forum

Modeling the Formation of Nanoparticles During Early Stages of Zeolite Growth

Lin Jin, Chemical Engineering, University of Massachusetts, Amherst, MA

Scott M. Auerbach, Chemistry, University of Massachusetts, Amherst, MA

Peter A. Monson, Univ. of Massachusetts, Amherst, MA

A significant amount of experimentation has been carried out to characterize the growth kinetics of zeolites, but the mechanism is still unclear. Since the key events in zeolite nucleation are not easily accessible by experimental methods, molecular modeling may be able to shed some light on how zeolites form. In the past decade, silica nanoparticles have been found to play an important role in the clear-solution synthesis of zeolites. Jorge et al. developed a simple lattice model to describe the formation of silica nanoparticles in the early stages of the clear-solution templated synthesis of silicalite-1, and obtained qualitative agreement with published experimental observations for some of the properties of the nanoparticles, including the core-shell structure [J. Am. Chem. Soc. 127, 14388(2005)]. Their model was based on simple cubic (SC) lattice, which does not model the four-coordinate bonding character of silica solids. In this work we are exploring two approaches to improving the lattice model by incorporating four-coordinate interactions. It is anticipated that this effort will lead to a more realistic structure for the nanoparticles and provide a basis for modeling the growth at later stages.

A four-coordinated tetrahedral network can be generated by imposing a second neighbor repulsion on a body-centered cubic (BCC) lattice . In this way we successfully generated nanoparticles with four-coordinated tetrahedral network. The nanoparticles possess a core-shell structure with silica in the core and template at the shell. More importantly, some template particles were able to penetrate the precursor nanoparticles, indicating that the nanoparticles generated by this model have some porosity. In addition to the SC and BCC models where a unit of silica is represented by a single site we have developed an atomistic Silicon-Oxygen lattice model. The Silica-Oxygen model represents silica as a rigid tetrahedron as one silicon atom at the center and four oxygen atoms (or hydroxyls) at the corners. We have developed a set of rules for resolving the configurations of these tetrahedra after condensation or hydrolysis reactions. This atomistic lattice model gives a quite realistic treatment of the structure of polymerizing silica.

See more of this Session: Templated Assembly of Inorganic Nanomaterials I
See more of this Group/Topical: Nanoscale Science and Engineering Forum

Synthesis of Silicalite-1 Nanocrystals in Three Dimensionally Ordered Mesoporous Carbon

Wei Fan, Department of Chemical Engineering and Materials Science, University of Minnesota, Minneapolis, MN

Pyung-Soo Lee, Department of Chemical Engineering and Materials Science, University of Minnesota, Minneapolis, MN

Sandeep Kumar, Department of Chemical Engineering and Materials Science, University of Minnesota, Minneapolis, MN

Won Cheol Yoo, Department of Chemistry, University of Minnesota, Minneapolis, MN

Xueyi Zhang, Department of Chemical Engineering and Materials Science, University of Minnesota, Minneapolis, MN

Andreas Stein, Department of Chemistry, University of Minnesota, Minneapolis, MN

R. Lee Penn, Department of Chemistry, University of Minnesota, Minneapolis, MN

Alon V. McCormick, Department of Chemical Engineering and Materials Science, University of Minnesota, Minneapolis, MN

Michael Tsapatsis, Department of Chemical Engineering and Materials Science, University of Minnesota, Minneapolis, MN

The sustained interest in the synthesis of zeolite nanocrystals is justified because they can serve as seeds for thin zeolite film growth and for preparation of polymer-zeolite nanocomposite membranes. In this study, silicalite-1 nanocrystals were synthesized inside three-dimensionally-ordered mesoporous (3DOm) carbon with a uniform pore size that can be tuned from 20 nm to 50 nm [1,2]. The 3DOm carbon was prepared using ordered arrays of silica nanoparticles (Lys-Sil) as a template. Electron Microscopy, X-ray diffraction and N_2 adsorption and desorption reveal that the silicalite-1 nanocrystals can reliably replicate the 3DOm porosity, resulting in the highly monodisperse silicalite-1 nanocrystals with a tunable size from 20 nm to 50 nm. Their application in the fabrication of zeolite thin film will be discussed along with implications in understanding zeolite nucleation and growth.

1. Davis, T. M.; Snyder, M. A.; Krohn, J. E.; Tsapatsis, M. Chem. Mater. 2006, 18, 5814.

2. Fan, W.; Synder, M. A.; Sandeep, K.; Lee, S. P.; Yoo, C. W.; McCormick, A. V.; Penn, L. R.; Stein, A.; Tsapatsis, M. Nat. Mater. 2008, 7, 984.

See more of this Session: Templated Assembly of Inorganic Nanomaterials I
See more of this Group/Topical: Nanoscale Science and Engineering Forum

Reverse Templating of Mesoporous Ceramics to Produce Highly Porous Structured Carbons From Sugars

J.E. St.Dennis, Chemical and Biomolecular Engineering, Tulane University, New Orleans, LA

Pradeep Venkataraman, Chemical and Biomolecular Engineering, Tulane University, New Orleans, LA

Bhanukiran Sunkara, Chemical and Biomolecular Engineering, Tulane University, New Orleans, LA

Vijay T. John, Chemical and Biomolecular Engineering, Tulane University, New Orleans, LA

Gary L. McPherson, Chemistry, Tulane University, New Orleans, LA

Jibao He, Coordinated Instrument Facility, Tulane University, New Orleans, LA

Camille Y. Jones, Department of Chemistry, Hamilton College, Clinton, NY

Steven J. Obrey, Chemistry Division, Los Alamos National Laboratory, Los Alamos, NM

Robert S. Currier, Chemistry Division, Los Alamos National Laboratory, Los Alamos, NM

Arijit Bose, Department of Chemical Engineering, University of Rhode Island, Kingston, RI

Novel carbon materials, such as nano-tubes, fullerenes and uni-dimensional structures, have received significant attention in recent years due to their applications in gas storage, catalysis, and in electrochemical devices (supercapacitors and electrodes). We describe here methods to use sugars as precursors to produce a variety of nanostructured carbons using the reverse template of porous titanias and silicas. These templates are prepared from surfactant mesophases that exhibit long range crystallinity and exist as stable gels. In particular, we focus on the 2-D hexagonal mesophase templates formed using the twin-tailed anionic surfactant bis(2-ethylhexyl) sodium sulfosuccinate (AOT) and the zwitterionic surfactant phosphatidylcholine (lecithin). These materials exhibit the characteristics of a rigid gel and maintain initial structural integrity when used as templates for the synthesis of ceramics such as silica and titania. The templated ceramics exhibit a hierarchical pore structure where 10 nm pores evolve to 200 nm pores. We then use these ceramics as templates by infiltrating carbon precursors, forming carbons through pyrolysis and then etching away the ceramic. The resultant carbon materials exhibit the inverse pore structure and long range order of the template. High resolution electron microscopy is used to understand the evolution and growth of these carbon structures. Variations of carbon structure and morphology develop upon changing the carbon precursor from sucrose to the cyclodextrins to a polysaccharide (carboxymethyl cellulose). The role of "threading" cyclodextrin to form templated carbons will be described.

See more of this Session: Templated Assembly of Inorganic Nanomaterials I
See more of this Group/Topical: Nanoscale Science and Engineering Forum

Aligned Carbon Nanotube Membranes with Enhance Fluid Flow and Active Gate Keeper Control

Mainak Majumder, University of Kentucky, Lexington KY, KY

Ji Wu, University of Kentucky, Lexington KY, KY

Karen Gerstandt, University of Kentucky, Lexington KY, KY

Bruce J. Hinds, Chemical and Materials Eng., University of Kentucky, Lexington KY, KY

Carbon nanotubes have three key attributes that make them of great interest for novel membrane applications 1) atomically flat graphite surface allows for ideal fluid slip boundary conditions 2) the cutting process to open CNTs inherently places functional chemistry at CNT core entrance and 3) CNT are electrically conductive allowing for electrochemical reactions and application of electric fields gradients at CNT tips. To explore the hypothesis of 'Gatekeeper' membrane selectivity, the entrances to CNT's cores were functionalized with aliphatic amines of different lengths, charged dye molecule and poly-peptides. A hindered diffusion model with a geometry of CNT tip functionalization, not along the length of CNT core, was consistent with the experimentally observed selectivities. Bio-chemical gating of CNTs is also seen by tethering desthiobiotin to CNT tips with the reversible binding to streptavidin. The complete ATP cycle (phosphylation/dephosphylation) can be performed on CNT tips with corresponding modulation of flux across CNT membrane. The functional density of tethered charge molecules can be substantially increased by the use of electrochemical grafting of diazonium salts. Voltage modulated membranes were realized with tethered molecules at the tips of CNTs being drawn into the CNT core at positive bias thus hindering/gating flux across the membrane. In general, the transport mechanisms through CNT membrane are a) ionic diffusion is near bulk expectation with no enhancement from CNT b) gas flow is enhanced by ~1-2 order of magnitude due to specular reflection off of flat graphitic surface c) and pressure driven flux of a variety of solvents (H_2O, hexane, decane ethanol, methanol) are 4-5 ORDERS OF MAGNITUDE FASTER than conventional Newtonian flow due to atomically flat graphite planes inducing nearly ideal slip conditions.

See more of this Session: Carbon Nanotubes III: Adsorption and Transport
See more of this Group/Topical: Nanoscale Science and Engineering Forum

Fast Fluid Flow and Electrolyte Transport in Carbon Nanotube Pores

Francesco Fornasiero, Lawrence Livermore National Laboratory, Livermore, CA

Sangil Kim, Lawrence Livermore National Laboratory, Livermore, CA

Jung Bin In, Mechanical Engineering, University of California, Berkeley, Berkeley, CA

Hyung Gyu Park, Lawrence Livermore National Laboratory, Livermore, CA

Jason K. Holt, Lawrence Livermore National Laboratory, Livermore, CA

Michael Stadermann, Lawrence Livermore National Laboratory, Livermore, CA

Costas P. Grigoropoulos, Mechanical Engineering, University of California, Berkeley, Berkeley, CA

Aleksandr Noy, Lawrence Livermore National Laboratory, Livermore, CA

Olgica Bakajin, Lawrence Livermore National Laboratory, Livermore, CA

Through advancing our understanding of novel fluid behavior at nanoscale, nanofluidic research is expected to significantly advance many technological areas such as chemical and biological sensing, drug delivery, lab-on-chip, and fluid separations. Robust nanofluidic platforms will allow us to develop a basic understanding of fluid flow under nanoscale confinement.

In our laboratory, we have developed a nanofluidic membrane with well-aligned, sub 2-nm carbon nanotubes as pores. We have used these membranes to study mass transport and selectivity through carbon nanotubes for both pure fluids and mixtures. Our measured gas and water transport rates confirm the molecular dynamic predictions of an exceptionally fast fluid flow in carbon nanotube pores. For example, the measured water flow exceeds values calculated from continuum hydrodynamics models by more than three orders of magnitude [1]. Our work on the mass-transport selectivity for single [2] and binary electrolyte solutions through carbon nanotubes suggest that, for dilute solutions, small ions can be selectively excluded by carbon nanotube pores when their open rim is enriched by charged groups. In some cases, ion exclusion can be as high as 98%. The selectivity mechanism for salt rejection/permeation through 1-2 nm carbon-nanotube pores appears to be dominated by electrostatic interactions between the ions and the charged carbon nanotube rims. The observed ion selectivity rules agree with the Donnan membrane equilibrium theory. Steric and hydrodynamic effects are less important [2]. For binary electrolyte solutions with common cation, we observed a negative rejection of the monovalent anion at the smallest mole fractions. The measured rejection of binary salt solutions as a function of the salt mole fraction is also consistent with a charge-based exclusion mechanism. Demonstrated ion selectivity, high multivalent ion rejection, and simultaneous ultrafast fluid transport make our carbon nanotube membranes promising platforms for future low-cost water purification processes.

This work performed under the auspices of the U.S. Department of Energy by Lawrence Livermore National Laboratory under Contract DE-AC52-07NA27344.

[1] J. Holt, H.G. Park, Y. Wang, M. Stadermann, A.B. Artyukhin, C.P. Grigoropoulos, A. Noy and O. Bakajin, "Fast Mass Transport Through Sub-2-Nanometer Carbon Nanotubes," Science, 312, 1034 (2006)

[2] F. Fornasiero, H.G. Park, J. Holt, M. Stadermann, C.P. Grigoropoulos, A. Noy and O. Bakajin, "Ion exclusion by sub-2-nm carbon nanotube pores," Proc. Natl. Acad. Sci. USA, 10.1073/pnas.0710437105 (2008)

See more of this Session: Carbon Nanotubes III: Adsorption and Transport
See more of this Group/Topical: Nanoscale Science and Engineering Forum

Pressure-Driven Water Transport through Single-Walled Carbon Nanotubes

John Thomas, Mechanical Engineering, Carnegie Mellon University, Pittsburgh, PA

Alan J. H. McGaughey, Mechanical Engineering, Carnegie Mellon University, Pittsburgh, PA

Water flow through carbon nanotubes (CNTs) with diameters ranging from 0.83 nm to 6.98 nm is examined using molecular dynamics simulation. A reflecting particle membrane is used to drive the flow and the relationship between the axial pressure gradient, CNT diameter, and volumetric flow rate is examined. The flow enhancement, defined as the ratio of the observed flow rate to that predicted from the no-slip Hagen-Poiseuille relation, is calculated for each CNT. In CNTs with diameters greater than 1.39 nm, flow can be accurately described in terms of continuum mechanics and the enhancement agrees with predictions from the slip-modified Hagen-Poiseuille flow relation. In CNTs with diameters smaller then 1.39 nm, we find that the liquid structure varies with CNT diameter and a continuum-based description of the fluid flow is inappropriate. The flow enhancement for these CNTs do not agree with predictions from the slip-modified Hagen-Poiseuille flow relation. They can, however, be correlated to the diameter-specific liquid structure. We also examine thermal transport in water/CNT composite systems.

See more of this Session: Carbon Nanotubes III: Adsorption and Transport
See more of this Group/Topical: Nanoscale Science and Engineering Forum

"Gated" Ion Transport through Dense Carbon Nanotube Membranes

Miao Yu, Chamical and Biological Engineering, University of Colorado at Boulder, Boulder, CO

"Gated" ion transport through hydrophobic protein domains in various cellular membranes was found widely, while these hydrophobic protein channels regulate ions transport through these membranes with both high selectivity and high flux. Hydrophobic and narrow carbon nanotubes (CNT) may serve as a simplified model to mimic these biological pores and thus reveal main properties of these hydrophobic nanochannels. Dense carbon nanotube (CNT) membranes with both 3-nm CNT and interstitial pores were prepared by a solvent-evaporation process. Ion diffusion through these membranes has an on/off behavior, which may be due to the sensitive water wettability of CNTs. At 298 K, ion diffusion through CNT membranes gradually stopped, but it increased dramatically when the temperature was increased 20 K or the membrane was subjected to ultrasound. Adsorption measurements show that water adsorption increased dramatically when the temperature was increased from 293 to 306 K, and this suggests that the adsorbed water forms a discontinuous phase at 293 K, but possibly a continuous layer at higher temperature in both the CNTs and the interstitial regions. When the ion diffusion channel was open, due to a continuous water phase, ions diffused through the CNT membranes at a rate similar to bulk diffusion in water.

See more of this Session: Carbon Nanotubes III: Adsorption and Transport
See more of this Group/Topical: Nanoscale Science and Engineering Forum

Gate Functionalized CNT Membranes for Desalination

Chandrashekar Shankar, Chemical Engineering, University of Pittsburgh, Pittsburgh, PA

Prashant Kumar, Chemical Engineering, University of Pittsburgh, Pittsburgh, PA

J. Karl Johnson, Dept.of Chemical and Petroleum Engineering, University of Pittsburgh, Pittsburgh, PA

The exceptionally high flux of water through CNTs shown in experiments and simulations suggests a novel method for desalination. However, to increase the viability of CNTs in desalination, greater selectivity is essential. Here, we show preliminary simulation studies of CNT membranes functionalized with gate groups at the CNT tip in order to achieve greater selectivity without sacrificing throughput. A variety of functional groups from simple carboxy (-COOH) groups to more complicated molecules such as 1,3-diaminopropane or 1,4-diaminobutane in various configurations are evaluated. The simulation results are then used in membrane design process.

See more of this Session: Carbon Nanotubes III: Adsorption and Transport
See more of this Group/Topical: Nanoscale Science and Engineering Forum

Diffusion Mechanisms of Fluids Confined in Carbon Nanotubes, Carbon Nanotube Bundles and Hierarchial Carbons: Single-File, Fickian and Dual-Mode Diffusion

Joshua D. Moore, Department of Chemical & Biomolecular Engineering and Center for High Performance Simulation (CHiPS), North Carolina State University, Raleigh, NC

Ying-Chun Liu, Department of Chemical & Biomolecular Engineering and Center for High Performance Simulation (CHiPS), North Carolina State University and, Raleigh, NC

Thomas J. Roussel, Department of Chemical & Biomolecular Engineering and Center for High Performance Simulation (CHiPS), North Carolina State University, Raleigh, NC

Jeremy C. Palmer, Department of Chemical & Biomolecular Engineering and Center for High Performance Simulation (CHiPS), North Carolina State University, Raleigh, NC

Keith E. Gubbins, Department of Chemical & Biomolecular Engineering and Center for High Performance Simulation (CHiPS), North Carolina State University, Raleigh, NC

Carbon nanotubes have received extensive interest in materials science and chemical engineering due to their possibilities of being used as host structures for gas separation; they can yield faster diffusion rates of fluid molecules compared to other widely used materials (e.g., zeolites). Several diffusion mechanisms are possible in narrow cylindrical pores. Among the most important are ballistic motion, Fickian diffusion, and single-file diffusion. Ballistic motion occurs for very short times, before the molecule has had a chance to collide with anything. For longer times the motion becomes either Fickian or single-file. If the pore is large enough, the molecules will diffuse in 3-dimensions as they would in a bulk fluid (Fickian diffusion), but if the pore diameter becomes small enough the diffusion will crossover from Fickian to single-file diffusion, where the molecules can no longer pass each other. Clearly single-file motion is much slower than Fickian, which is much slower than ballistic.

We use a combination of Grand Canonical Monte Carlo and Molecular Dynamics simulations to investigate the adsorption and self-diffusion properties of pure component Lennard-Jones Ar and mixtures of Lennard-Jones fluids (e.g., Ar, Kr, Xe, Ne) in isolated carbon nanotubes and carbon nanotube bundles for a range of densities and nanotube diameters.

In the case of the diffusion of fluids in severe confinement, the fluids are inevitably trapped in or close to a state of Knudsen diffusion where the fluid-wall interactions are important. For pure component fluids, the carbon nanotube transition diameter is a function of the fluid-wall interaction diameter. A single transition diameter of the carbon nanotube can be extracted for each fluid. For mixtures, dual diffusion mechanisms can occur in isolated carbon nanotubes, in which one component exhibits Fickian diffusion while the other shows single-file diffusion. We show that for carbon nanotube bundles and hierarchical carbons a dual diffusion mechanism can also occur, with 3-dimensional Fickian diffusion occurring inside the larger pores and single-file diffusion occurring simultaneously in smaller pores or interstices. Both results have implications for separations and for applications of porous materials with bimodal pore size distributions as catalysts and electrodes.

See more of this Session: Carbon Nanotubes III: Adsorption and Transport
See more of this Group/Topical: Nanoscale Science and Engineering Forum

Magnetically Driven Mixing within a Microarray Geometry Using Functionalized Magnetic Nanoparticles

Paul E. Laibinis, Department of Chemical and Biomolecular Engineering, Vanderbilt University, Nashville, TN

Sandip Agarwal, Rice University, Houston, TX

Microarrays have found widespread use for probing genomic and proteinomic functions. These systems consist of a purposefully spotted microscopic slide of different biological samples (the microarray), a sub-milliliter solution volume containing target molecules of interest, and a coverslip to contain and spread the solution over the microarray surface. Incubation times for the target solution with the microarray can be as long as a day to allow binding of target molecules to probes on the array surface. These times are limited by the reliance on diffusional processes for transport within this thin-film geometry. We have pursued approaches for providing solution movement and mixing within this confined geometry enlisting functionalized magnetite nanoparticles that are included within the target solution and an externally sweeping magnetic field that together provide mixing within the microarray chamber. The effects of particle size, concentration, magnetic path, and movement velocity have each been investigated, resulting in an optimization of accelerated mixing within the chamber. The particles have been prepared to incorporate surface functional groups that make them non-interacting with functionalities present on the DNA microarray surface or toward DNA target molecules in solution, thereby providing a component that is readily removed from the system prior to reading of a microarray. A goal of this work has been to reduce the timescale for performing a microarray from roughly a day to a few hours. Our experiments demonstrate the ability for both accelerating timescales as well as for improving the quality and uniformity of data sets by use of the developed mixing strategy.

See more of this Session: Magnetic Nanoparticles in Biotechnology and Biomedicine III
See more of this Group/Topical: Nanoscale Science and Engineering Forum

Nucleic Acid Extraction with Chemically Stable Metallic Nanomagnets

Robert N. Grass, Department of Chemistry and Applied Biosciences, ETH Zurich, 8093 Zurich, Switzerland

Stefan Huber, Department of Chemistry and Applied Biosciences, ETH Zurich, 8093 Zurich, Switzerland

Magnetic particles have been used commercially for many years for the extraction of nucleic acids from various biological samples. This talk will give a short overview of silica and polymer coated magnetic particles for nucleic acid extraction and focus on the trend towards decreasing particle sizes, down to the 10-30 nm size range. Problems of small particle sizes are discussed in terms of particle stability, aggregation, separation velocity and selectivity. Our recent advances in using specifically designed carbon coated metallic nanoparticles for the extraction of nucleic acids from biological media will show how the individual problems can be addressed and how the novel and highly substrate sensitive magnetic separation agents can be synthesized. Image: Extraction of a standard DNA ladder from biological samples with specifically designed metallic nanoparticles. Line M represents the standard. The particles in line 1, 2 and 3 were treated with water (1), strong acid (2) and strong base (3) prior to the extraction of the length standard from a biological fluid and depict the high chemical stability of the particles. Ref: Grass, Athanassiou and Stark, Angew. Chem. Int. Ed. 2007.

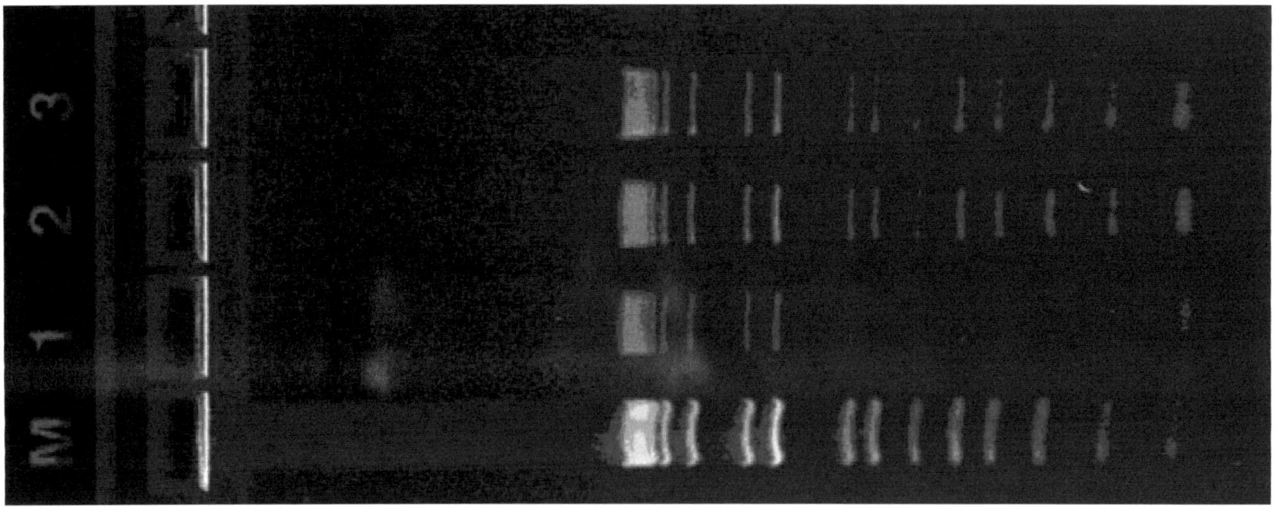

See more of this Session: Magnetic Nanoparticles in Biotechnology and Biomedicine III
See more of this Group/Topical: Nanoscale Science and Engineering Forum

Combining Fluorescent Imaging and Magnetic Force in Single Molecule Studies

Gang Ruan, Department of Chemical and Biomolecular Engineering, The Ohio State University, Columbus, OH

Dhananjay Thakur, Chemical and Biomolecular Engineering and Biophysics, the Ohio State University, Columbus, OH

Greg Vieira, Department of Physics, Ohio State University, Columbus, OH

Thomas Henighan, Department of Physics, Ohio State University, Columbus, OH

Aaron Chen, Department of Physics, Ohio State University, Columbus, OH

R. Sooryakumar, Department of Physics, The Ohio State University, Columbus, OH

Jessica O. Winter, Chemical and Biomolecular Engineering, The Ohio State University, Columbus, OH

Studying single nanoparticles and molecules has led to a number of important discoveries in biology and nanoscience. In particular, single quantum dots (semiconductor nanocrystals) have been successfully used to image and track biological processes at single molecule level in vitro and in vivo. On the other hand, magnetic tweezers have been used to apply force on micron-sized magnetic beads. Combining the fluorescence of quantum dots and the force of magnetic tweezers, we have achieved the first concurrent magnetic manipulation and fluorescent tracking of single nanoparticles (smaller than 100 nm).

Quantum dots and superparamagnetic iron oxide nanoparticles were first loaded into nanocontainers with diameters adjustable from 10 to 100 nm, forming nanocomposites with both fluorescence and superparamagnetism properties. Magnetic tweezers were then used to apply force on the composite nanoparticles, while the movement of these nanoparticles was tracked by a fluorescent microscope. It was found that the composite nanoparticles had fluorescent intermittency (i.e. blinking), indicating that they were single nanoparticles (rather than large aggregates of nanoparticles). The ability of concurrent magnetic manipulation and fluorescent tracking of single sub-100 nm nanoparticles could have broad impact in many areas, including biomechanics and nanofabrication.

See more of this Session: Magnetic Nanoparticles in Biotechnology and Biomedicine III
See more of this Group/Topical: Nanoscale Science and Engineering Forum

Magnetic Nanoparticles/Glucose Oxidase Mediated Cascade Catalysis for Cancer Marker Rapid Detection

Lizeng Gao, Chemical and Petroleum Engineering, University of Pittsburgh, Pittsburgh, PA

Jiamin Wu, Chemical and Petroleum Engineering, University of Pittsburgh, Pittsburgh, PA

Di Gao, Chemical and Petroleum Engineering, University of Pittsburgh, Pittsburgh, PA

Many technologies have been developed for rapid detection of cancer marker, such as typical ELISA, biochips and novel nanobiosensors. These technologies always take many steps, for example, antigen/antibody binding and washing step by step, which will take long time and make the test become burdensome. So it is imperative to develop novel detection system for rapid diagnosis of disease, especially for point of care or home test. Here we reported a novel method for cancer marker detection based on cascade enzymatic activity of magnetic nanoparticles and glucose oxidase. Our previous work found that magnetic nanoparticles have peroxidase like catalysis and catalyze color reaction in H_2O_2/TMB system which is usually used in ELISA. The fundmental strategy is to combine catalytic magnetic nanoparticles and glucose oxidase which can produce H_2O_2 by catalyzing glucose. The mechanism is that adjacent effect between magnetic nanoarticles and glucose oxidase accelerates the color reaction, which means that magnetic nanoparticles mixed with glucose oxidase can perform color reaction only when there is an intermediate that can connects two moieties together. Cancer marker CA125 hase been tested by this method. Two CA125 specific antibodies are crosslinked onto magnetic nanoparticles and glucose oxidase repecitively. For the detection we just need to mix CA125 with antibody coupled magnetic nanoparticles and glucose oxidase at one-step. The color signal produced by cascade catalysis showed clear relation with the concentration of CA125, which means the new system can be used to rapidly detect cancer marker. The system has two advantages compared to classic bioassay. One is that it is very simple and just need to mix the antigen into reaction buffer, which especially suits for point of care or home test. The other is that the reaction is fast because the binding between antigen and antibody is achieved in homogenous liquid phase. The new method has promising potential in rapid diagnosis for cancer markers and other diseases.

See more of this Session: Magnetic Nanoparticles in Biotechnology and Biomedicine III
See more of this Group/Topical: Nanoscale Science and Engineering Forum

Biodegradable Magnetite Stent for Implant-Assisted Magnetic Drug Targeting

Jan O. Mangual, Department of Chemical Engineering, University of South Carolina, Columbia, SC

Armin D. Ebner, Department of Chemical Engineering, University of South Carolina, Columbia, SC

James Ritter, Department of Chemical Engineering, University of South Carolina, Columbia, SC

A biodegradable magnetic nanocomposite stent was developed as the implant for implant assisted-magnetic drug targeting (IA-MDT). The composite consisted of a poly(D,L-lactic-co-glycolic) acid copolymer with dispersed iron oxide nanopowder as the magnetic component. Polymer composites with various iron oxide loadings were melt extruded into fibers and coiled to simulate a ferromagnetic stent implant. An in vitro setup was used to study the capability of this implant to capture 100 nm magnetite particles as simulated magnetic drug carrier particles (MDCPs) under physiological flow conditions.

Degradation and dynamic mechanical thermal analysis were conducted on the fiber samples to characterize the effect of the iron oxide on the polymer composites properties. The addition of iron oxide was found to increase the glass transition temperature (Tg) of the polymer fiber. It also increases the mechanical properties of the fiber composite. Increasing the iron oxide loading in the polymer composite lead to an increase in the capture of the MDCPs. In contrast, reducing the concentration of the MDCPs in suspension and increasing the fluid velocity both had negative impacts on the capture efficiency of the MDCPs. An overview of these findings will be given during this presenatation.

See more of this Session: Magnetic Nanoparticles in Biotechnology and Biomedicine III
See more of this Group/Topical: Nanoscale Science and Engineering Forum

Spin-Polarized Calculations of Hydrated Magnetite Surfaces: Implications for Biomedical Applications

Sándor Á. Kovács, Energy, Environmental and Chemical Engineering, Washington University in St. Louis, St. Louis, MO

Cynthia S. Lo, Energy, Environmental and Chemical Engineering, Washington University in St. Louis, St. Louis, MO

Magnetite (Fe_3O_4) is a commonly used material for biomedical applications, such as bioseparations, drug delivery and diagnostics, including magnetic resonance imaging. In each of these cases, the magnetite-water interface is expected to play a key role in controlling surface chemistry and reactivity, yet the structure of this interface remains elusive. In this study, the surface structure and energetics of both clean and hydrated Fe_3O_4 (1 0 0) and Fe_3O_4 (1 1 1) are calculated using spin-polarized density functional theory and are compared with experimental characterizations. The energetics of the clean (1 0 0) surface are measured against the (1 1 1) surface in order to obtain a fundamental understanding of water ordering at the interface. Water is both molecularly adsorbed and heterolytically dissociated on the two surfaces, and the energetics of these hydrated surfaces are examined. The implications of water ordering at magnetite surfaces will be discussed in the context of surface reactivity for the aforementioned biomedical applications.

See more of this Session: Magnetic Nanoparticles in Biotechnology and Biomedicine III
See more of this Group/Topical: Nanoscale Science and Engineering Forum

Cyclodextrin Conjugated Magnetic Nanoparticles as a Solid-Phase Artificial Chaperone for Protein Refolding

A. Z. M. Badruddoza, Department of Chemical & Biomolecular Engineering, National University of Singapore, Singapore, Singapore

M. S. Uddin, Department of Chemical & Biomolecular Engineering, National University of Singapore, Singapore, Singapore

K. Hidajat, Department of Chemical & Biomolecular Engineering, National University of Singapore, Singapore, Singapore

A solid-phase artificial chaperone (CD-APES-MNPs) was fabricated for protein refolding by grafting mono-tosyl-â-cyclodextrin (TsCD) onto (3-aminopropyl) triethoxysilane (APES) modified magnetic nanoparticles. For confirmation of the grafting of â-CD onto nano-sized magnetic particles, characterizations were carried out using Fourier Transform Infra-Red Spectroscopy (FTIR), Transmission Electron Microscopy (TEM), X-ray Photoelectron Spectroscopy (XPS), Vibrating Sample Magnetometer (VSM), X-ray Diffraction (XRD) and elemental analyses.

The use of CD-APES-MNPs as a solid-phase artificial chaperone to assist protein refolding in vitro was demonstrated using carbonic anhydrase from bovine (CA) and lysozyme as model proteins.Our refolding results indicate that the as-synthesized magnetic nanoparticles can greatly affect the refolding yields of thermally denatured CA and urea denatured lysozyme. Under the optimal refolding environment, the maximum refolding yield obtained was 89% for 0.029 mg/ml of CA and 70% for 0.05 mg/ml lysozyme proteins which were as the same level as that using liquid-phase artificial chaperone-assisted refolding. Additionally, the stripping agents can be easily separated from the refolded-active protein solution by magnetic decantation and can be re-used after recycling. Our results reveal that CD-APES-MNPs is a suitable and efficient host matrix for general use in solid-phase artificial chaperone-assisted refolding.

See more of this Session: Magnetic Nanoparticles in Biotechnology and Biomedicine III
See more of this Group/Topical: Nanoscale Science and Engineering Forum

Permanent and Patterned Adjustment of the Surface Potential of Graphene-Like Carbon through Chemical Functionalization

Fabian M. Koehler, Departement of Chemistry and Applied Biosciences, ETH Zurich, Zurich, Switzerland

Norman A. Luechinger, Departement of Chemistry and Applied Biosciences, ETH Zurich, Zurich, Switzerland

Dominik Ziegler, Department of Mechanical Engineering, ETH Zurich, Zurich, Switzerland

Evagelos K. Athanassiou, Department of Chemistry and Applied Biosciences, ETH Zurich, 8093 Zurich, Switzerland

Robert N. Grass, Department of Chemistry and Applied Biosciences, ETH Zurich, 8093 Zurich, Switzerland

Antonella Rossi, Department of Material Science, ETH Zurich, Zurich, Switzerland

Christofer Hierold, Department of Mechanical Engineering, ETH Zurich, Zurich, Switzerland

Andreas Stemmer, Department of Mechanical Engineering, ETH Zurich, Zurich, Switzerland

Wendelin J. Stark, Department of Chemistry and Applied Biosciences, ETH Zurich, 8093 Zurich, Switzerland

The potential use of graphene in electronics requires reliable and pattern-controlled methods to inject or remove electron density from the two-dimensional carbon honeycomb lattices. Combining well established radical chemistry at ambient conditions and classical lithography, we find that the structure of graphene layers can be permanently altered through covalent chemical functionalization. We further demonstrate how the classic Hammett concept and the Linear Free Enthalpy Relationship from organic chemistry can be used to predict the surface potential shifts in graphene-like carbon surfaces. Our study (1) reveals an astonishingly simple yet accurate method to adjust the surface potential in graphene layers. Next to the direct application of such patterning in device fabrication, the here shown covalent attachment introduces a third dimension in the two-dimensional graphene base. Such functionalization can ultimately lead to covalent attachment of molecular electronics, circuitry incorporated chemical sensors or actuators and offers an alternative approach to address the contacting problem in the nm range. The possibility to print a potential pattern onto a graphene sheet was only conceptually shown at present, but this approach offers a permanent polarization of the graphene sheets as in the case of gate-induced changes in conductivity. Latter will be of crucial importance when using a larger graphene sheet as a starting material in device fabrication (3D analogy: native Si wafer), which is then doped and partially oxidized (3D analogy: insulating silicon oxide). One might envision that for graphene, doping or oxidation to an insulating state could therefore be achieved through simple chemical processing and classical or dip-pen lithography.

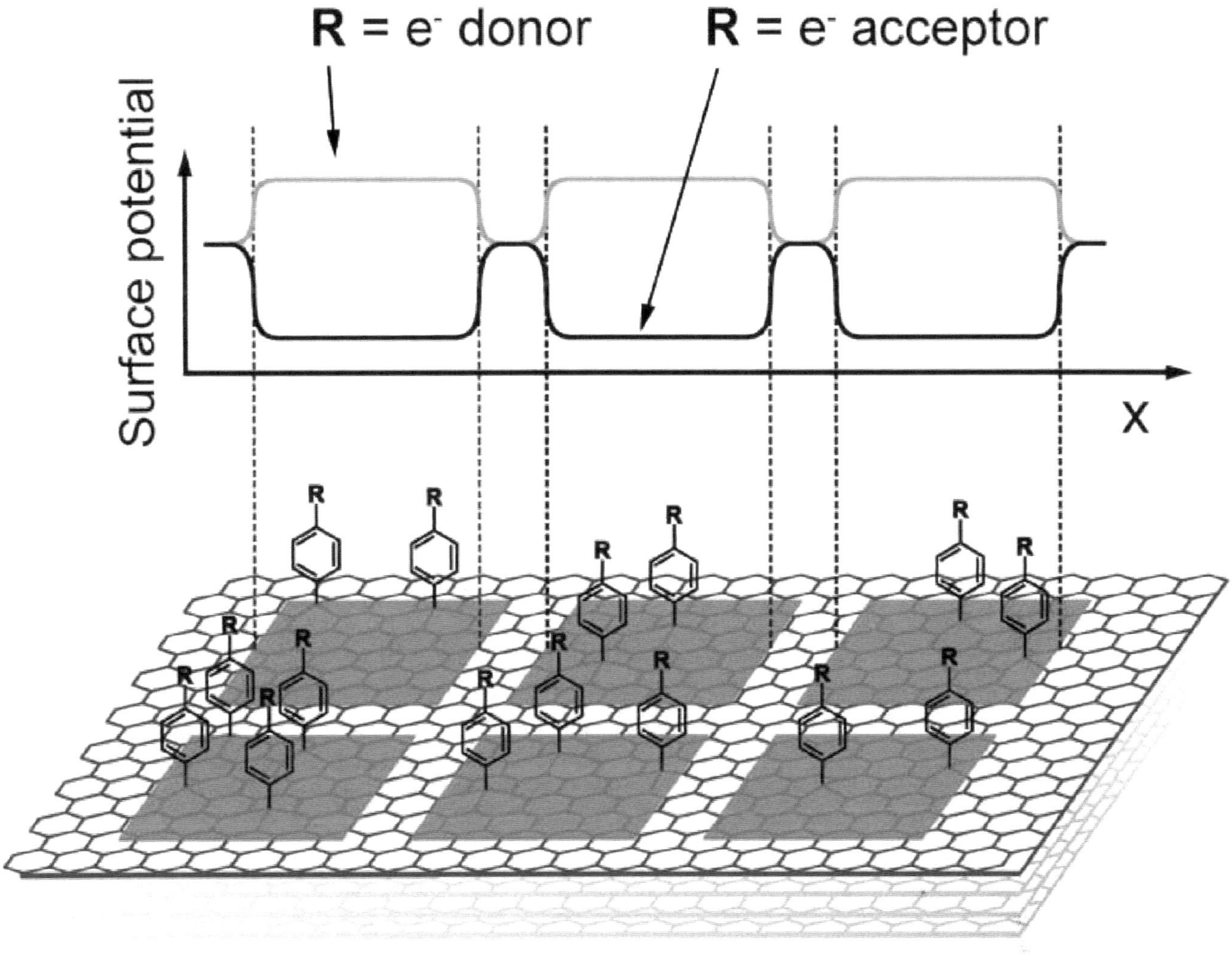

See more of this Session: Nanoelectronic Materials II
See more of this Group/Topical: Nanoscale Science and Engineering Forum

Separating Carbon Nanotubes: Connecting Single Molecule Electrical Measurements to Ensemble Spectroscopic Properties

Woo-Jae Kim, Department of Chemical Engineering, Massachusetts Institute of Technology, Cambridge, MA

Nitish Nair, Department of Chemical Engineering, Massachusetts Institute of Technology, Cambridge, MA

Chang Young Lee, Chemical Engineering, Massachusetts Institute of Technology, Cambridge, MA

Michael Strano, Chemical Engineering, Massachusetts Institute of Technology, Cambridge, MA

The separation and sorting of single-walled carbon nanotubes (SWNTs) is one of the most significant challenges in this rapidly progressing field. A variety of methods including density gradient centrifugation1 and free solution electrophoresis2 invariably utilize Raman, photoabsorption, or photoluminescent spectroscopy to benchmark the resulting separation. However, it is not clear how to relate these qualitative numbers to absolute quantities of individual SWNTs. Moreover, for nanoelectronic applications, separation and purity standards need to be studied using single molecule electrical measurements, and ensemble spectroscopic

assessments are thus far uninformative for this purpose. There is also recent evidence that solution phase processing of the types employed for the above separation methods induces undesirable defects on SWNTs that adversely affect their electrical performance. However, there is currently no quantitative standard by wh ich the degree of defect generation on individual SWNTs from solution phase processing is accessed. In this presentation, we directly compare, for the first time, the ensemble spectroscopic measurements to a systematic and statistically rigorous counting of individual metallic and semiconducting SWNT devices using a high throughput electrical probe station. The comparison allows us for the first time to report an accurate extinction coefficient ratio for metallic and semiconducting SWNTs from HiPco and laser oven preparations. This parameter should greatly aid in the analytical chemistry of separation methods for SWNTs. The systematic counting of metallic and semiconducting types from solution also allows us to examine the variances associated with device properties and therefore provide the first measure of potential defect generation during processing.

See more of this Session: Nanoelectronic Materials II
See more of this Group/Topical: Nanoscale Science and Engineering Forum

Solvent Extraction and Size Effects On the Electrical and Optical Properties of P3HT Nanostructures

Steven D. Bearden, Jr., Joseph Cannon, Ahmed Minhas, Peter Venema, Scott A. Gold
Louisiana Tech University, Institute for Micromanufacturing, Ruston, LA, USA 71272

Abstract

The effect of Soxhlet extraction procedures on the production and properties of P3HT-based semiconducting polymer nanotubes via template wetting nanofabrication will be examined. The extraction process is examined in terms of the Flory-Huggins theory. Electronic properties studied using current voltage characteristics of nanotube diode devices along with corresponding charge transport models while Uv-vis spectroscopy was used to examine the optical properties of various fractions. Comparison of the fractions with the as received bulk material is made. This analysis will help to further understand and qualify the directed assembly of polymer nanotubes constructed by template wetting.

Introduction

Since the introduction of polythiophene-based electronic devices in 1986[1], advances in synthesis procedures[2] have led to dramatic improvements in molecular ordering and device performance for polythiophene systems[3]. While various semiconducting thiophene derivatives have demonstrated higher mobilities, poly(3-hexylthiophene) (P3HT) has proven to be the most widely studied thiophene material. This is largely because of its solubility in common organic solvents and its exceptionally high carrier mobility among soluble materials[3]. The high mobilities observed originate from the fact that the material forms well ordered films in which high degrees of intermolecular π-orbital overlapping is present[3].

Synthesis of high quality materials is of the utmost importance[2], as impurities present in the material can have significant, adverse effects on device performance. Impurities in organic electronic materials are often introduced during synthesis, and can act to limit molecular ordering, trap free carriers, and accelerate device degradation[4]. Key issues with synthetic processes include broad molecular weight distributions and high metallic content present in the resultant material.[5] Typical synthetic procedures react halogenated 3-hexylthiophene monomers, with some form of metallic, heterogeneous catalyst to produce the polymer.[2] This leads to the presence of undesirable metallic and ionic impurities within the synthesized polymer.

Purification processes based on various extraction techniques have been developed and utilized to produce P3HT material with very low metallic content and narrow molecular weight distributions.[5] Trznadel et al. demonstrated Soxhlet extraction as an effective means of purifying bulk P3HT after synthesis.[5] Several narrow molecular weight fractions were collected when following an extraction sequence using five different solvents.[5] The acetone and chloroform fractions were noted to contain large amounts of metallic impurities and small oligomers.[5] Utilizing this Soxhlet extraction procedure, Samitsu demonstrated that P3HT synthesized by the Rieke process (as obtained from Sigma-Aldrich) could be separated into 2 major fractions: one extracted from the bulk with dichloromethane with an average molecular weight of approximately 15,000 g/mol, and another extracted with tetrahydrofuran whose average molecular weight was approximately 110,000 g/mol [133]. The dichloromethane fraction accounted for ~12% of the mass of the original bulk polymer, and the tetrahydrofuran fraction accounted for ~84% [133]. The remaining four percent included both the acetone and hexane fractions, but the metallic content of these fractions was not commented upon [133].

The effect of Soxhlet extraction procedures on the production and properties of P3HT-based semiconducting polymer nanotubes is examined here. The extraction process is examined in terms of the Flory-Huggins theory and the electronic and optical properties of various fractions are examined and compared with the bulk material before purification.

Experimental

Regioregular poly(3-hexylthiophene-2,5-diyl), or P3HT, with >98.5 % head-to-tail linkages (Rieke Metals Inc.) was obtained from Sigma-Aldrich. High purity acetone (AC), chloroform (CF), dichloromethane (DC), hexane (HE), and tetrahydrofuran (THF) were also obtained from Sigma-Aldrich, and were used as received. Purification procedures were performed using a ChemGlass micro Soxhlet extractor (25 ml capacity). Commercially available anodic porous aluminum oxide membranes (Whatman Co.) were used as templates. Templates were 60 μm thick with monodisperse pore diameters of 100 nm and 200 nm, with porosities of 10^{10} pores per cm^2 and 10^9 pores per cm^2, respectively.

Soxhlet extraction procedures were performed following a variation of the procedures described elsewhere,5 which called for extraction until the filtrate appeared colorless in each of five solvents: AC, HE, DC, THF, and CF in that order. Flory-Huggins solubility methods were also used to examine the extraction process. Solubility parameters for P3HT were calculated as a function of molecular weight for three possible end group combinations, and solvent interaction parameters were calculated for each of the extraction solvents. This is done to provide insight into the possible contents of each fraction.

Electronic devices are created using template wetting techniques. Wetting solutions were created using the unfiltered polymer (UNF), dichloromethane fraction (DC), and tetrahydrofuran fraction (THF) by dissolving in chloroform at a concentration of 2 mg/ml. Templates with both 100 nm and 200 nm nominal pore diameters were wetted with 100 μl droplets, and solvent was allowed to evaporate for several hours. Residual polymer was removed from template surfaces by means of non-reactive ion etching with a 100 W He plasma, at 200 mtorr for 10 minutes on each side. Electronic contacts were created by depositing 150 nm Au layers through a shadow mask via electron beam evaporation.

Spectroscopic analysis was used to provide information about the π-electron system present in the conjugated, organic materials.6 UV-vis spectroscopy was used to determine optical absorbance spectra and to approximate optical band gaps of the polymers in the template. Samples were examined immediately after He plasma etching, with no further modification.

Results and Discussion

Extraction procedures are often developed and analyzed by examining the relationships between the dispersion, polarity, hydrogen bonding solubility, δ, and fractional, Fr, parameters.[7] Here, we use these procedures, along with the two methods of calculating solvent interaction to access the effects of molecular weight and molecular termination groups on solubility in the solvents used in the extraction procedure. Combinations of methyl and Br termination groups for the P3HT molecules are examined here. Due to lack of complete empirical data for group contributions to cohesive energies and molar attraction constants, the effects of iodine and MgBr terminations are not examined. The various methods of calculating the solvent interaction parameter:

$$\chi_{H,1} = \frac{v_i}{RT}(\delta_{t,i} - \delta_{t,j})^2 \qquad [1]$$

and

$$\chi_{H,2} = \frac{v_i}{RT}\{(\delta_{d,i} - \delta_{d,j})^2 + (\delta_{p,i} - \delta_{p,j})^2 + (\delta_{h,i} - \delta_{h,j})^2 \qquad [2]$$

produce significantly different results for the molecular weight dependence of solubility at room temperature. Little affect was actually observed due to changes in end group, thus only the Me-(3HT)n-Me data is examined here. From the $\chi_{H,1}$ calculation, we expect that P3HT is most soluble in dichloromethane, regardless of molecular weight, and that the material is more soluble in acetone than in tetrahydrofuran. Neither of these predictions is true, as the material is indeed soluble in tetrahydrofuran, and virtually insoluble in acetone. However, the predictions obtained from $\chi_{H,2}$ calculations are in much better agreement with experiment. Solubility is predicted to be best in dichloromethane at low molecular weights, and better in tetrahydrofuran for higher molecular weights. This is in good agreement with extraction procedures performed here, and observations made elsewhere as well.[5,8]

Weight percentages of material extracted for each fraction are recorded and in compared with those reported in the literature. Our experimental results are in good agreement with those of Samitsu[8] for SA-P3HT, but significantly higher levels of material were extracted in the AC and HE fractions for ADS-P3HT. Due to the limited solubility predicted for P3HT in these solvents, the extracted material for these fractions is assumed to be made up of compounds other than P3HT. These contaminants are likely residual catalyst material and defective oligomers, as observed by Trznadel.[5]

Current densities measured in Au/P3HT/Au hole-only, electronic devices yielded consistent results from device to device, and substrate to substrate. Electronic devices created from the DC fraction exhibited consistently higher current densities than those observed in either the THF fraction or the unfiltered material, regardless of pore size or vendor. Devices created from the THF fraction also demonstrated higher current densities when compared with the unfiltered material. The mobility enhancement for the DC fraction was strongest in P3HT, but larger standard deviations of the mobility were also observed for this material, indicative of lower purity in the bulk material. The DC fraction is known to consist of moderate molecular weight material,[8] and likely contains the most narrow molecular weight distribution of all fractions.[5] Reduced molecular weight variation should lead to improved ordering of the material, as interstitial defects created by mixing of dissimilar molecular lengths disrupts long range order[4], thus the devices fabricated with the DC fraction should be most ordered. Mobilities calculated for the THF fractions are virtually identical to those of the unfiltered material. This supports the hypothesis that the DC fraction should be more ordered due to narrower molecular weight distributions, because the THF fraction has been observed to contain a very broad range of molecular weights[5]. Further, the THF fraction likely contains any metallic impurities not removed in the AC or HE extractions.

The Uv-vis optical absorbance spectrums were also consistent. The onset of the absorption was used to determine a rough optical bandgap. The DC devices showed a higher energy onset than unfiltered or THF. The reduction in the molecular weight can be seen in the reduction in energy gap. The lower absorbance corresponds to a reduction in effective conjugation length. Thus, a shorter polymer length of the DC fraction polymer may increase the mobility, but the intrachain interactions are more dominant than the interchain interactions. The THF fraction onset is at a slightly lower energy. This is thought to be due to the removal of the impurities as well as the removal of the lower molecular weight polymers that will leave only longer chains with longer conjugation lengths.

Conclusions

Soxhlet extraction has been demonstrated as an effective means of purifying bulk P3HT material. The extraction process has been accurately predicted following a variation of Flory-Huggins theory for polymer solutions. Extracted P3HT was used to successfully create semiconducting polymer nanotubes, and electronic devices based on these nanotubes. Slight improvements in hole mobility were observed for the dichloromethane fraction in all sample groups, and is likely due to the high degree of purity in this fraction. As well as, a slight improvement in the optical bandgap of the THF fraction likely due to the removal of short chain polymers.

The prediction scheme for Soxhlet extraction processes following[7] adequately described the extraction process used in this work, when analyzed in conjunction with molecular weight dependent solvent interaction parameters calculated following [2]. Predictions made using the standard solvent interaction expression [1], proved inaccurate, suggesting that [2] may be a more accurate approach to calculation of solubility for the purposes of developing extraction procedures. The extraction procedure produced significant quantities of material for each of the DC and THF fractions, with extracted mass percentages in good agreement with other work for P3HT[8].

References

(1) Tsumura, A.; Koezuka, H.; Ando, T. *Applied Physics Letters* **1986**, *49*, 1210-1212.
(2) Chen, T. A.; Wu, X.; Rieke, R. D. *Journal of the American Chemical Society* **1995**, *117*, 233.
(3) Sirringhaus, H.; Brown, P. J.; Friend, R. H.; Nielsen, M. M.; Bechgaard, K.; Langeveld-Voss, B. M. W.; Spiering, A. J. H.; Janssen, R. A. J.; Meijer, E. W.; Herwig, P.; de Leeuw, D. M. *Nature* **1999**, *401*, 685-688.
(4) Forrest, S. R. *Nature* **2004**, *428*, 911-918.
(5) Trznadel, M.; Pron, A.; Zagorska, M.; Chrzaszcz, R.; Pielichowski, J. *Macromolecules* **1998**, *31*, 5051-5058.
(6) Davids, P. S.; Campbell, I. H.; Smith, D. L. *Journal of Applied Physics* **1997**, *82*, 6319-6325.
(7) Fitzpatrick, L. J.; Dean, J. R. *Analytical Chemistry* **2001**, *74*, 74-79.
(8) Samitsu, S.; Shimomura, T.; Heike, S.; Hashizume, T.; Ito, K. *Macromolecules* **2008**, *41*, 8000-8010.

Covalent Molecular Assembly of Gold Nanoparticles for Non-Volatile Memory Applications

M.P. Srinivasan, Department of Chemical and Biomolecular Engineering, National University of Singapore, Singapore, Singapore

Raju Kumar Gupta, Chem and Biomol Engg, NUS, Singapore, Singapore

P.S. Lee, School of Materials Science and Engineering, Nanyang Technological University, Singapore, Singapore

D.Y. Kusuma, of Materials Science and Engineering, Nanyang Technological University, Singapore, Singapore

The synthesis of metal nanoparticles has been the focus of extensive research over the past decade due to their intriguing size and shape-dependent properties and, more importantly, their applications as building blocks for nanoelectronic devices. Controlled assembly of nanoparticles into ordered architectures on substrates is a promising path to develop miniaturized electronic and optical devices. The stability of nanostructures containing organic thin films has always been one of the critical issues in fabricating electronic devices. An avenue to solve this issue is to assemble nanoparticles covalently on to a desired surface to enhance stability of electronic devices for real-world applications.

In this work, we have synthesized thiol-functionalized gold nanoparticles (AuNPs) with a narrow size distribution (diameter °5 nm) as confirmed by Transmission electron microscopy (TEM). The Au NPs were immobilized on a SiO2 substrate using a functionalized polymer as a surface modifier. Microscopic and spectroscopic techniques were used to characterize the AuNPs and their morphology beofre and after immobilization. In addition, the electrical characterization of covalently bound Au NPs as a charge trapping layer in non-volatile memory is also investigated by means of a metal-insulator semiconductor (MIS) device. The present synthetic route to covalently immobilized gold nanoparticles system will be a step towards realization for the nanoparticle-based electronic devices and related applications.

See more of this Session: Nanoelectronic Materials II
See more of this Group/Topical: Nanoscale Science and Engineering Forum

Electrochemical Synthesis of Ultra-Long Sb2Te3 Nanoribbons for Phase Change Memory

Hyunsung Jung, Chemical engineering, University of California, Riverside, Riverside, CA

Youngwoo Rheem, Chemical engineering, University of California, Riverside, Riverside, CA

Nicha Chartuprayoon, Chemical engineering, University of California, Riverside, Riverside, CA

Nosang V Myung, Chemical engineering, University of California, Riverside, Riverside, CA

Phase change memory is one of the promising candidates for the next generation nonvolatile memory because of faster operating rate and high storage density. Chalcogenides such as Sb2Te3, Ge2Te¬3 and their ternary alloys have been extensively studied as phase transition materials because of the ability to transition between amorphous and crystalline phase at moderate temperature and different electrical properties at crystalline and amorphous phase. In conventional phase change memory, chalcogenides were deposited using top-down vacuum processes. In this work, SbxTey nanoribbons with the controllable dimensions were synthesized on pre-determined locations by novel Lithographically Patterned Nanowire Electrodeposition (LPNE) method. The composition of SbxTey nanoribbons was tuned to modulate the electrical properties including FET properties. To determine size dependent transition temperature, the I-V characteristics of Sb2Te3 nanoribbons with different thickness and width were measured in both the SET (crystalline) and RESET (amorphous) states.

See more of this Session: Nanoelectronic Materials II
See more of this Group/Topical: Nanoscale Science and Engineering Forum

Structurally Controlled Bio-Hybrid Materials Based On Unidirectional Association of Anisotropic Microparticles with Human Endothelial Cells

Joerg Lahann, Chemical Engineering, University of Michigan, Ann Arbor, MI

Mutsumi Yoshida, Chemical Engineering, University of Michigan, Ann Arbor, Ann Arbor, MI

Kyungho Roh, University of Michigan, Ann Arbor, MI

Suparna Mandal, Chemical Engineering, University of Michigan, Ann Arbor, Ann Arbor, MI

In recent decades, biomedical researchers have increasingly realized the importance of local interactions between synthetic materials and cells for their macroscopic functionality. To date, a range of different approaches towards custom-tailored biomaterials exist, such as hydrogels, microfabricated biomaterials, or electrospun fiber scaffolds. In addition, so-called bio-hybrid materials have been proposed, where cells and synthetic building blocks form structurally and functionally integrated architectures. While bio-hybrid materials have so far been mostly realized as two-dimensional cell sheets, a next generation of integrated biomaterials may be contemplated, where both synthetic building blocks, such as micro- or nanoparticles, and cells become integral, structural elements that ultimately form the actual biomaterial through directional self-assembly in all three dimensions.

We will present initial work towards a novel type of microstructured bio-hybrid material comprised of anisotropic polymer particles with spatially-controlled affinity towards human endothelial cells. These anisotropic particles consist of two distinct compartments that can exhibit vastly different functions. Here, functional bipolarity was realized by selective surface modification to create polymer particles with two biologically distinguishable hemispheres: one exhibiting high binding affinity for human endothelial cells and the other being essentially resistant towards cell binding. Bicompartmental particles are synthesized by simultaneous electrohydrodynamic co-jetting of two different solutions of polyacrylamide/ polyacrylic acid co-polymers. While the base polymers were generally the same for both hemispheres, different functional groups were selectively incorporated into compartments to enable subsequent surface modifications. For this purpose, acetylene-modified PAAm-co-AA was prepared using a method similar to the one used for biotin modification of PAAm-co-AA. Free acid groups were activated in the presence of EDC/sulfo-NHS and were reacted with propargyl amine at room temperature for 4 hrs and confirmed by proton NMR. Moreover, fluorescein isothiocyanate (FITC)- or rhodamine-conjugated dextran were selectively added to the jetting solutions to confirm particle anisotropy or to monitor unidirectional association between particles and cells using fluorescence imaging. When self-assembled with human endothelial cells, these particles exhibit a strongly bipolar behavior, where one compartment exhibits binding affinity towards human endothelial cells, while the other compartment was essentially non-binding. In this design, particles not only bound to the top of the human endothelial cells, but associated all around the perimeter of cells forming a single particle lining and effectively isolating cells.

See more of this Session: Nanostructured Biomimetic and Biohybrid Materials and Devices
See more of this Group/Topical: Nanoscale Science and Engineering Forum

Self-Assembly of Semiconductor Nanoparticles Mimicking Biomolecules

Sudhanshu Srivastava, Chemical Engineering, University of Michigan, Ann Arbor, MI

Kevin Critchley, Chemical Engineering, University of Michigan, Ann Arbor, MI

Aaron Santos, Chemical Engineering, University of Michigan, Ann Arbor, MI

Ki-Sub Kim, Chemical Engineering, University of Michigan, Ann Arbor, MI

Paul Podsiadlo, Chemical Engineering, University of Michigan, Ann Arbor, MI

Kai Sun, Materials Science and Engineering, University of Michigan, Ann Arbor, MI

Sharon C. Glotzer, Chemical Engineering, University of Michigan, Ann Arbor, MI

Nicholas A. Kotov, Chemical Engineering, University of Michigan, Ann Arbor, MI

Self-assembly of inorganic nanocolloids has produced a variety of shapes and structures including 1D, 2D and 3D architectures. The ability to utilize novel synthetic routes to assemble nanoparticles (NPs) into complex morphologies for specific functions is a huge challenge in the field of nanotechnology. The assembly is often a quite difficult multistage process involving structure-directing ligands such as organic or biomolecular ligands attached to nanocolloids. Fabricating nanoscale structures similar to alpha-helix or beta-sheets using NP-NP assembly is a unique method for mimicking the biomolecular structures. This approach opens the door for replicating complex biological topologies and functions using inorganic NP assemblies. There is a need to understand and rigorously demonstrate the possibilities of significantly increased complexity of the self-organized structures made from NPs. This study demonstrates the assembly of CdTe NPs into nanowires, nanosheets, or helical nanoribbons depending on the amount and type of capping group used. Computer simulations were also carried out to understand and predict the packing of tetrahedral CdTe NPs in the superstructures. This study also demonstrates the idea of crafting natural structures by self-assembly of NPs to be employed as machinery with advanced physical responses.

See more of this Session: Nanostructured Biomimetic and Biohybrid Materials and Devices
See more of this Group/Topical: Nanoscale Science and Engineering Forum

Environmentally Responsive Gold Nanorod-Polypeptide Assemblies

Huang-Chiao Huang, Department of Chemical Engineering, Arizona State University, Tempe, AZ

Kaushal Rege, Chemical Engineering, Arizona State University, Tempe, AZ

Environmentally responsive nanoassemblies based on polypeptides and nanoparticles are promising in a number of biological/biomedical applications. We demonstrate the formation of gold nanorod (GNR)-elastin-like polypeptide (ELP) nanoassemblies whose optical response can be remotely controlled based on near-infrared (NIR) light exposure. In this study, cysteine-containing ELPs (C2ELP) were self -assembled on gold nanorods (GNRs) leading to GNR- C2ELP nanoassemblies. Exposure of GNR- C2ELP assemblies to NIR laser resulted in the heating of GNRs due to surface plasmon resonance. Heat transfer from the GNRs resulted in a temperature increase of the self-assembled C2ELP above its transition temperature (Tt), leading to a detectable phase transition and aggregation of the GNR- C2ELP assemblies. This phase transition was quantified using an optical readout, based on an increase in the optical density of the GNR- C2ELP dispersion. This optical response was seen only in the case of GNR- C2ELP nanoassemblies. While a photothermal response was observed in the case of a dispersion consisting of GNRs alone, no optical response could be seen due to the absence of the C2ELP. In addition C2ELP solutions demonstrated neither a photothermal nor an optical response. The reproducibility and reversibility of optical response of GNR-C2ELP nanoassemblies was demonstrated across multiple cycles following exposure and removal of the laser irradiation. In summary, our results indicate that polypeptides can be interfaced with GNRs resulting in optically responsive nanoassemblies. Current work involves an investigation into the optical and photothermal properties of nanoassemblies using higher cysteine-containing ELP homologues. We anticipate that these nanoassemblies can be useful for sensing, tissue engineering, and drug delivery applications.

See more of this Session: Nanostructured Biomimetic and Biohybrid Materials and Devices
See more of this Group/Topical: Nanoscale Science and Engineering Forum

Investigation of Palladium Sorption On the Hydroxyl, Sulfhydryl, and Amine Functionality to Improve the Palladium Deposition On the Genetically Engineered Tobacco Mosaic Virus

Jung-Sun Lim, Chemical Engineering, Purdue University, West Lafayette, IN

Sang-Yup Lee, Chemical Engineering, Yonsei University, Seoul, South Korea

James N. Culver, University of Maryland Biotechnology Institute, College Park, MD

Michael T. Harris, Chemical Engineering, Purdue University, West Lafayette, IN

Inorganic mineralization is used in metal/semiconductor coating on the biologically driven templates. With the aid of genetic engineering, the structure of the biotemplate or the surface functionality is potentially designable to prepare nanostructured inorganics for nano devices. The tobacco mosaic virus (TMV) has served as a model template to synthesize metal nanorods utilizing its unique tubular rod-shape (4 nm/18 nm of inner/external diameter with 300 nm longitudinal length). Also, the genetic engineering of TMV is well developed to modify it with active surface functionalities such as sulfhydry groups. There have been reports of biotemplated metal nanowires with gold, silver, palladium, and platinum with various biotemplates including TMV. However, further progresses of biotemplated metal nanowire synthesis are limited by few experimental observations elucidating the precursor interaction with the template surface.

The binding of pd(II) ion on the different functional groups on genetically modified tobacco mosaic virus (TMV1Cys) is studied through pd(II) sorption on the silica nanoparticles functionalized with hydroxyl, sulfhydryl, and amine surface groups. Pd(II) uptake/desorption on the surface functionalized silica is measured and the resulting Pd(II) staining on the functionalized silica will be further characterized with X-ray photoelectron spectrometer (XPS) and transmission electron microscopy (TEM).

See more of this Session: Nanostructured Biomimetic and Biohybrid Materials and Devices
See more of this Group/Topical: Nanoscale Science and Engineering Forum

Effect of Surface Attachment Characteristics On Photoactivity of Photosystem I Assembly On Thiol-Activated Au Substrates

Dibyendu Mukherjee, Department of Chemical and Biomolecular Engineering, Sustainable Energy Education and Research Center (SEERC), University of Tennessee, Knoxville, TN

Michael Vaughn, Department of Biochemistry, Cellular and Molecular Biology, Sustainable Energy Education and Research Center (SEERC), University of Tennessee, Knoxville, TN

Barry D. Bruce, Department of Biochemistry, Cellular and Molecular Biology, Sustainable Energy Education and Research Center (SEERC), University of Tennessee, Knoxville, TN

Bamin Khomami, Department of Chemical and Biomolecular Engineering, Sustainable Energy Education and Research Center (SEERC), University of Tennessee, Knoxville, TN

Cyanobacterial Photosystem I (PSI), a supra-molecular protein complex (MW~300 kDa, disk shaped with diameter ~30 nm & height ~9 nm), is a biological photodiode responsible for driving natural photosynthesis mechanism. Upon exposure to light, PSI undergoes charge separation enabling unidirectional electron flow. This makes PSI a promising candidate for future research in the fabrication of solid-state bio-electronic, hybrid photovoltaic devices, or bio-sensors. But, a first step in the rational design of aforementioned devices and sensors involves in arranging these protein complexes as a uniform homogeneous monolayer on chemically/ physically decorated self-assembled monolayer (SAM) substrates with proper directionality to allow efficient electron capture pathways. Previous studies have suggested that PSI preferentially attaches onto hydroxyl (OH)-terminated alkanethiols with the electron vectors pointing outward. But, our present studies indicate that experimental parameters like solution concentration and deposition protocol can dictate the attachment dynamics of PSI onto alkanethiolate SAM/ Au substrates resulting in complex structural arrangements which can either hinder or facilitate the photo-electrochemistry of such systems.

In this work, morphology and orientations of PSI assembly on OH-terminated alkanethiolate SAM/Au substrates are studied with: 1) gravity-driven and 2) electric field assisted deposition techniques to elucidate the surface attachment dynamics and properties of these protein complexes. Atomic force microscopy (AFM) images and spectroscopic ellipsometry (SE) data indicate the formation of a large number of columnar PSI aggregates on the substrates for gravity-driven deposition from high PSI density aqueous buffer solutions. These larger structures dissipate into a monolayer at lower concentrations indicating formation of Brownian diffusion mediated aggregates in the solution phase. However, in electric-field assisted deposition, electrophoretic mobility of PSI can be tuned based on the field strength to prevent such aggregation thereby allowing a more uniform deposition. Experimental studies of solution phase aggregation dynamics via dynamic light scattering and analytical ultracentrifugation also support these observations, thereby providing an unprecedented level of understanding into the complex dynamics of protein-protein and/or protein-surface interactions. Furthermore, Quartz Crystal Microbalance (QCM) experiments help us study PSI adsorption/ desorption kinetics with different surface treatments for optimal monolayer formation. PSI orientation from QCM based surface arrangements and packing density data, validated by fluorescence polarization spectroscopy to track emission anisotropy from fluorescently tagged PSI molecules are also discussed. Such studies could play significant role in monitoring and tailoring surface attachment of PSI to specific SAM substrates and hence, in the rational design of future bio-electronic devices.

See more of this Session: Nanostructured Biomimetic and Biohybrid Materials and Devices
See more of this Group/Topical: Nanoscale Science and Engineering Forum

Ultrathin, Free Standing Biomimetic Membranes From the Vapor Phase:

Ramaswamy Sreenivasan, Chemical Engineering, Massachusetts Institute of Technology, Cambridge, MA

Erik Bassett, Tissue Engineering Lab, Massachusetts General Hospital, Cambridge, MA

David Hoganson, Tissue Engineering Lab, Massachusetts General Hospital, Cambridge, MA

Joseph Vacanti, Tissue Engineering Lab, Massachusetts General Hospital, Cambridge, MA

Karen Gleason, Chemical Engineering, Massachusetts Institute of Technology, Cambridge, MA

Decomposition of a gas phase initiator into free radicals to facilitate polymerization of a monomer/monomers is the cornerstone of the initiated chemical vapor deposition (iCVD) process that distinguishes it from other types of vapor deposition of polymers. iCVD has been successfully used earlier in our group to deposit a wide variety of conformal coatings with the desired functional groups for varying applications. In this work, for the first time, we have used initiated iCVD to fabricate free standing ultra thin biomimetic membranes. Ultrathin membranes as thin as 100 nm were made by this method. iCVD thus is capable of producing membranes with thicknesses thinner than typical solution based methods such as spin coating. The resulting membranes were characterized for thickness, mechanical robustness and permeability. Various chemistries resulted in membranes with varying mechanical robustness. Potential applications in microfluidic devices will be discussed. Three strategies of device fabrication using iCVD free standing membranes will be explained. Selecting the right chemistry to impart mechanical strength to the free standing membrane while serving its function (gas permeation, liquid barrier etc.) is key to successful applications. Optimizing the handling of the ultra thin free standing membrane during the steps of device fabrication is of paramount importance. This work promises more future research into integrating iCVD into the process flow of biomedical device fabrication entailing biomimetic coatings and free standing membranes.

See more of this Session: Nanostructured Biomimetic and Biohybrid Materials and Devices
See more of this Group/Topical: Nanoscale Science and Engineering Forum

Fluorescence Quenching of CdSe/ZnS Nanocrystals near Copper Nanoparticles in Aqueous Solution

Sanchari Chowdhury, Chemical and Biomedical Engineering, University of South Florida, Tampa, FL

Venkat Bhethanabotla, Department of Chemical and Biomedical Engineering, University of South Florida, Tampa, FL

Rajan Sen, Civil Engineering, University of South Florida, Tampa, FL

Fluorescence quenching of CdSe/ZnS nanocrystals near copper nanoparticles in aqueous solution

Quenching of fluorescence of vicinal fluorophores of nanoparticles is efficiently utilized for many different applications such as improvement of homogeneous and competitive fluorescence immunoassay, optical detection of DNA hybridization, competitive hybridization assay and in optoelectronics[1-3]. Metal nanoparticles are known to both quench and enhance fluorescence depending on the optical properties of nanoparticles, fluorophore-particle separation distance, molecular dipole orientation with respect to particle surface and size of the nanoparticles. The presence of nanoparticles close to the luminophores can create new nonradiative channels due to light absorption inside the metal, thus quenching the emission of luminophores [4]. If the probe molecules are very close to the nanoparticles (typically, less than 5 nm), luminescence emission is quenched due to Forster transfer of energy from the excited state of the molecule to the surface plasmons of the metal surface. Metal quenched fluorescence has been studied mostly using gold nanoparticles[1, 2, 5].

In this present study, we investigate the fluorescence quenching of CdSe/Zns nanocrystals in the presence of copper nanoparticles. Copper nanoparticles coated with stabilizing polyvinylpyrrolidone (PVP) polymer, quench the fluorescence of CdSe/Zns nanocrystals coated with mercaptoundecanoic ligands. Stable copper nanoparticle colloid was synthesized by reducing copper salt precursor using ascorbic acid in the presence of PVP. The fluorescence quenching of the CdSe/Zns nanocrystals follows the Stern-Volmer relations against concentration of the copper nanoparticles in water with a detection limit of 5 nanomol. The mechanism of the fluorescence quenching on the copper nanoparticles surfaces has been treated in the light of electromagnetic interaction between nanoparticles and fluorophores which influences both radiative and non-radiative decay rate of fluorophores. We presented theoretical calculation using Gersten-Nitzan (GN) model[6] to provide insight into the influence of the Cu nanoparticles on radiative and non-radiative decay rates of luminophore molecules at their close proximity.

References:

(1) Zai-Sheng, W.; Jian-Hui, J.; FU, L.; Guo-Li, S.; Ru-Qin, Y. *Anal. Biochem.* **2006**, *353*, 22-29.

(2) Dulkeith, E.; Ringler, M.; Klar, T. A.; Feldmann, J.; MunozJavier, A.; Parak, W. J. *Nano Lett.* **2005**, *5*, 585-589.

(3) Ao, L.; Gao, F.; Pan, B.; He, R.; Cui, D. *Anal. Chem.* **2006**, *78*, 1104-1106.

(4) Azoulay, J.; Débarre, A.; Richard, A.; Tchénio, P. *Europhys. Lett.* **2000**, *51*, 374-380.

(5) Kato, N.; Caruso, F. *J. Phys. Chem. B* **2005**, *109*, 19604-19612.

(6) Mertens, H.; Koenderink, A. F.; Polman, A. *Phys. Rev. B* **2007**, *76*, 115123-115121-115112.

See more of this Session: Sensors and Bio-Imaging Contrast Agents at the Cellular Level
See more of this Group/Topical: Nanoscale Science and Engineering Forum

Fluorescence Level Manipulation by Gold Nano Particles

Jianting Wang, Chemical Engineering, University of Louisville, Louisville, KY

Martin O'Toole, Chemical Engineering, University of Louisville, Louisville, KY

Samuel Achilefu, Department of Radiology, Washington University in St. Louis, St. Louis, MO

Jacek Jasinski, Chemical Engineering, University of Louisville, Louisville, KY

Kyung A. Kang, Chemical Engineering, University of Louisville, Louisville, KY

Fluorephore mediated molecular imaging is becoming a leading methodology in medical diagnostic and biotechnology. The ability of fluorescence alteration can be highly beneficial for improving sensitivity and specificity: artificial enhancement in the quantum yield can be highly beneficial to improve the sensitivity. Likewise, if fluorescence can be quenched when the molecule of interest reacts with the target, it can be very useful and also minimizing non-specific signal.

We have been studying alteration of fluorescence using nano metal particles. Gold nano particles (GNPs) are known to generate strong surface plasmon fields when light is applied and this field can effectively be used to quench or enhance fluorescence. GNP is of our interest because of its chemical inertness and ability for easy surface modification. The level of the fluorescence change by GNPs can be from complete quenching to extensive enhancement, depending upon the strength of plasmon field where the fluorophore is placed. The field strength depends upon the particle size and the distance from the particle surface. For a system where the metal type, particle size, and the fluorophore are fixed, the distance between particle and fluorophore can be controlled by placing a molecular spacer at a desired length between the two.

We have manipulated the distance between an NIR fluorophore (Indocyanine Green based) and a GNP, by adding various number of polymer layers of known thickness on the surface of a GNP. The NIR fluorophore was immobilized on the outer layer of the polymer. The distance was also altered by placing bio-compatible molecular spacers at various lengths. The fluorescence was, then, correlated with the distance between the GNP and the NIR fluorophore.

This fluorescence alteration mechanism can be highly beneficially utilized for fluorescent bioimaging with higher contrast. By combining both fluorescence enhancement and quenching in one entity, it is also possible to develop molecular-beacon like, contrast imaging agent with excellent specificity and sensitivity.

The authors acknowledge the financial support from U.S. Army Breast Cancer Program.

See more of this Session: Sensors and Bio-Imaging Contrast Agents at the Cellular Level
See more of this Group/Topical: Nanoscale Science and Engineering Forum

Antibody-Conjugated Gold Nanoclusters (Nanoroses) for Targeted Cancer Cellular Imaging and Therapy

Li Leo Ma, Chemical Engineering, University of Texas at Austin, Austin, TX

Justina O. Tam, Biomedical Engineering, University of Texas at Austin, Austin, TX

Jinze Qiu, Biomedical Engineering, University of Texas at Austin, Austin, TX

Tianyi Wang, Biomedical Engineering, University of Texas at Austin, Austin, TX

James T. Jenkins, Division of Cardiology, Department of Medicine, University of Texas Health Science Center at San Antonio, San Antonio, TX

Geoffrey D. Clarke, Division of Cardiology, Department of Medicine, University of Texas Health Science Center at San Antonio, San Antonio, TX

Konstantin Sokolov, Departments of Biomedical Engineering and Imaging Physics, M.D. Anderson Cancer Center, Houston, TX

Marc D. Feldman, Division of Cardiology, Department of Medicine, University of Texas Health Science Center at San Antonio, San Antonio, TX

Thomas E. Milner, Biomedical Engineering, University of Texas at Austin, Austin, TX

Keith P. Johnston, Chemical Engineering, University of Texas at Austin, Austin, TX

The ability of 30 to 100 nm nanoparticles to target and modulate the biology of specific cell types can enable major advancements in cellular imaging and therapy in cancer and atherosclerosis. A key challenge is to combine targeting, imaging, and therapeutic functionality into small, yet stable nanoparticles. We report ~30 nm stable uniformly-sized near infrared (NIR) active, superparamagnetic nanoclusters formed by kinetically controlled self-assembly of gold-coated iron oxide nanoparticles. The controlled assembly of nanocomposite particles into clusters with short primary particle spacing produces collective optical response of the electrons that shift absorbance into the NIR. The epidermal growth factor receptor (EGFR) is a protein tyrosine kinase over expressed on many types of tumor cells, including breast, ovarian, bladder, and prostatic carcinoma. Anti-EGFR conjugated gold nanoroses show efficient binding to the EGFR in cancer cell lines, which facilitates NIR imaging contrast enhancement and therapy. Because of the thin gold shells on the particles, the r2 spin-spin magnetic relaxivity is 200 mM-1s-1, an order of magnitude larger than observed for typical iron oxide particles. The high relaxivity is utilized for magnetic resonance imaging (MRI) in a mouse model. The increased capacity of anti-EGFR conjugated gold nanoroses has the potential to activate signal transduction pathways that regulate cell proliferation, and therefore to inhibit in vivo tumor growth. Antibody conjugated nanoclusters with optical, magnetic and therapeutic functionality, designed by assembly of nanoparticle building blocks, can provide broad opportunities for targeted cancer cellular imaging, therapy, and combined imaging and therapy.

See more of this Session: Sensors and Bio-Imaging Contrast Agents at the Cellular Level
See more of this Group/Topical: Nanoscale Science and Engineering Forum

Gold Nanoparticle-Bovine Serum Albumin Complexes for Imaging and Analysis of Macromolecule Drug Delivery through Murine Endothelial Monolayers

Hao Shen, Department of Chemical Engineering, University of Illinois at Chicago, Chicago, IL

Richard D. Minshall, Department of Pharmacology and Anesthesiology, University of Illinois at Chicago, Chicago, IL

Ying Liu, Department of Chemical Engineering and Biopharmaceutical Sciences, University of Illinois at Chicago, Chicago, IL

Gold nanoparticles (GNP) have been widely used for bioimaging and biomedical sciences because of their unique optical and electronic properties related to their sizes and shapes. Here, we focus on the use of a gold-BSA complex as an imaging-contrast agent to understand the effect of size and surface properties on the pathways of particle transport through a monolayer of mouse endothelial cells.

Transport from the blood into the surrounding tissues has direct implications for understanding healthy metabolic function, disease states involving transmigration of certain classes of viruses, inflammatory cells, and metastatic cancer cells, and tissue delivery of drugs. Transport across the endothelial barrier occurs by two main pathways - the paracellular pathway through the intercellular junctions and the transcellular pathway through endothelial cells mediated by transcytosis via caveolae. We have prepared gold-BSA complexes by Flash NanoPrecipitation (FNP) to directly image and elucidate the hypothesized mechanisms of junctional and vesicular transport across the endothelial barrier. BSA has been used to convert the hydrophobic GNP suspension to a hydrophilic suspension. The hydrodynamic diameter of the gold-BSA complexes varied from 5 nm to 50 nm for various ratios of GNP and BSA from 1:10 to 50:1. The sizes and structures of the gold-BSA complexes were measured by dynamic light scattering (DLS) and confirmed by transmission electron microscopy (TEM).

Compared to hydrophilic polymers, albumin was chosen because it is effectively internalized via caveolae in endothelial cells. However, when conjugated with gold NPs, most proteins including BSA undergo conformational changes at the binding boundary with GNP, which may or may not result in the loss of biological activity. Confirmation that BSA retained its native conformation was conducted using UV-Vis, fluorescence, and circular dichroism (CD) spectroscopy and nuclear magnetic resonance (NMR).

See more of this Session: Sensors and Bio-Imaging Contrast Agents at the Cellular Level
See more of this Group/Topical: Nanoscale Science and Engineering Forum

High-Temperature Stable Metal-Silica Core-Shell and Yolk-Shell Materials with Exceptional Dimensional Control

Lu Zhang Whaley, Department of Chemical Engineering, University of Pittsburgh, Pittsburgh, PA

Rongwen Lu, State Key Laboratory of Fine Chemicals, Dalian University of Technology and University of Pittsburgh, Dalian, China

Michelle Najera, Department of Chemical Engineering, University of Pittsburgh and National Energy Technology Laboratory, Pittsburgh, PA

Goetz Veser, Department of Chemical Engineering, University of Pittsburgh and National Energy Technology Laboratory, Pittsburgh, PA

The controlled synthesis of nanomaterials holds great promise for the design of highly efficient catalysts. However, the low stability of nanoparticles poses a severe limitation on the use of such materials in many industrial applications. This limitation is motivating efforts to stabilize nanoparticles via embedding or encapsulation in nanostructured matrix materials. The main challenges thereby are the control over key material dimensions and accessibility to the nanoparticle via porosity of the matrix.

We are presenting results of different types of metal-silica core-shell nanomaterials with highly controllable size and shape. The materials are synthesized in a one-pot reverse microemulsion-templated approach. By adjusting the synthesis parameters, the structure of the core-shell materials can be changed from solid (but porous) core-shell structures to spherical "yolk-shell" structure with a pronounced cavity, to strongly elongated, rod-shaped structures with high aspect ratios. Furthermore, the thickness of the silica shell can be adjusted with nanometer-size precision. In this way, we have been able to synthesize nickel-, copper-, iron- and palladium-based metal-silica core-shell materials.

Electron microscopy (TEM) demonstrates overall core-shell particles sizes of 20 nm - 50 nm with adjustable silica wall thickness of 3-20 nm, encapsulating metal cores with size from sub-nanometer clusters to nanoparticles with ~10 nm diameter. The porosity of the silica walls is confirmed via CO chemisorptions and complete oxidation and reduction of the metal nanoparticle in TPO/TPR. Remarkably, the materials are thermal stable up to at least ~800°C, without significant changes in morphology.

See more of this Session: Templated Assembly of Inorganic Nanomaterials II
See more of this Group/Topical: Nanoscale Science and Engineering Forum

Templated Gold Nanoparticle Assembly On Two-Dimensional Protein Crystals

Matthew M. Shindel, Department of Chemical Engineering and Materials Science, University of California, Irvine, CA

Daniel R. Mumm, Department of Chemical Engineering and Materials Science, University of California, Irvine, CA

Szu-Wen Wang, Department of Chemical Engineering and Materials Science, University of California, Irvine, CA

Nature has evolved the ability to efficiently construct complex, nanoscale architectures in self-assembling biological systems, and this provides synthetic nanotechnology with insightful paradigms and functional fabrication platforms. Two-dimensional crystals of the protein streptavidin exhibit tunable crystalline lattice structures and the ability to control interactions between colloidal nanoparticles and the protein monolayer through electrostatic and specific receptor-ligand mechanisms. Furthermore, these highly-ordered protein monolayers can be grown at both the air/water interface and on a solid substrate. Using the C222 crystal form of streptavidin, we demonstrate that this protein system can be manipulated to template the assembly of ordered gold nanoparticle arrays. We examine experimental conditions (e.g., pH, ionic strength, nanoparticle size) which promote the assembly process to yield nanoparticle arrays with longer-range order. Our work shows that the streptavidin protein system is a novel platform for the "bottom-up" fabrication of highly-organized arrays of inorganic nanomaterials.

See more of this Session: Templated Assembly of Inorganic Nanomaterials II
See more of this Group/Topical: Nanoscale Science and Engineering Forum

In vitro enzymatic synthesis of phytochelatin-capped CdS nanocrystals

Semiconductor nanocrystals, also known as quantum dots (QDs), have been widely used as fluorescent dyes in biological application. To synthesize biocompatible quantum dots, *E. coli* was designed to produce phytochelatin synthase (PCS), an enzyme which catalyzes the synthesis of a naturally occurring metal-chelating peptide phytochelatin (PC) upon the addition of glutathione (GSH). Presence of a hexa-histidine tail enabled the direct immobilization of PCS onto Ni-resins. Addition of inorganic salt precursors and GSH resulted in the room-temperature synthesis of CdS nanocrystals using PC as the capping agents. The synthesized nanocrystals exhibited an absorption peak at 305nm and a significantly blue-shifted fluorescent peak around 380nm compared with GSH capped CdS QDs. High-resolution transmission electron microscopy confirmed the formation of quantum sized CdS nanocraystals. These PC capped CdS QDs showed good stability. Moreover, by varying the enzymatic reaction time, different compositions of PCs (PC2 and PC3) can be synthesized, forming CdS nanocrystals with tunable size. Since PC is known to bind a wide range of heavy metal ions such as Zn, Pb, and As, this approach may be generalized to guide the in vitro self assembly of a wide range of semiconductor nanocrystals with different compositions and size.

Synthesis of Doped Zinc Selenide Quantum Dots and Core-Shell Structures in Microemulsion Templates

Tracy Heckler, Chemical Engineering, University of Massachusetts, Amherst, MA

Qi (Grace) Qiu, Chemical Engineering, University of Massachusetts, Amherst, MA

Jun Wang, Chemical Engineering, University of Massachusetts, Amherst, MA

T. J. Mountziaris, Chemical Engineering, University of Massachusetts, Amherst, MA

The doping of II-VI Quantum Dots (QDs) with transition metals is attracting significant attention because of the possibility of tuning their emission wavelength to longer wavelengths than the corresponding bulk gap emission and for the development of magneto-optical QDs for spintronic applications.

In this presentation we will discuss the synthesis of ZnSe:Mn QDs in a microemulsion template, their extraction from the template, surface modification, and stabilization in water. The synthesis route is based on the templated growth of ZnSe nanocrystals reported by Karanikolos, et al. [1-3]. The template consists of a microemulsion having p-xylene as the continuous phase, water as the dispersed phase, and a poly(ethylene oxide)-poly(propylene oxide)-poly(ethylene oxide) (PEO-PPO-PEO) block copolymer as the surfactant. This template is very stable and has very slow droplet-droplet coalescence kinetics, thus eliminating particle aggregation and allowing precise control of particle size. The precursors for Zn and Mn were acetate salts of these metals that were dissolved in the aqueous dispersed phase. The concentration of Zn acetate in the dispersed phase determines the final size of the QDs and the concentration of Mn acetate determines the dopant concentration in the nanocrystal. The Se precursor was hydrogen selenide gas diluted in hydrogen and was bubbled through the microemulsion.

The fluorescence spectra of the resulting QDs exhibit a blue-violet peak attributed to ZnSe gap emission and a yellow-orange peak at 585nm attributed to an electronic transition of Mn^{2+}. The evolution of the intensity and ratio of these emission peaks was studied as a function of dopant salt concentration in the precursor solution and time. The results indicate that ZnSe QDs are formed first and Mn slowly diffuses into the ZnSe lattice.

Core-shell structures were synthesized by sequential addition of precursors. The capping of ZnSe:Mn QDs with ZnSe has been shown to increase the Mn^{2+} peak emission.

To make the doped ZnSe QDs useful for multi-color tagging of biomolecules, a procedure was developed that allows QD removal from the microemulsion template, capping with functional hydrophilic ligands, and stabilization in an aqueous solution.

The microemulsion-based synthesis technique is easy to scale up and employs less expensive chemicals compared to the most common QD synthesis technique that involves injection of organometallic precursors in a hot coordinating solvent.

References

1. G. N. Karanikolos, et al., Langmuir 20, 550 (2004).

2. G.N. Karanikolos, et al., Nanotechnology, 16, 2372 (2005)

3. G.N. Karanikolos, et al., Nanotechnology, 17, 3121(2006).

See more of this Session: Templated Assembly of Inorganic Nanomaterials II
See more of this Group/Topical: Nanoscale Science and Engineering Forum

One-pot Synthesis of Gold Nano/Micro Structures in PEG-SDS Solution

Yueping REN, Yun FANG, Jia LIU*

School of Chemical and Material Engineering, Jiangnan University, Wuxi, Jiangsu, China

Abstract

In PEG-SDS solution, nano/micro gold particles were synthesized through a one-pot route at atmosphere. Gold nanoplates were the dominant products at 25°C, the size of which increased from 500 nm to 1μm as the concentration of $HAuCl_4$ ranges from 0.2 mM to 1.0 mM. When the reaction temperature was ranges from 15°C to 60°C, the products were quasi-spheres (15 °C), nanoplates (25 °C), and irregular plates (40 °C and 60 °C), respectively.

Introduction

Noble metal nanostructures have been widely investigated because of their peculiar properties and potential applications. In recent years, anisotropic gold (Au) nanostructures have drawn increasing attention due to their unique optical properties and potential use in areas such as catalysis[1], biosensors[2], and surface-enhanced Raman scattering[3]. These properties are known to strongly depend on their sizes and shapes. Therefore, shape-controlled synthesis of anisotropic Au nanostructures has become one of the research focuses in current nanoscience and nanotechnology. Numerous approaches have been developed to fabricate Au nanoparticles with different shapes such as rods[4], wires[5], belts[6], plates[7], mutipods[8] and polyhedra[9]. Among these shapes, two-dimensional (2D) nanostructures have attracted considerable attention because of their high absorption in the NIR region which makes them promising candidates for hyperthermia of cancer cells, optical coatings of architectural materials and heat absorption in special equipments[10]. Up to now, surfactants, and polymers are usually employed as capping materials to direct an anisotropic crystal growth and has achieved some progress[6]. However, to purposively and effectively control the shapes of the nanostructures is still a difficult task. According to the studies on the interaction between surfactant and polymer in aqueous solution, certain anionic surfactant SDS and some water-soluble nonionic polymer (PEG, polyvinylpyrrolidone (PVP)) self assembles into necklace-like soft clusters in aqueous solutions once the concentration of surfactant exceeded a critical value[11]. The noble structure of the necklace-like soft cluster makes it an ideal soft template for the preparation of inorganic nanoparticles. Our group is focused on the synthesis of metal or metal oxide nanoparticles in surfactant-polymer necklace-like soft cluster. In the past few years, using SDS-PEG and similar SDS-PVP necklace-like soft clusters aqueous solution as soft templates at ambient temperature and atmosphere, we have successfully prepared Au nanorings[12], Au multipod-like[13] and ultrafine Ni particles[14] with unique surface morphology.

---

* Corresponding author: yunfang@126.com. Supported by the National Natural Science Foundation of China (20871059).

Recently, simple and environmentally benign strategies have been motivated in chemical synthesis of nanomaterials. One of the typical examples was that reductive polymer began to be used both as capping agent and reductant to prepare nanostructures in aqueous solution. For instance, Hoppe *et al* prepared ribbon-like and branched gold nano- and microstructures by simple heating of diluted aqueous solutions of PVP and $HAuCl_4$[15] and Xia *et al* fabricated Au nanotadpoles, nanokites and microplates by reducing aqueous $HAuCl_4$ with PVP in aqueous solutions at 100 ^{o}C[16]. However, these reductions mentioned above were all performed at high temperatures which may be not conducive to energy conservation. In this paper, Au nano/micro plates were fabricated and templated by SDS-PEG necklace-like soft clusters at 25 ^{o}C and in this route, PEG act both as reducing agent and one of the composites of the soft templates.

Experimental

$HAuCl_4$ and PEG (M_w=20000) were purchased from Sinopharm Chemical Reagent Co. Ltd. SDS was purchased from Acros Organics Co., USA. All reagents were used as received without further purification. Ultra pure water was used through out to prepare $HAuCl_4$ and PEG-SDS aqueous solutions. In a typical synthesis, an aqueous solution of 9.712 $mmol \cdot L^{-1}$ $HAuCl_4$ (1.03 mL) was added to the solution containing PEG and SDS. The volume was adjusted to 50 mL and the final concentrations of $HAuCl_4$, PEG and SDS in the reaction solution were 0.2 $mmol \cdot L^{-1}$, 5 $mmol \cdot L^{-1}$ and 3.5 $mmol \cdot L^{-1}$ respectively. Solutions were immerged into a water bath which was 25 ^{o}C and kept away from ambient light to avoid photoreactions of the Au complex. Then the reaction mixture was stirred at 25 ^{o}C for 8 h. SEM measurements were made on a HITACHI S-4700 field-emission scanning electron microscope (FESEM), operated at 15 kV. TEM images and the corresponding selected area electron diffraction (SAED) pattern were obtained on a Tecnai 12 transmission electron microscope (TEM).

Results and Discussion

Characterization of Au nanoplates

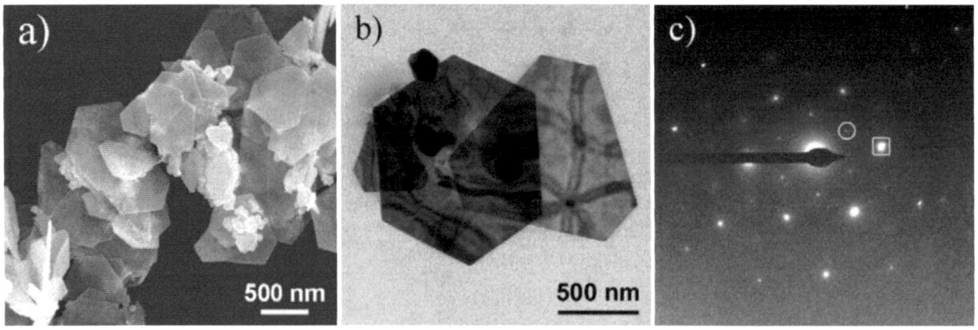

Figure 1 a) SEM, b) TEM images and c) SAED pattern of Au nanoplates.

Figure 1 shows that the products are mainly Au nanoplates ~500 nm in size. The corresponding selected-area electron diffraction (SAED) pattern of the products (Figure 1c) obtained by directing the incident electron beam perpendicular to a single nanoplate is a hexagonal symmetry diffraction spot pattern, demonstrating that the hexagonal nanoplates are single crystals. The spots can be identified from this diffraction pattern: the strongest spots (in a box) can be indexed to the {220} reflections and the inner spots (in a circle) with the weakest intensity correspond to the forbidden 1/3{422} reflections which indicate the presence of twin planes in the {111} plane perpendicular to the electron beam. These results indicate that the top and bottom faces of both Au triangular and hexagonal microplates are enclosed by the {111} planes[16].

Factors influencing the formation of nanostructures

$HAuCl_4$ concentration TEM images (Figure 2) clearly indicates that the $HAuCl_4$ concentration is found to be of paramount importance to the final shape of the products. When the $HAuCl_4$ concentration was 0.1 mmol·L^{-1}, the color of the reaction solution changed into deep purple 2 hours later and the products were mainly large irregular structures together with spheres as shown in Figure 2a. We increased the $HAuCl_4$ concentration to 0.2 mmol·L^{-1} and the products were mainly nanoplates ~500 nm in size as illustrated in Figure 1a-b. As Figure 2b shows, the Au nanoplates increase to ~1μm in size when $HAuCl_4$ concentration was increased to 0.3 mmol·L^{-1}. Large nanoplates ~2 μm in size (see Fig. 2c) together with equiaxial Au particles are obtained at the $HAuCl_4$ concentration of 0.5 mmol·L^{-1}. Therefore, the $HAuCl_4$ concentration is crucial for the formation of Au nanoplates and the increase of $HAuCl_4$ results in the enlargement of Au nanoplates which is due to a faster selectively growth along the lateral facets of the nanoplates.

Figure 2 SEM images of the samples synthesized with different $HAuCl_4$ concentrations: a) 0.1 mmol·L^{-1}; b) 0.3 mmol·L^{-1}; c) 0.5 mmol·L^{-1}.

Reaction temperature Figure 3 shows the morphology of the products fabricated at different temperatures. Regular hexagonal nanoplates were obtained at 25 °C as illustrated in Figure 1. We reduced the reaction temperature to 15 °C, equiaxial Au particles became the dominant products (see Fig. 3a). When the reaction temperature was kept at 40 °C, we found that the angles of some plates began to roundout and some antenniform salient occurred on the edge of them. Furthermore, the products obtained at 60 °C were plates whose pungent angles had completely disappeared and several different nanoplates fused together to form large irregular structures.

Figure 3 SEM and TEM images of the samples synthesized at different temperatures: (a) 15 °C, (b) 40 °C and (c) 60 °C.

In conclusion, Au nanoplates were successfully synthesized templated by SDS-PEG necklace-like soft clusters through a simple, mild, one-pot route. PEG acted both as reducing agent and one of the cluster components. $HAuCl_4$ concentration and reaction temperature were demonstrated influence factors to the growth of Au nanostructures. With the increase of $HAuCl_4$ concentration from 0.2 $mmol \cdot L^{-1}$ to 0.5 $mmol \cdot L^{-1}$, the size of Au plates enlarged from ~500 nm to ~2 μm. At lower temperature, the products were equiaxial Au particles. However, the fusion between different Au nanoplates was demonstrate when the temperature beyond 40 °C.

References

1. a) Nadagouda N. Mallikarjuna and Rajender S. Varma (2007), "Microwave-Assisted Shape-Controlled Bulk Synthesis of Noble Nanocrystals and Their Catalytic Properties", *Cryst. Growth Des.* July, pp. 686-690; b) Claire M. Cobley, Dean J. Campbell, and Younan Xia (2008), "Tailoring the Optical and Catalytic Properties of Gold-Silver Nanoboxes and Nanocages by Introducing Palladium", *Adv. Mater.* January, pp. 748-752.
2. a) Maya Zayats, Ronan Baron, Inna Popov, and Itamar Willner (2004), "Biocatalytic Growth of Au Nanoparticles: From Mechanistic Aspects to Biosensors Design" *Nano Lett.*, December, pp. 21-25; b) Weian Zhao, William Chiuman, Jeffrey C. F. Lam, Simon A. McManus, Wei Chen, Yuguo Cui, Robert Pelton, Michael A. Brook, and Yingfu Li, (2008), "DNA Aptamer Folding on Gold Nanoparticles: From Colloid Chemistry to Biosensors" *J. Am. Chem. Soc.* February, pp. 3610-3618.
3. Hak Soo Kim, Seung Joon Lee, Nam Hoon Kim and Jae Keun Yoon (2003), "Adsorption Characteristics of 1,4-Phenylene Diisocyanide on Gold Nanoparticles: Infrared and Raman Spectroscopy Study" *Langmuir*, July, pp. 6701-6710.
4. Danielle K. Smith and Brian A. Korgel (2008), "The Importance of the CTAB Surfactant on the Colloidal Seed-Mediated Synthesis of Gold Nanorods", *Langmuir*, January, pp. 644-649.
5. Lihua Pei, Koichi Mori, Motonari Adachi (2004), "Formation Process of Two-Dimensional Networked Gold Nanowires by Citrate Reduction of $AuCl_4$- and the Shape Stabilization", *Langmuir*, July, pp. 7837-7843.
6. N. N. Zhao, Y. Wei, N. J. Sun, Q. Chen, J. W. Bai, L. P. Zhou, Y. Qin, M. X. Li, and L. M. Qi (2008),

"Controlled Synthesis of Gold Nanobelts and Nanocombs in Aqueous Mixed Surfactant Solutions", *Langmuir*, January, pp. 991-998.

7. S. Shiv Shankar, Akhilesh Rai, Absar Ahmad, and Murali Sastry (2005), "Controlling the Optical Properties of Lemongrass Extract Synthesized Gold Nanotriangles and Potential Application in Infrared-Absorbing Optical Coatings", *Chem. Mater.* 2005, January, pp. 566-572.
8. Sihai Chen, Zhong Lin Wang, John Ballato, Stephen H. Foulger, and David L. Carroll (2003), "Monopod, Bipod, Tripod, and Tetrapod Gold Nanocrystals", *J. Am. Chem. Soc.*, November, pp. 16186-16187.
9. Cuncheng Li, Kevin L. Shuford, Minghai Chen, Eun Je Lee and Sung Oh Cho (2008), "A Facile Polyol Route to Uniform Gold Octahedra with Tailorable Size and Their Optical Properties", *ACS Nano*, August, pp. 1760-1769.
10. a) Caixia Kan, Xiaoguang Zhu, and Guanghou Wang (2006), "Single-Crystalline Gold Microplates: Synthesis, Characterization, and Thermal Stability", *J. Phys. Chem. B*, February, pp. 4651-4656; b) Cuncheng Li, Weiping Cai, Bingqiang Cao, Fengqiang Sun, Yue Li, Caixia Kan, and Lide Zhang, L. D. Zhang (2006), "Mass Synthesis of Large, Single-Crystal Au Nanosheets Based on a Polyol Process", *Adv. Funct. Mater.*, November, pp. 83-90.
11. a) R. Bury, B. Desmazières, C. Treiner (1998), "Interactions between poly(vinylpyrrolidone) and ionic surfactants at various solid/water interfaces: a calorimetric investigation", *Colloids and Surfaces A: Physicochemical and Engineering Aspects*, March, pp. 113-124; b) Saburo Shimabayashi, Tadayuki Uno, Masayuki Nakagaki (1999), "Formation of a surface complex between polymer and surfactant and its effect on the dispersion of solid particles" *Colloids and Surfaces A: Physicochemical and Engineering Aspects*, June, pp. 283-295.
12. Chunrong Wang, Yun Fang, Jiewen Feng (2007), "Synthesis of Au Nanorings Using SDS-PEG Cluster as the Soft Template under Mild Conditions", *Acta Chim. Sin.*, Febrary, pp. 1177-1180.
13. Chunrong Wang, Yun Fang, Bo Li (2007), "Self-Assembly of Multipod-like Gold Nanoparticles in the Soft Template of SDS-PVP Clusters", *Acta Phys. Chim. Sin.*, October, 183-186.
14. Xiaoyun Ruan, Yun Fang, Ye Fan (2008), "Preparation of Ultrafine Nickel Powders in SDS-PVP Aqueous Solution"*Acta Phys. Chim. Sin.*, April, pp. 1513-1518.
15. Iva´n Pardin˜as-Blanco, Cristina E. Hoppe, Yolanda Pin˜eiro-Redondo, M. Arturo Lo´pez-Quintela, and Jose´ Rivas (2008), "Formation of Gold Branched Plates in Diluted Solutions of Poly (vinylpyrrolidone) and Their Use for the Fabrication of Near-Infrared-Absorbing Films and Coatings", *Langmuir*, December, pp. 983-990.
16. Byungkwon Lim, Pedro H. C. Camargo, and Younan Xia (2008), "Mechanistic Study of the Synthesis of Au Nanotadpoles, Nanokites, and Microplates by Reducing Aqueous $HAuCl_4$ with Poly (vinyl pyrrolidone)" *Langmuir*, August, pp. 10437-10442.

Graphene-Based Materials

Rodney Ruoff, Mechanical Engineering and the Texas Materials Institute, The University of Texas at Austin, Austin, TX

Our top-down approaches [1,2] inspired physicists to study individual layers of graphite obtained by micromechanical exfoliation, but our current approaches include growth on metal substrates and judicious use of isotopic labeling (13C vs 12C) to study the kinetics and mechanisms of deposition of large-area graphene and few layer graphene on metal substrates. This talk will focus on our suggested path for obtaining large area growth of high quality graphene in ways compatible with methods of the semiconductor industry. In addition, I will present highlights of published work on polymer and ceramic matrix composites with graphene as filler, on ultracapacitors based on graphene, on paper-like materials based on graphene, and on the use of 13C-labeled graphite (and 12C-pure graphite and graphene) in a variety of areas. Support and prior support of our work by the state of Texas, The University of Texas, DARPA-iMINT, DARPA-CERA, and previously by the NASA and the NSF, is appreciated.

1. Lu XK, Yu MF, Huang H, and Ruoff RS, Tailoring graphite with the goal of achieving single sheets, Nanotechnology, 10, 269-272 (1999).

2. Lu XK, Huang H, Nemchuk N, and Ruoff RS, Patterning of highly oriented pyrolytic graphite by oxygen plasma etching, Applied Physics Letters, 75, 193-195 (1999).

See also papers on http://bucky-central.me.utexas.edu/publications.htm such as #139,146, 150,155, 160, 164, 166, 168, 169, 174, 179-182, 184-192, etc.

Prior to joining The University of Texas at Austin as a Cockrell Family Regents Chair in Mechanical Engineering, Rod Ruoff served as Director of the Biologically Inspired Materials Institute at Northwestern University. He has been a 'Visiting Chair Professor' at Sungkyunkwan University in South Korea. He received his B.S. in Chemistry from the U. of Texas (Austin) and Ph.D. from the University of Illinois-Urbana. He was a Fulbright Fellow at the Max Planck Institute-Goettingen, Germany. From '89-'90, he was a Postdoctoral Fellow at the IBM T. J. Watson Research Center in New York. Prior to joining Northwestern in 2000, he was a Staff Scientist at the Molecular Physics Laboratory of SRI International and Associate Professor of Physics at Washington University. His research activities include global environment and energy; synthesis and physical/chemical properties of nanostructures and composites; nanorobotics, NEMS, and developing new tools for biomedical research. Ruoff has published 192 refereed journal articles in the fields of chemistry, physics, mechanics, & materials science at the time of submission of this abstract.

See more of this Session: Carbon Nanotubes IV
See more of this Group/Topical: Nanoscale Science and Engineering Forum

Novel Chemical Schemes to Functionalize Graphene without Introducing Defects: Avenue for Building Sensitive Graphene Sensors

Kabeer Jasuja, Chemical Engineering, Kansas State University, Manhattan, KS

Nihar Mohanty, Chemical Engineering, Kansas State University, Manhattan, KS

Vikas Berry, Chemical Engineering, Kansas State University, Manhattan, KS

In this talk, we will present novel approaches for interfacing and functionalizing graphene sheets with biological and physical molecules via non-covalent bond to retain the electrical properties of graphene while introducing molecule-recognition specificity. Graphene has received tremendous research attention over recent years which can largely be attributed to its unique electrical & mechanical properties with a surface construct that is atomic-scale thick and yet is multi micron-scale large. While the atomic thickness imparts electronic confinement, graphene's large area provides excellent potential for detecting molecules via surface interaction. However, the current surface modification schemes introduce several defects in graphene by removing the pi-electrons and significantly degenerate graphene's superior properties. Here, non-covalent interfacing on graphene will be demonstrated as an effective method to facilitate defect-free functionalization to retain graphene's superior properties. Further The resultant change in electrical properties of graphene will also be illustrated. We will show the conjugate's detection sensitivity for DNA & protein sensing. We envision that by leveraging the non-covalent chemistry on graphene, a novel class of defect free graphene-hybrids can be fabricated with tunable physico-chemical and electrical properties for advanced applications in biological/chemical sensing, nanoelectronics, optoelectronics and catalysis.

See more of this Session: Carbon Nanotubes IV
See more of this Group/Topical: Nanoscale Science and Engineering Forum

Structure of Carbon Nanocrystals Nucleated by Hydrogen-Induced Intershell C-C Bonding in Multi-Walled Carbon Nanotubes

Andre R. Muniz, Department of Chemical Engineering, University of Massachusetts, Amherst, MA

Tejinder Singh, Department of Chemical Engineering, University of Massachusetts, Amherst, MA

Dimitrios Maroudas, Department of Chemical Engineering, University of Massachusetts, Amherst, MA

It has been observed that the exposure of multi-walled carbon nanotubes (MWCNTs) to atomic hydrogen from a hydrogen plasma results in the formation of nanocrystalline carbon phases (e.g., cubic diamond nanocrystals) embedded into the nanotubes. Motivated by these experimental findings, in this presentation, we investigate the formation of inter-shell sp^3 C-C bonds in MWCNTs induced by atomic hydrogen and carry out a comprehensive theoretical analysis of the crystalline structures that can be nucleated through such C-C bonding.

In this context, we report (i) pathways of H interactions with the graphene walls of the MWCNTs that induce inter-shell C-C bonding, (ii) a detailed analysis of the atomic and electronic structure of the resulting nanocrystal nuclei, and (iii) a systematic study of the nucleated nanocrystal structure as a function of the chiralities and relative alignment of the adjacent graphene walls of MWCNTs that are bridged by the inter-shell C-C bonding. The analysis is based on a synergistic combination of classical molecular-dynamics (MD) simulations with first-principles density functional theory (DFT) calculations. The MD simulations are used for large-scale structural relaxation in the isothermal-isobaric ensemble. In the MD simulations, the interatomic interactions are described according to the Adaptive Interatomic Reactive Empirical Bond Order (AIREBO) potential. The DFT calculations are performed within the generalized gradient approximation (GGA) and employ plane-wave basis sets, ultrasoft pseudopotentials, and supercell models. They are used for atomic and electronic structure computations, as well as for analysis of optimal reaction pathways for the interactions of H with the MWCNT graphene walls in conjunction with the climbing-image nudged elastic band method.

We show that the resulting MWCNT structures with inter-shell C-C bonds are stable and demonstrate that the atomic clusters in the vicinity of this inter-shell C-C bonding provide seeds for the nucleation of crystalline carbon phases embedded into the MWCNTs. We demonstrate that various such crystalline phases can be generated, including cubic and hexagonal diamond, as well as new tetragonal and monoclinic carbon phases. The structure of these crystalline seeds depends on the chiralities and on the relative alignment between the adjacent concentric walls of the MWCNTs; this relationship is explored systematically and quantified.

See more of this Session: Carbon Nanotubes IV
See more of this Group/Topical: Nanoscale Science and Engineering Forum

Brownian Dynamics Simulations of Carbon Nanotubes Breaking During Sonication

Guido Pagani, Chemical and Biomolecular Engineering, Rice University, Houston, TX

Micah J. Green, Chemical and Biomolecular Engineering, Rice University, Houston, TX

Philippe Poulin, Centre de Recherche Paul Pascal, CNRS Bordeaux, Pessac, France

Matteo Pasquali, Chemical and Biomolecular Engineering, Rice University, Houston, TX

The extraordinary mechanical, thermal, and electrical properties of carbon nanotubes (CNTs) are well-known; however, the processing of CNTs into high-performance macroscopic materials is hampered by CNTs' poor solubility in virtually all common aqueous and organic solvents. CNTs naturally tend to form insoluble bundles and ropes due to van der Waals interactions. One of the most common approaches to liquid-phase dispersion is ultrasonication, which breaks the bundles apart. Stabilizing agents such as surfactants and polymers are then used to prevent the CNTs from reaggregating. However, a number of experiments indicate that the average length of CNTs in solution decreases with sonication time; this is due to CNT breaking due to the extreme flow field around cavitating bubbles.

We couple Brownian Dynamics simulations of CNTs with a model for the formation and collapse of bubbles during sonication. These simulations show that CNTs are prone to breakage when they become oriented in a direction tangent to the bubble wall during bubble growth; when the bubble collapses, such CNTs become compressed and can break. This contradicts previously-suggested mechanisms for CNT shortening. The simulations suggest phase diagram for nanotube breakage as a function of length and stiffness; we compare simulation results to experimental data for the scaling of average length as a function of sonication time.

See more of this Session: Carbon Nanotubes IV
See more of this Group/Topical: Nanoscale Science and Engineering Forum

Addressing HiPco Single-Walled Carbon Nanotube Bundles with Optically Patterned Virtual-Electrodes

Peter Pauzauskie, Chemical Sciences Division, Lawrence Livermore National Laboratory, Livermore, CA

Arash Jamshidi, Electrical Engineering, University of California, Berkeley, Berkeley, CA

Hsan-Yin Hsu, Electrical Engineering, University of California, Berkeley, Berkeley, CA

Ted Laurence, Condensed Matter and Materials Division, Lawrence Livermore National Laboratory, Livermore, CA

Joseph M. Zaug, Chemical Sciences Division, Lawrence Livermore National Laboratory, Livermore, CA

Joe H. Satcher, Jr., Chemical Sciences Division, Lawrence Livermore National Laboratory, Livermore, CA

Ming C. Wu, Electrical Engineering, University of California, Berkeley, Berkeley, CA

Solution phase processing of carbon nanotube materials requires massive parallelism for high throughput unit operations combined with specificity aimed at discriminating between semiconducting and metallic fractions commonly found after nanotube synthesis. As shown recently with nanowires and multiwalled carbon nanotubes [1, 2], optoelectronic tweezers (OET) offer a unique combination of real-time, massively-parallel manipulation of nanostructures with the ability to distinguish particles based on their dielectric constant and conductivity alone. In this work we present the ability of OET to trap and translate bundles of HiPco single-walled carbon nanotubes (SWCNTs). Fluorescent labeling enables visualization of SWCNT solution-phase dynamics in conjunction with time-correlated single-photon counting and video microscopy. High-speed digital thermography as well as anti-Stokes Raman spectroscopy are used to characterize potential heating profiles of SWCNT ensembles trapped within discrete OET virtual-electrodes. Future work will focus on extending OET to semiconducting and metallic SWCNTs materials enriched with density-gradient ultracentrifugation.

[1] Jamshidi, A.*; Pauzauskie, P.J.*; et al., Nature Photonics (2008)

[2] Pauzauskie, P.J.*; Jamshidi, A.*; et al., Applied Physics Letters, in press (2009)

* - equal contribution

See more of this Session: Carbon Nanotubes IV
See more of this Group/Topical: Nanoscale Science and Engineering Forum

Multi-Angle Depolarized Dynamic Light Scattering of Short Functionalized Single Walled Carbon Nanotubes

Abhishek M. Shetty, Chemical Engineering, University of Michigan, Ann Arbor, MI

Georgina M.H Wilkins, Chemical Engineering, University of Michigan, Ann Arbor, MI

Jagjit Nanda, Research and Advanced Engineering, Dearborn, MI

Michael J. Solomon, Chemical Engineering, University of Michigan, Ann Arbor, MI

Precise characterization of single-walled carbon nanotube (SWCNT) length and diameter is essential in a number of areas, particularly for optimizing and predicting their electrical, optical and rheological properties. Current characterization methods, such as AFM and TEM, are time intensive, and potentially suffer from serious systematic errors due to the effects of drying and surface deposition. We introduce the method of multi-angle depolarized dynamic light scattering (MA-DDLS), which is a solution-based technique to characterize the length and diameter of single-walled carbon nanotubes (SWCNTs). We have used covalently functionalized SWCNT with poly(amino benzene sulfonic acid) (PABS) and poly(ethylene glycol) (PEG) for this study. MA-DDLS yields simultaneous characterization of the mean translational and rotational diffusivities of dilute solutions of SWCNTs. Using an anisotropic rigid rod model, we uniquely determine the length and diameter of the SWCNTs from the independent measurements of rotational and translational diffusion. The multi-angle depolarized light scattering technique is found to be a fast, non-invasive and reproducible method for identifying the average length and diameter of SWCNTs in solution. Furthermore this method can be performed in situ and is compatible with continuous separation methods such as size-exclusion chromatography.

See more of this Session: Carbon Nanotubes IV
See more of this Group/Topical: Nanoscale Science and Engineering Forum

Nanocomposite Fibrous Mats for Strain Sensing Applications

Russell E. Gorga, Fiber and Polymer Science Dept., North Carolina State University, Raleigh, NC

Laura I. Clarke, Physics, North Carolina State University, Raleigh, NC

Electrospinning is a process to develop fibers with diameters at the nanometer range. Recently much interest has been generated in this technique due to the large number of potential applications such as tissue engineering, drug delivery, and filtration. Our effort has been to produce nanocomposite fibers embedded with carbon nanotubes for for sensitive strain-sensing applications. Here, strategies to obtain unique morphologies (including core-sheath structures) of multi-wall carbon nanotube reinforced nanofibrous mats are discussed. The morphologies of these materials are characterized using electron microscopy. The mechanical properties (accounting for the porosity of the mat) and the electrical properties (including percolation behavior) are measured. Strain sensing capability is demonstrated.

See more of this Session: Nanoscale Structure in Polymers II: Nanostructured Polymeric Materials
See more of this Group/Topical: Nanoscale Science and Engineering Forum

Self Sterilizing Polymer Surfaces Based On Silver/Tricalcium Phosphate Nanoparticles

Oliver D. Schneider, Department of Chemistry and Applied Biosciences, ETH Zurich, 8093 Zurich, Switzerland

Stefan Loher, Department of Chemistry and Applied Biosciences, ETH Zurich, 8093 Zurich, Switzerland

Wendelin J. Stark, Department of Chemistry and Applied Biosciences, ETH Zurich, 8093 Zurich, Switzerland

The prevention of microbial surface contamination is a very important issue in today's health care system, in food and pharmaceutical production. Unwanted contamination events cause costly production time out, serious health problems and infections particularly in hospitals. The present contribution shows the development of a novel, transparent plastic film, which can kill bacteria (figure). This strongly antibacterial polymer surface contains 20-50 nm tricalcium phosphate nanoparticles which themselves serve as carriers for 1-2 nm silver particles [1]. Since bacteria need calcium and phosphate for their metabolism, the silver gets released when the calcium phosphate particles are taken up by the micro-organism as nutrition. The antibacterial effect of this triggered mechanism lead to an up to three orders of magnitude higher bacteria killing efficiency compared to conventionally used silver systems. Clinically relevant pathogens such as P. aeruginosa, C. albicans, E. coli were chosen to investigate the antibacterial performance. Figure: Silver nanoparticles are only released from the transparent polymer films when required or stimulated by the presence of microorganism. Scanning transmission electron micrograph of Ag-TCP showing the small 1-2nm silver particles. References: [1] S. Loher, O. D. Schneider, T. Maienfisch, S. Bokorny, W. J. Stark, Small, 2008, 4(6), 824-832.

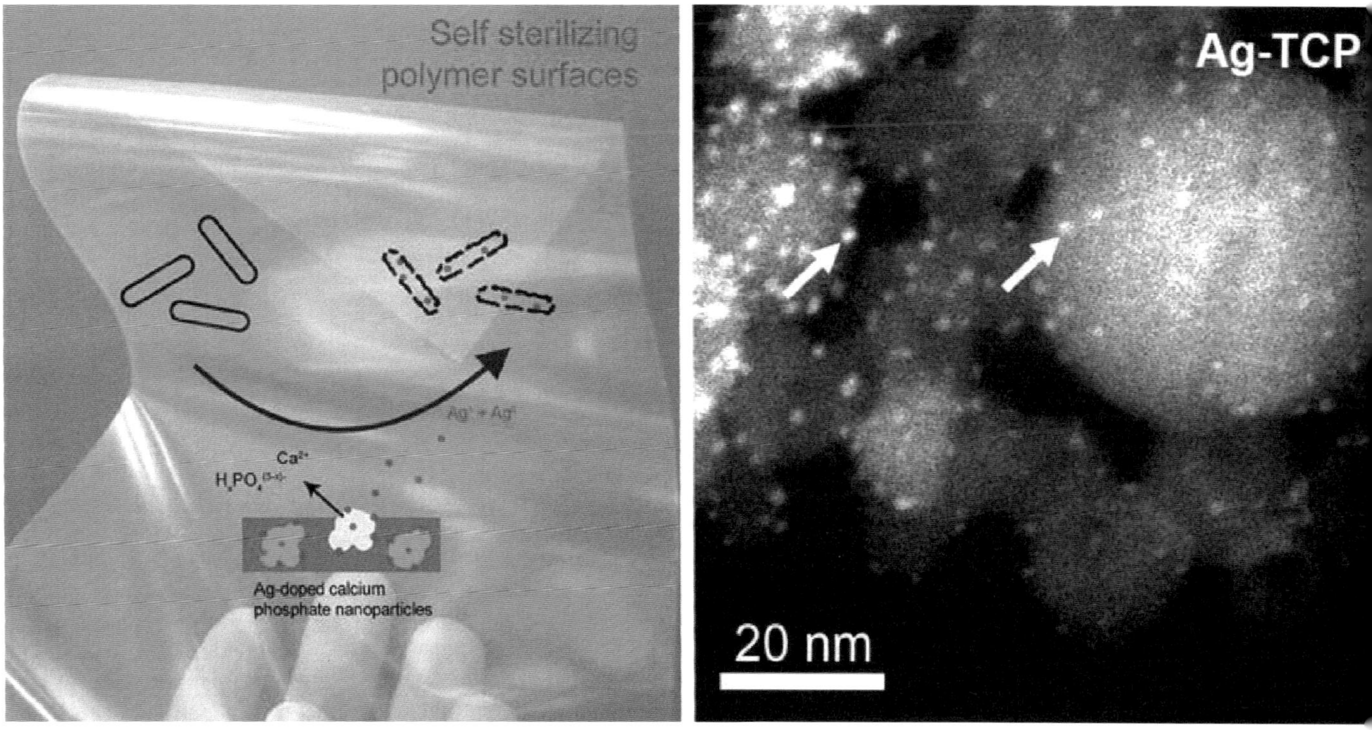

See more of this Session: Nanoscale Structure in Polymers II: Nanostructured Polymeric Materials
See more of this Group/Topical: Nanoscale Science and Engineering Forum

Structure and Deformation Response of Rod-Containing Nanocomposites

Gregory N. Toepperwein, Department of Chemical and Biological Engienering, University of Wisconsin-Madison, Madison, WI

Juan J. DePablo, University of Wisconsin-Madison, Madison, WI

Addition of nanoparticles, such as metallic nanoparticles or carbon nanotubes, into a polymer drastically affects the mechanical properties of the resultant nanocomposite. This includes elastic modulus, a function of localrdering, and hardening modulus, a function of chain topology. Adequate experimental analysis is frustrated by challenges associated with sample preparation and particle dispersion, prompting this work's use of Monte Carlo of molecular dynamics simulation of course-grained models to provide a accurate nano-scale view of these nanocomposite systems. Studies into the nature of these changes have examined a large parameter space of variables including particle size, effective surface interactions, concentration, and thermal history. However, comparatively few works have examined the role of nanoparticle geometry. To that end we have examined the effect of introducing various length nanorods into polymeric materials via course-grained simulation, and applied the plethora of well-developed techniques that have previously been used to study spherical-particle nanocomposite systems. This work first examines the organization of nanorods at different concentrations and comments on resulting alignment behavior of polymer as it changes with particle size. Second, deformation of nanocomposites reveals changes in the elastic behavior, strain-softening behavior, and hardening modulus as a function of rod length. Third, recently developed algorithms that allow for the extraction of primitive path from simulation configurations are used to describe how particles affect the underlying polymer entanglement network, as well as how that network evolves due to particle diffusion. Fourth, this work studies how bridged polymer-particle network topology depends on nanorod length and orientation, including during deformation. The primary result of these efforts is a characterization of how anisotropic rods fundamentally alter polymer behavior on the nanoscale resulting in magnified macroscopic observables.

See more of this Session: Nanoscale Structure in Polymers II: Nanostructured Polymeric Materials
See more of this Group/Topical: Nanoscale Science and Engineering Forum

Fluorescent Probe as Model Solute to Study the Phase Behavior in Polymeric Nanoparticles

Varun Kumar, Chemical Engineering, Princeton University, Princeton, NJ

Robert K. Prud'homme, Chemical Engineering, Princeton University, Princeton, NJ

Nanostructured polymeric particles are effectively used as a carrier for various therapeutics [1]. The hydrophobic solute molecules are encapsulated into the hydrophobic polymeric core using different formulation techniques [2]. Higher drug loading in the polymeric core results in the formation of drug crystals and hence phase separation [3]. The release of drug molecules from the nanoparticle depends on the extent of dispersion, and is diffusion and dissolution controlled at low and higher drug loading respectively [4]. The solubility of pharmaceutical materials, and hence their bioavaiolability, also depend on the physical state. Hancock et al. [5] demonstrated the higher aqueous solubility for amorphous Indomethacin drug as compared to its crystalline counterpart.

In the present work, pyrene is used as a model fluorescent probe to study the phase behavior in the polymeric core. The emission spectra of pyrene displays distinct peaks of monomer and excimer molecules [6]. The fluorescence quenching of monomer and excimer content were studied for a range of pyrene loading using various core forming polymer co-solutes. The distribution of pyrene in nanoparticle core was found to be dynamic and controlled by the miscibility between the probe and co-solutes.

Hostasol yellow, another fluorescent dye with larger molar volume compared to pyrene, was also used for the fluorescence study. Even though the fluorescence quenching was observed at higher dye loading, the spectra remains stable unlike pyrene. The Flory- Huggins theory of mixing [7] has been used for qualitative estimation of dye dispersity in various polymer co-solutes.

1. Allen, C.; Maysinger, D.; Eisenberg, A. Colloids and Surfaces B: Biointerfaces, 1999. 16(1-4):p.3.

2. Kumar, V.; Prud'homme, RK. Journal of Pharmaceutical Sciences, 2008. 97:4904-4914.

3. Gref, R. et al. Science, 1994. 263(5153): p. 1600-1603.

4. Polakovic, M. et al. Journal of Controlled Release, 1999. 60(2-3): p. 169.

5. Hancock, B.C. and Parks, M. Pharmaceutical Research, 2000. 17(4): p. 397-404.

6. Birks, J.B.; Dyson, DJ.; Munro, IH. Proceedings of the Royal Society of London. Series A, Mathematical and Physical Sciences, 1963. 275(1363): p. 575.

7. Rubinstein, M. and Colby, RH. Polymer Physics. 2003.

See more of this Session: Nanoscale Structure in Polymers II: Nanostructured Polymeric Materials
See more of this Group/Topical: Nanoscale Science and Engineering Forum

Focused Ion Beam Imaging of Highly Dispersed Benzocyclobutene-Maleimide/Layered Silicate Thermoset Nanocomposites

Edwin Hampton, Tuskegee University, Tuskegee, AL

Mitra Yoonessi, Structure and Materials Division, at NASA Glenn Research Center, Cleveland, OH

Derrick R. Dean, Department of Materials Science and Engineering, University of Alabama at Birmingham, Birmingham, AL

Loon-Seng Tan, Materials & Manufacturing Directorate, U.S. Air Force Research Laboratory, Wright-Patterson AFB, OH

Hilmar Koerner, Materials and Manufacturing Directorate, Air Force Research Laboratory, Wpafb, OH

Robert Wheeler, Wright Patterson Air Force Research Laboratory, Wpafb, OH

Richard A. Vaia, Materials and Manufacturing Directorate, Wright Patterson Air Force Research Laboratory, Wpafb, OH

The advancement of thermoset based nanocomposites depends upon the development of fabrication methods that will provide extensive dispersion of nanoparticles within the thermoset and the development of controllable interfacial chemistry. A general methodology has been used to achieve high dispersion in a high temperature benzocyclobutene-maleimide (BCB-MI) thermosetting polymer. A modified approach, which directly addresses the nanoclay dispersion difficulties of thermosets, resulted in a highly disordered thermoset nanocomposite with little or no agglomeration of individual clay platelets, and most importantly no visible remnants of the initial clay tactoid structure. Specifically, imidazolium based reactive and nonreactive surfactants were used to modify montmorillonite, thus providing the necessary thermal stability as well as the ability to covalently connect the nanoparticle surfactant to the BCB-MI matrix. This 'traditional' organoclay served as the starting point to generate a lightly cross-linked corona of BCB-MI around the montmorillonite through a surface catalyzed solution procedure. The subsequent hybrid montmorillonite was then added to neat BCB-MI resulting in a highly disorderly thermoset nanocomposite, as determined by small angle x-ray diffraction and transmission electron microscopy. The extent of coupling between the organic modifier and matrix (reactive and nonreactive surfactants) had a substantial impact on the mechanical properties and thermal properties. Three dimensional visualization of the extent of the disordered structure of the tactoids and nanoclay layers were studied by slice imaging using focused ion beam (FIB) dual-beam SEM. This novel approach provided understanding of the clay nanostructure in the polymer matrix three dimensionally beyond the traditional two dimensional projection imaging.

See more of this Session: Nanoscale Structure in Polymers II: Nanostructured Polymeric Materials
See more of this Group/Topical: Nanoscale Science and Engineering Forum

Glass Transition Temperature of Isolated Polymer Chains Dispersed within a Bulk Phase: Novel Characterization by Fluorescence

Robert W. Sandoval, Department of Chemical and Biological Engineering, Northwestern University, Evanston, IL

John M. Torkelson, Chemical and Biological Engineering and Materials Science & Engineering, Northwestern University, Evanston, IL

Since the publication of "Self-concentration and effective glass transitions in (miscible) polymer blends" (Macromolecules 2000, 33, 5278) by Lodge and McLeish in the year 2000, many dozens of experimental studies have focused on characterizing the alpha-relaxation dynamics and glass transition temperatures (Tgs) of particular components in miscible polymer blends. In numerous cases, these studies have demonstrated that the two polymer components in miscible polymer-A/polymer-B blends can exhibit substantially different cooperative segmental dynamics if the Tgs of the two homopolymers are sufficiently different in magnitude. However, the lack of instrumental sensitivity via conventional techniques to a minor component in highly asymmetric blends (≤ ~1 wt% minor component) has hindered the direct characterization of Tg associated with trace levels of one polymer dispersed in a sea of a second polymer. Such measurements are useful to test the validity of the Lodge-McLeish model in the limits of infinite dilution, both in blends that are miscible at all compositions and in blends that are normally immiscible except near the limit of infinite dilution of one component. Such measurements are also useful to provide information regarding the length scales over which the cooperative segmental mobility of a single polymer-A chain may be impacted by perturbations to its environment consisting of bulk polymer-B, e.g., by the presence some hundreds of nanometers away of other polymer-A chains as the concentration of polymer-A is slowly increased from infinite dilution.

In response to this challenge, we have developed a novel fluorescence technique that utilizes the intrinsic fluorescence of styrene repeat units to determine the Tg of a styrene-containing minor component dispersed homogeneously near trace levels in a polymer blend (e.g., at ≤ 2 wt % where the blend is miscible and the minor component is dissolved in the bulk phase). In particular, we have applied this technique to determine the Tg of polystyrene (PS) chains near infinite dilution in other polymers via the intersection of the rubber-state and glassy-state temperature dependences of fluorescence intensity. This technique also allows us to confirm that the PS chains are individually dispersed in the second polymer and not segregated or in significant contact with other PS chains because single PS chains isolated in a second polymer exhibit only monomer fluorescence of the phenyl rings while segregated PS chains exhibit both excimer and monomer fluorescence. When dispersed at 0.1 wt% within poly(methyl methacrylate) (PMMA) (neat, bulk Tg = 125 °C), polystyrene (PS) (neat, bulk Tg = 100 °C) exhibits a Tg of 118 °C. However, when PS chains are dispersed at 0.1 wt% within poly(t-butyl acrylate) (PtBA) (neat bulk Tg = 45 °C), the PS Tg is reduced to 64 °C. This demonstrates that the bulk environment can have dramatic effects on the chain dynamics of the individually dispersed PS chains. This technique has also been extended to block copolymer/homopolymer blends. This study is the first to allow direct Tg measurements of isolated polymer chains dispersed within a bulk phase and also allows for a critical test of the Lodge-McLeish model at highly asymmetric blend ratios.

See more of this Session: Nanoscale Structure in Polymers II: Nanostructured Polymeric Materials
See more of this Group/Topical: Nanoscale Science and Engineering Forum

Mesoscale Simulation of Nanoscale Segregation in Polyelectrolyte Membranes

Aleksey Vishnyakov, Rutgers University, Piscataway, NJ

Shuang Yang, Chemical and Biochemical Engineering, Rutgers University, Piscataway, NJ

Alexander V. Neimark, Department of Chemical and Biochemical Engineering, Rutgers University, Piscataway, NJ

Nanostructured polyelectrolyte membranes (PEM) are widely used as perm-selective diffusion barriers in fuel cell technologies and electrochemical processing. PEMs are also suitable for protection against chemical warfare agents. PEMs are made of nanostructured polymers that contain hydrophilic and hydrophobic fragments. For example, in Nafion polymer (Du Pont) hydrophilic sidechains terminated by sulfonate groups are attached to a hydrophobic perfluorocarbon backbone. Upon hydration, PEM segregates into hydrophobic and hydrophilic subphases. The former is made by the hydrophobic organic backbone, while the latter comprises water, counterions and ionic groups of the polymer, forming a dynamic network of channels in the hydrophobic organic matrix. With respect to transport and adsorption of chemicals, solvated PEM represents a nanoporous material. Water diffuses easily through the network of pores comprised of hydrophilic channels, while more hydrophobic chemical agents get trapped in the essentially immobile hydrophobic subphase.

Mesoscale simulations of nanoscale segregation in PEM provide a powerful tool for optimization of the polymer structure and membrane composition. In this work, we describe two different approaches to mesoscale modeling of PEM, using Nafion and sulfonated polystyrene-polyolefinepolystyrene triblock copolymers as examples. First, we developed a dissipative particle dynamics scheme suitable for PEM. The polymers were presented as sequences of beads, either linear (triblocks) or branched (Nafion) linked by harmonic bonds. Weak harmonic angles were introduced for rigidity. We applied two methods for the DPD parameter derivation from (1) radial distribution functions obtained in atomistic molecular dynamics simulations of smaller fragments and (2) thermodynamic properties of reference solutions. The second approach to mesoscale modeling of PEM is based on self-consistent field (SCF) theory, which we extended to complex systems that include a charged triblock copolymer, water, and polar chemical agents. SCF is parameterized from available experiments coupled with ab-initio and thermodynamic modeling of reference solutions. With these techniques we explored nanoscale segregarion in Nafion (DPD only) and sSEBS (DPD, SCF). We predicted the morphology of hydrated Nafion with equivalent weight between 1200 and 1800D and 4-17 wt % water content, observing a transition between separated hydrophilic clusters and a continuous 3D network of hydrophilic channels. No particular segregation morphology in Nafion was detected. In sSEBS, the morphology was more regular, transiting from reverse micelles at lower hydration to lamellae and even reverse micelles as water and DMMP content increased. We discuss the influence of polymer composition and hydration on the shape and connectivity of hydrophilic channels. Using this information on the membrane morphology, we perform lattice Monte Carlo simulations of water and chemical agent transport through PEMs and compare our results with the experiments.

See more of this Session: Nanoscale Structure in Polymers II: Nanostructured Polymeric Materials
See more of this Group/Topical: Nanoscale Science and Engineering Forum

Regioselective Competitive Adsorption of Water and Organic Vapor Mixtures On Heterogeneous Bundles of Single-Walled Carbon Nanotube: Experimental and Simulation Study

Sandeep Agnihotri, Civil and Environmental Engineering, University of Tennessee, Knoxville, TN

J. P. B. Mota, Departamento de Química, Faculdade de Ciências e Tecnologia - Universidade Nova de Lisboa, 2829-516 Caparica, Portugal

Pyoungchung Kim, Civil and Environmental Engineering, University of Tennessee, Knoxville, Knoxville, TN

Liangcheng Yang, Civil and Environmental Engineering, University of Tennessee, Knoxville, TN

Sequential adsorption of water and organic vapor mixtures onto single-walled carbon nanotube (SWNT) bundles is studied experimentally and by grand canonical Monte Carlo (GCMC) simulation to elucidate the distinct interactions between the select adsorbates and the nanoporous structure of SWNTs. First, we developed a practical method for adsorption modeling of heterogeneity of SWNTs. Our method integrates experimental analysis with GCMC simulation of simple molecules, such as nitrogen, at 77 K. Using this method it is possible to estimate adsorption inside the nanotubes, adsorption on the external surface of the bundles, and adsorptive contributions from the impurities present in samples (Agnihotri et. al., J Phys Chem C, 2007). Using this methodology, we conducted a study on competitive adsorption of water and organic vapors to elucidate the distinct interactions between the select adsorbates and the nanoporous structure of SWNTs (Agnihotri et al, Langmuir, 2008). Experimental adsorption isotherms on SWNT bundles for hexane, methyl ethyl ketone, cyclohexane and toluene individually mixed in carrier gases that were nearly saturated with water vapor are compared with the GCMC simulated isotherms for hexane (chosen as a representative organic molecule) on only the external surface of the heterogeneous SWNT bundles. From the nearly perfect overlap between the experimental and simulated isotherms, it is concluded that until near saturation only the internal pore volume of pristine SWNT bundles fills with water. The adsorption of water vapor on the peripheral surface of the bundles remains insignificant, if not negligible, in comparison to the adsorption of water in the internal volume of the bundles. This is in contrast with the adsorption of pure hexane, which exhibits appreciable adsorption both inside the bundles as well as on their external surface. It is also suggested that during competitive adsorption, water molecules take precedence over small non-polar and polar organic molecules for adsorption inside SWNTs and leave unoccupied the hydrophobic external surface of the bundles for other more compatible adsorbates.

See more of this Session: Carbon Nanotubes V: Adsorption and Transport
See more of this Group/Topical: Nanoscale Science and Engineering Forum

Ferroelectric Water Inside Single-Walled Carbon Nanotubes

Yutaka Maniwa, Department of Physics, Faculty of Science, Tokyo Metropolitan University, Hachioji, Japan

In this paper, we report novel ferroelectric properties of a new form of ice inside single-walled carbon nanotubes (SWCNTs). The ices are called "ice nanotubes" (ice NTs) and they consist of polygonal water rings stacked one-dimensionally along the SWCNT axis. We performed molecular dynamics (MD) calculations for the ice NTs under an external electric field and in a temperature range between 100 K and 350 K, using TIP3P water model. First, it is revealed that the present MD simulations in a zero-external electric field reproduce reported x-ray diffraction experiments. From a comparison, it is strongly suggested that ice-NTs consisting of more than 20 polygon units are indeed formed inside real SWCNTs. The calculated polarization indicates significant fluctuation with time and space around the ice-NT phase transition temperature . The fluctuation was more remarkable for thinner SWCNTs and for parallel polarization with respect to the SWCNT axis. Therefore it could be ascribed to quasi-one-dimensional fluctuations in the ice-NT structures. With reduced temperatures well below , a ferro- or antiferro-electric single domain is formed inside the SWCNTs; although the polarization fluctuates between negative and positive polarity with the same magnitude in the intermediate temperature region, it finally freezes along one direction of the SWCNT axis. The polarization as a function of the external field indicates novel hysteresis behavior that cannot be observed in the bulk: The n-gonal ice NT, where n=5, 6, or 7, has (n+1)-polarized structures with different polarizations. The results predict that ice-NTs can be the world's smallest ferroelectrics with different polarized states, as small as 2 nm in diameter and shorter than 4 nm in length. The present findings suggest potential applications of SWCNTs encapsulating dielectric materials to fabricate the smallest ferroelectric devices.

Reference:

F. Mikami, K. Matsuda, H. Kataura and Y. Maniwa, Dielectric Properties of Water inside Single-Walled Carbon Nanotubes, ACS Nano, (2009) DOI: 10.1021/nn900221t

See more of this Session: Carbon Nanotubes V: Adsorption and Transport
See more of this Group/Topical: Nanoscale Science and Engineering Forum

A Nano-Syringe for the Transport of Thin Water Films Across Carbon Nanotube Membranes

Jose L. Rivera, Department of Chemical & Biomolecular Engineering, Vanderbilt University, Nashville, TN

Francis W. Starr, Physics Department, Wesleyan University, Middletown, CT

The controlled flow of water molecules at the nanoscale is an initial step to many fluidic processes in nanotechnology. Here we show how thin films of water can be drawn through a nano-syringe built from a carbon nanotube membrane and a "plunger". By increasing the speed of withdrawal of the plunger, we can obtain molecular transport through the membrane at flux rates exceeding 10^{25} molecules cm^{-2} s^{-1}. Above a threshold speed around 0.25 nm/ns, molecules cannot fill the chamber created by the plunger motion as fast as the chamber expands, and the resulting flux rate drops. By considering hydrophobic or hydrophilic plungers, we unexpectedly find that the nature of the water-plunger interactions does not affect the flux rate or the threshold plunger speed. While the water structure near the plunger surface differs significantly for different plunger interactions, the failure of the film away from the plunger surface is responsible for loss of transport. As a result, the surface interactions play a limited role in controlling the flux.

See more of this Session: Carbon Nanotubes V: Adsorption and Transport
See more of this Group/Topical: Nanoscale Science and Engineering Forum

Nonpolar Microenvironments around Single-Walled Carbon Nanotubes

Kirk J. Ziegler, Department of Chemical Engineering, University of Florida, Gainesville, FL

Randy Wang, University of Florida, Gainesville, FL

Wei-Chiang Chen, University of Florida, Gainesville, FL

Single-walled carbon nanotubes (SWNTs) have a unique near-infrared fluorescence associated with their (n,m) type. Here we show that mixing aqueous suspensions with immiscible solvents results in solvatochromic shifts associated with environmental changes surrounding the SWNTs. These shifts suggest the hydrophobic region of the micelle swells with the organic solvent when mixed. We use these new microenvironments around SWNTs to investigate the photophysical properties of SWNTs in 16 nonpolar solvents and mixtures. The solvatochromic shifts follow the expected behavior from a polarizable solute in a polarizable solvent. The PL intensity is shown to be very sensitive to small concentrations of polar components. The solvatochromic shifts of the aqueous SWNT suspensions are reversible once the solvent evaporates. However, some surfactant-solvent systems show permanent changes to the NIR fluorescence emission intensity after exposure to the organic solvent. The intensity of some large diameter SWNT (n,m) types increase by more than 175%. These differences are attributed to surfactant reorganization, which can improve nanotube coverage, resulting in decreased exposure to quenching mechanisms from the aqueous phase.

See more of this Session: Carbon Nanotubes V: Adsorption and Transport
See more of this Group/Topical: Nanoscale Science and Engineering Forum

Structure and Phase Behaviors of Confined Fluids in Single-Walled Carbon Nanotubes

HuanCong Huang, Nanyang Technological University, Singapore, Singapore

Jayant K. Singh, Chemical Engineering, Indian Institute of Technology, Kanpur, Ka, India

Sang Kyu Kwak, Division of Chemical and Biomolecular Engineering, Nanyang Technological University, Singapore, Singapore

Single-walled Carbon Nanotubes (SWCNTs) have become one of the most potential materials in fabrication of nano-fluidic devices since their discovery, due not only to their unusual electrical and mechanical but also structural properties. SWCNTs are rolled up graphite into cylindrical shape providing a hollow interior that can be served as a vessel for filling of fluid molecules, which delimit applications in capillary condensation, gas adsorption and transport phenomena. To understand the behaviors of confined fluids in SWCNTs, one can consider three dominant factors to structurize fluid molecules within cylindrical constraint; interaction forces of carbon atoms-fluid molecules and fluid-fluid molecules, and radius of SWCNTs depending on surface chirality. The preliminary study indicates high sensitivity of those factors on variation of fluid and solid properties as well as the first order phase transition. Many studies have been conducted to investigate the behaviors of confined fluids in SWCNTs, yet the structure and phase behaviors are still little explored in terms of solidification of fluids. We aim to investigate the effects of those factors on confined fluid properties via molecular simulation methodology. Different molecular models such as hard sphere, square-well, and water under different chiral structures of SWCNTs are to be explored to elucidate, in particular, the influencing degree of those factors on the vicinity of the coexisting region of solid and fluid.

See more of this Session: Carbon Nanotubes V: Adsorption and Transport
See more of this Group/Topical: Nanoscale Science and Engineering Forum

ROLE OF AGGREGATION IN THE ADSORPTION BEHAVIORS OF CARBON NANOTUBES

Shujuan Zhang, Ting Shao, Sule Kaplan, and Tanju Karanfil*

* *Department of Environmental Engineering and Earth Sciences, Clemson University, Anderson, SC 29625, USA*
shuz@clemson.edu; tshao@clemson.edu; sulek@clemson.edu; tkaranf@clemson.edu

INTRODUCTION

Carbon nanotubes (CNTs), as newly emerging nanomaterials, have been examined for various applications in many fields due to their unique structures and outstanding mechanical, optical, and electronic properties. However, due to their nanosize, CNTs can enter into cells and cause damage to plants, animals and humans (Oberdorster et al. 2005). In addition, owing to their strong adsorption affinities for organic chemicals, the fate and transport of organic contaminants can be significantly changed by the release of CNTs into the environment (Colvin 2003). On the other hand, due to the strong van der Waals forces along the length axis, CNTs are prone to aggregate. Consequently, the availability of adsorption sites for organic chemicals is highly dependent on the aggregation of CNTs. Therefore, a fundamental understanding of the aggregation and adsorption properties of CNTs is critical for assessing the environmental risk of these nanomaterials and for evaluating their potential use in water/wastewater treatment and environmental remediation applications.

EXPERIMENTAL

A single-walled carbon nanotube (SWNT) and a multi-walled carbon nanotube (MWNT) were used as received. Two synthetic organic chemicals (SOCs), phenanthrene (PNT) and biphenyl (BP), were employed as liquid phase adsorbates. Nitrogen adsorption at 77 K was performed with a physisorption analyzer (Micromeritics ASAP 2010) to characterize the surface area and pore volume by using the Brunaur-Emmett-Teller (BET) model and *t*-plot method. Constant carbon dose aqueous phase isotherm experiments were conducted using the completely mixed batch reactors (255 mL glass bottles with Teflon-lined screw caps). The isothermal data were analyzed with the Freundlich equation.

RESULTS

The nitrogen adsorption isotherms of both the SWNT and the MWNT (not shown) can be divided into four phases, indicative of multi-stage adsorption, which can be attributed to micropore filling, monolayer formation, capillary condensation in mesopores and macropores, respectively. The surface area and pore volume obtained from the nitrogen adsorption isotherms by using the BET and *t*-plot are sumarized in Table 1 as S_{BET} (m^2/g) and V_t (cm^3/g). By assuming that (1) the length of the C-C bonds in the curved graphene sheets of CNTs is the same as that in the planar sheet (i.e. $d_{C\text{-}C} = 0.1421$ nm), (2) the MWNTs are composed of concentric walls and the inter-wall distance is $d_{w\text{-}w} = 0.34$ nm, and (3) the aspect ratio of CNTs is sufficiently high to neglect the area and volume of the tip ends, the theoretical surface area (S) and pore volume (V) of the CNTs were calculated and are listed in Table 1 (Detailed information about the calculation is provided in our previous report) (Zhang et al., 2009). The significantly higher V_t than V values of CNTs indicates that besides the adsorption in inner cavities, a large portion of the nitrogen adsorption comes from the nanospaces generated from aggregation of CNTs, such as the interstitial channels and peripheral grooves. In terms of surface area, the aggregation of CNTs led to a great reduction of S_{BET} for the SWNT but only a slight reduction for the MWNT. The slighter reduction in surface area but larger increase in pore volume of the MWNT than the SWNT induced by aggregation demonstrate that the aggregation structures of the MWNT were much looser than those of the SWNT. There was nearly no sieving effect in the adsorption of nitrogen by the MWNT.

Table 1. Theoretical and experimental determined structure parameters of CNTs and their adsorption capacities to SOCs

CNT	Theoretical		N_2		PNT			BP		
	S	V	S_{BET}	V_t	K_{FS}	S_{BET}/S_o	V_t/V_o	K_{FS}	S_{BET}/S_o	V_t/V_o
SWNT	1420	0.288	486	0.739	297	1.1	3	273	1.1	3
MWNT	214	0.056	192	0.765	60	1.9	12	90	1.2	7

For the liquid phase adsorption, the adsorption capacities were obtained from the solubility normalized Freundlich constant ($K_{FS,}$ mg/g). The surface area and pore volume occupied by the SOCs (S_o and V_o) were calculated by dividing the K_{FS} with the molecular cross-section area and density of the SOCs. As shown in Table 1, the ratios of S_{BET} to S_o for the two SOCs were larger than 1, suggesting that the surface coverage of the CNTs by the SOCs was lower than that by the N_2, which can be attributed to the more severe steric hindrance in the adsorption of SOCs than that of N_2, because the molecular sizes of the two SOCs were significantly larger than that of N_2. Another possibility for the low surface coverage of CNTs by SOCs is that the aggregation structures of CNTs in aqeous solution might be different from those in bulk phase. For both of the CNTs the V_t/V_o ratios were much higher than the S_{BET}/S_o ratios and the V_t/V_o ratios of the MWNT were much larger than those of the SWNT. As aforementioned, aggregation of CNTs led to a decrease of surface area but a significant increase of pore volume, especially for the MWNT. A large portion of the nitrogen adsorption came from capillary condensation in meso- and macropores trapped in the aggregates of CNTs. Unlike what happened in nitrogen adsorption, no capillary condensation occurred during the liquid phase SOC adsorption. Therefore, most of the nanospaces in the aggregates of CNTs that can be filled with N_2 could not be efficiently occupied by the SOCs. The adsorption of SOCs was controlled by the available surface area rather than the pore volume. The increased pore volume of CNTs due to aggregation exerted a positive effect on the adsorption of N_2 but had no positive effect on adsorption of SOCs in aqueous solutions. Since aggregation reduced surface area of CNTs, it is an unfavorable factor in their liquid phase adsorption.

CONCLUSIONS

The theoretical calculation and experimental determination performed in this study demonstrate that aggregation of CNTs played a crucial role in their adsorption. Aggregation of CNTs exerted a positive effect on nitrogen adsorption due to increased pore volume, while a negative effect was observed for the adsorption of SOCs in aqueous solutions. Approaches to minimize aggregation of CNTs may enhance their liquid phase adsorption capabilities.

ACKNOWLEDGEMENTS

This work was partly supported by a research grant from National Science Foundation (CBET 0730694). However, the manuscript has not been subjected to the peer and policy review of the agency and therefore does not necessarily reflect its views.

REFERENCES:

Oberdorster, G., E. Oberdorster & J. Oberdorster (2005) Nanotoxicology: An emerging discipline evolving from studies of ultrafine particles. *Environmental Health Perspectives*, 113 (7), 823-839.

Colvin, V.L. (2003) The potential environmental impact of engineered nanomaterials. *Nature Biotechnology*, 21 (10), 1166-1170.

Zhang, S.J., T. Shao, S.S.K. Bekaroglu, and T. Karanfil (2009) The impacts of aggregation and surface chemistry of carbon nanotubes on the adsorption of synthetic organic compounds. *Environ. Sci. Technol.* 43 (15), 5719-5725.

Dissolution of Long Carbon Nanotubes in Superacids

Nicholas G. Parra-Vasquez, Chemical and Biomolecular Engineering, Rice University, Houston, TX

Natnael Behabtu, Chemical and Biomolecular Engineering, Rice University, Houston, TX

Micah J. Green, Chemical and Biomolecular Engineering, Rice University, Houston, TX

Colin C. Young, Chemical and Biomolecular Engineering, Rice University, Houston, TX

Cary L. Pint, Smalley Institute, Rice University, Houston, TX

Virginia Davis, Chemical Engineering, Auburn University, Auburn, AL

Ellina Kesselman, Chemical Engineering, Technion-Israel Institute of Technology, Haifa 32000, Israel

Judith Schmidt, Chemical Engineering, Technion-Israel Institute of Technology, Haifa 32000, Israel

Yachin Cohen, Chemical Engineering, Technion-Israel Institute of Technology, Haifa 32000, Israel

Robert H. Hauge, Smalley Institute, Rice University, Houston, TX

Yeshayahu Talmon, Chemical Engineering, Technion-Israel Institute of Technology, Haifa 32000, Israel

Matteo Pasquali, Chemical and Biomolecular Engineering, Rice University, Houston, TX

Controlling the phase behavior of single-walled carbon nanotubes (SWNTs) in fluids is critical for the production of SWNT-based macroscopic materials. Yet, dissolution has been a major hurdle for researchers over the past decade. Here we show that SWNTs form true thermodynamic molecular solutions in superacids. In particular, SWNTs dissolve spontaneously in chlorosulfonic acid as isotropic solutions concentrations a thousand fold higher than those previously found in other solvents; at higher concentrations, they form a liquid crystal. We also show that such superacids can even be used to dissolve long (hundreds of microns) SWNTs and multi-walled carbon nanotubes; these solutions can also form liquid crystals. Dissolution of such large nanotubes had previously been thought impossible. The liquid crystalline phase is readily processed into fibers and sheets, whose morphology is controlled by the starting fluid phase.

See more of this Session: Carbon Nanotubes V: Adsorption and Transport
See more of this Group/Topical: Nanoscale Science and Engineering Forum

Fuel Cell Electrocatalysts Obtained From Bimodal Nano-Porous Templates

Svitlana Pylypenko, University of New Mexico, Albuquerque, NM

Tim S. Olson, University of New Mexico Chemical Engineering, Albuquerque, NM

Dimiter Petsev, Chemical and Nuclear Engineering, University of New Mexico, Albuquerque, NM

Plamen Atanassov, Department of Chemical Engineering, University of New Mexico, Albuquerque, NM

The development of hierarchically-structured electrocatalysts and their supports can effectively address some of the current limitations of fuel cells. It is highly desirable to decrease Pt loading, while increase the obtainable power densities and improve durability of fuel cells. One way to increase the current density is through designing of the electrocatalyst not only with high surface area but also with high amount of accessible three-phase sites. Porous structures can effectively minimize transport limitations, thus increasing the accessibility of the active sites by gas and electrolyte phases. The durability requirements of the fuel cells can be met through improvement of the corrosion stability of the current supports and design of novel corrosion resistant supports. The procedure discussed here allows to create materials that will address both, mass-transport and durability limitations of the fuel cells.

Hierarchical bimodal porous silica templates were obtained through oil/water/surfactant microemulsion templating. The first type of pores is in the range of 10-40 nm and is determined by dimensions of microemulsion droplets. The second type of pores is around 5 nm and is determined by micellar dimensions.

Silica was used as a template to make a range of materials. First, carbon supported platinum electrocatalysts will be discussed. The role of the small nanopores here is to lock Pt particle size in the range of 3-5 nm. Next, non-noble electrocatalysts based on pyridinic and cobalt precursors will be discussed. Here, small nanopores of the silica template play role of sites for creation of Me-Nx centers. Large nanopores are channels necessary for complete infiltration of the template with precursor materials and formation of the catalyst and/or catalyst support backbone. Synthesis conditions, such as amount of precursors, temperature of pyrolysis and conditions of removal of the silica template play an important role in formation of porous structure of electrocatalyst. Ability to control these conditions enables to create electrocatalytic materials with better transport properties.

See more of this Session: Nanomaterials for Energy Production and Fuel Cells II
See more of this Group/Topical: Topical 5: Nanomaterials for Energy Applications

Synthesis of Pt-Modified Tungsten Monocarbide and Its Activity towards the Hydrogen Evolution Reaction

Daniel V. Esposito, Chemical Engineering, University of Delaware, Newark, DE

Kevin D. Dobson, Institute of Energy Conversion, University of Delaware, Newark, DE

Brian E. McCandless, Institute of Energy Conversion, University of Delaware, Newark, DE

Robert W. Birkmire, Institute of Energy Conversion, University of Delaware, Newark, DE

Jingguang G. Chen, Chemical Engineering, University of Delaware, Newark, DE

Hydrogen is a promising energy carrier for the future because it represents a means for carbon-free energy storage and can be produced from biomass or water using renewable energy. When produced from water in an electrolyzer or photoelectrochemical cell (PEC), H_2(g) is evolved at the cathode via the hydrogen evolution reaction (HER) while water is split at the anode to produce O_2(g). For electrolysis in acidic solutions, the best HER catalysts are precious metals such as Pt, Rh, and Pd [1], but their prohibitively high costs and limited supplies are major hurdles for the widespread production of H_2 by this means. One approach for overcoming this challenge is to disperse the active precious metal catalyst in sub-monolayer or monolayer coverages on a low-cost, stable, supporting catalyst.

In this work, we investigate the performance of tungsten monocarbide (WC), modified by sub-monolayer amounts of Pt, as a HER catalyst. WC is an attractive material for this purpose because it has shown an appreciable activity for the HER [2], is known for its electrocatalytic stability in acidic solutions [3], and has demonstrated synergistic effects when combined with Pt as an electrocatalyst [4]. In recent years, several research groups have reported high HER activity for Pt-modified tungsten carbides in acidic solutions [5],[6], but these carbides are phase mixtures of WC, WC_{1-x}, and W_2C. One of the goals of our research is to distinguish between the effectiveness of these various carbide phases as HER catalysts. Initial studies have investigated the activity of phase-pure WC and Pt-WC thin film electrodes for the HER in H_2SO_4. The composition and phase purity of these films has been verified using XPS and GIXRD, and their activity towards the HER investigated using linear sweep voltammetry (LSV) and chronoamperometry. The LSV measurements show a significant improvement in WC HER activity upon modification by one monolayer of Pt, and post-electrochemistry XPS measurements indicate no loss of Pt from the WC surface. Current work is focused on investigating the effect of different coverages of Pt on the HER activity of WC and its comparison to polycrystalline Pt foils.

References

[1] H. Kita, Journal of the Electrochemical Society, 113 (1966) 1095.

[2] D.V. Sokolsky, V.S. Palanker, and E.N. Baybatyrov, Electrochemica Acta, 20 (1975) 71.

[3] M.B. Zellner and J.G. Chen, Catalysis Today, 99 (2005) 299.

[4] E.C. Weigert, A.L. Stottlemyer, and J.G. Chen, Journal of Physical Chemistry C, 111 (2007) 14617.

[5] M. Wu and M. Nie, Journal of Power Sources, 166 (2007) 310.

[6] D.J. Ham, R. Ganesan, and J.S. Lee, Int. J. of Hydrogen Energy, 33 (2008) 6865.

See more of this Session: Nanomaterials for Energy Production and Fuel Cells II
See more of this Group/Topical: Topical 5: Nanomaterials for Energy Applications

Novel Quaternary Platinum-Ru-Ni-i Alloy Based Anode Electrocatalysts for Direct Methanol Fuel Cell

Karan Kadakia, Chemical Engineering, University of Pittsburgh, Pittsburgh, PA

Prashanth Jampani Hanumantha, Chemical Engineering, University of Pittsburgh, Pittsburgh, PA

Nicolaus L. Rock, Mechanical Engineering & Materials Science, University of Pittsburgh, Pittsburgh, PA

Moni Kanchan Datta, Mechanical Engineering & Materials Science, University of Pittsburgh, Pittsburgh, PA

Prashant Kumta, Chemical Engineering, University of Pittsburgh, Pittsburgh, PA

Direct methanol fuel cells (DMFC) are of interest as suitable power sources for consumer electronic devices, remote and auxiliary power units for transportation. The main advantage of DMFC is the use of liquid methanol fuel rather than gaseous hydrogen in proton exchange membrane fuel cells (PEMFC). The ability to store and handle liquid better than gaseous hydrogen makes it a viable energy source for mobile and portable applications. The energy density of methanol and the amount of energy contained in a given volume of methanol is an order of magnitude greater than even compressed hydrogen. However, the currently achieved power densities of DMFC are still significantly below PEM and other fuel cell types. This is mainly related to the relatively sluggish kinetics of the electrochemical reactions occurring on the catalyst layers at the electrode surface combined with the methanol crossover problem to the cathode compartment. There is therefore a critical need to research and develop new and improved anode and cathode electro-catalysts with higher catalytic activity in order to enhance the efficiency and power density of DMFC, while lowering the precious metal loadings to realize the much desired cost savings to match the performance.

The nominal composition Pt-50at.% Ru is considered the most suitable anode catalyst composition yielding the best catalytic activity for use in DMFCs. Improved anode catalysts can be obtained by synthesizing high surface area Pt-Ru based alloy catalysts combined with engineered proper alloy design by modifying the chemical composition of the alloy, and doping or alloying with other elements onto the catalyst structure. Ternary and quaternary alloys with transition metals have been explored as possible alternatives to pure Pt-Ru electro-catalyst. The addition of other non-noble transition metals could help potentially reduced the cost while providing similar electrocatalytic activity. It has also been proposed that addition of other alloying elements could reduce materials loss and particle ripening during cycling. Ni-Zr and Ni-Ti have previously been shown to be electrochemically active and stable under anode conditions. Recently, sputter deposited Pt-Ru-Ni-Zr thin films have been shown to exhibit promising catalytic activity. In the present work, we report bulk synthesis of nanocrystalline Pt-Ru-Ni-Ti based quaternary alloys, for use as DMFC anodes, by a facile wet chemical synthesis route. X-ray results show complete alloying and formation of phase pure Pt(Ru,Ni,Ti) solid solution, and also nanocrystalline nature of the alloy. Alloy materials have been shown to have moderate BET surface areas (~70m2/g). Electrochemical characterization has been performed on a three electrode cell using a solution containing 1 M methanol and 0.5 M sulfuric acid as an electrolyte while also serving as a source of the fuel. Cyclic voltammetry and polarization results indicate that the nano-crystalline Pt-Ru-Ni-Ti alloy of suitable compositions possessing good chemical homogeneity, reveal promising catalytic activity, demonstrating the potential of the Pt-Ru-Ni-Ti based electro-catalysts for direct methanol fuel cells. Results of these studies will be presented and discussed.

See more of this Session: Nanomaterials for Energy Production and Fuel Cells II
See more of this Group/Topical: Topical 5: Nanomaterials for Energy Applications

Improving Oxygen Reduction Efficiency in Direct Methanol Fuel Cells through Structured Catalyst Design

Jinhua Yang, Department of Chemical and Biomolecular Engineering, National University of Singapore, Singapore, Singapore

Jing Xu, Department of Chemical and Biomolecular Engineering, National University of Singapore, Singapore, Singapore

Mark Saeys, Department of Chemical and Biomolecular Engineering, National University of Singapore, Singapore, Singapore

Jim Yang Lee, Department of Chemical and Biomolecular Engineering, National University of Singapore, Singapore, Singapore

We have adopted a rational approach to the design of fuel-tolerant oxygen reduction catalysts for direct methanol fuel cells (DMFCs). Our methodology is to use structured catalysts with different core and shell components. These catalysts may be prepared by simple galvanic replacement reactions which can be scaled up for volume production. Density functional theory (DFT) calculations of changes in the d-band centers of the surface catalytic metals; and the strengths of adsorbed species are used to gain some insights into the relationship between surface structure and catalytic activity of these heterogeneously structured catalysts. We found that carbon-supported core-shell PdM@PdPt catalysts (where M=Ni, Co, Fe and Cr) can provide us with the greatest latitude in activity and selectivity tuning. A volcano-curve was obtained where maximum ORR activities occur around the compositions PdCo@PdPt/C and PdFe@PdPt/C. It is believed that the Pt d-band centers in these catalysts are at values representing a good balance between O-O bond dissociation and the removal of adsorbed OH groups on the catalyst surface. Compared with commercial Pt/C catalysts, the PdM@PdPt catalysts utilize less platinum, and yet display favorable catalytic outcomes in terms of ORR activity and selectivity (methanol tolerance).

See more of this Session: Nanomaterials for Energy Production and Fuel Cells II
See more of this Group/Topical: Topical 5: Nanomaterials for Energy Applications

PdFe Nanorods as Highly Active Electrocatalysts for Ethanol Oxidation in Alkaline

Kamonwad Yangyuenthanasan, Materials Science and Engineering, Michigan Technological University, Houghton, MI

Zhiyong Zhang, Chemical Engineering, Michigan Technological University, Houghton, MI

Wenzhen Li, Chemical Engineering, Michigan Technological University, Houghton, MI

Palladium (Pd) and its alloys have been demonstrated promising electrocatalytic activity towards ethanol oxidation reaction (EOR) in alkaline [1]. Recently, we have successfully synthesized PdFe-NR with high oxygen reduction reaction (ORR) in acid electrolyte (both in three-electrode-cell and single proton exchange membrane fuel cell PEMFC) [2], based on a modified synthesis approach that was firstly reported by Sun et al. [3]. In this presentation we report a convenient organic phase reduction synthesis approach to small diameter PdFe nanorods (PdFe-NR), which show very high EOR activity in alkaline.

The synthesis of PdFe-NR was obtained through chemical reduction of Pd(acac)2 and thermo-decomposition of Fe(CO)5 in oleyamine (OAm) at 160oC. The EOR activity of PdFe-NR was tested in a conventional three-electrode-cell by recording their cyclic voltammetry (CV) curves. The electrolyte was 0.1M NaOH + 0.5 M CH3CH2OH. It is found that the EOR activity of the PdFe-NR-carbon nanotubes (CNTs) (PdFe-NR deposited on CNT covered glassy carbon electrode GCE) exhibits higher mass activity than Pt/C and PtRu/C, although its onset potential of EOR remains lower than PtRu/C. TEM images show that the PdFe-NR has a mean diameter of 2.5 nm and length of 23 nm. Other Pd-based nanocatalysts, such as CNT supported PdNi, PdCo, etc, and more characterizations, such as XRD, TGA, XPS, etc are currently under working in our lab to achieve better EOR activity and attain an insight to the activity enhancement.

References:

[1]Shen et al, Electrochemical Communications, 2006, 8, 184-188.

[2]Li et al, Electrochemical Communications, 2009, doi:10.1016/j.elecom.2009.03.046.

[3]Wang et al. Angewandte Chemie-International Edition, 2007, 46, 6333-6335. ->

See more of this Session: Nanomaterials for Energy Production and Fuel Cells II
See more of this Group/Topical: Topical 5: Nanomaterials for Energy Applications

Nitrogen Substituted Carbon Nanotubes as An Electrode Support for Pt in Formic Acid Electrooxidation

Sankaran Murugesan, Chemical & Metallurgical Engineering, University of Nevado at Reno, Reno, NV

Kirby Myers, Chemical & Metallurgical Engineering, University of Nevado at Reno, Reno, NV

Ravi Subramanian, Chemical & Metallurgical Engineering, University of Nevado at Reno, Reno, NV

Electrode supports play an important role in effective anchoring and dispersion of Pt metal in electrooxidation of fuels such as methanol and formic acid. The electrode supports reduces the sintering of Pt particles and show high influence in stable operation of fuel cells. At present, commercially Vulcan XC-72 carbon has been used as an electrode support due its properties such as high specific conductivity and low double layer charge compared to other carbon electrode supports. Recently, carbon in nanomaterials such nanotubes, nanofibres, and nanosheets have attracted great interest. Such electrode supports demonstrate different structural, physical, and electronic properties by the arrangement of carbon network. The surface properties of carbon nanotubes, for example, can be altered by chemical treatment or substitutional doping of heteroatoms. This work focuses on the effect of nitrogen substitution in carbon nanotubes (NCNT) as a support for Pt anchoring in formic acid electrooxidation. Synthesis of nitrogen containing carbon nanotubes and how they can help in anchoring Pt in preferential orientation is presented. The results of NCNTs for formic acid electrooxidation have been compared with the other commercial electro catalyst (HiSpec 3000) and commercial support Vulcan-XC 72. The results of this work play a significant role in the development of fuel cell electrodes.

See more of this Session: Nanomaterials for Energy Production and Fuel Cells II
See more of this Group/Topical: Topical 5: Nanomaterials for Energy Applications

Electrically Conductive and Super-Hydrophilic Bipolar Plate Coatings Prepared From Mixed Aqueous Suspensions of Graphene Sheets and Silica Nanospheres

Feng Wang, Chemical and Materials Engineering, University of Cincinnati, Cincinnati, OH

Indrajit Dutta, Chemical and Materials Engineering, University of Cincinnati, Cincinnati, OH

Ruidong Yang, Chemical and Materials Engineering, University of Cincinnati, Cincinnati, OH

Mubarak Alazemi, Chemical and Materials Engineering, University of Cincinnati, Cincinnati, OH

Anastasios Angelopoulos, Chemical and Materials Engineering, University of Cincinnati, Cincinnati, OH

Interfacial chemical domain size and distribution can have a significant influence on the electronic properties of composite coatings. In this paper, we demonstrate how the theoretical predictions of the Gouy-Chapman model for electrostatic interactions between nanoparticles may be used to control heterogeneous chemical domain size and distribution at interfaces. Specifically, we demonstrate through high resolution SEM, TEM, and AFM imaging how the ionic strength, dielectric constant, and pH of suspending media may be systematically altered to vary the three-dimensional distribution of hydrophobic graphene sheets (in the form of ~5 nm thick graphite platelets) and hydrophilic silica nanospheres (~60 nm diameter) simultaneously and irreversibly deposited onto metallic surfaces from mixed suspension utilizing a polymeric binder. The process is used to alter the water wetting properties of bipolar plate materials employed in proton exchange membrane (PEM) fuel cells while maintaining low electrical contact resistance. Such composite coatings are shown to yield advancing contact angles of not more than 10 degrees while maintaining an electrical contact resistance below 5 mΩ-cm^2 for ~100 nm adsorbed layer thickness. This result represents an order of magnitude reduction in contact resistance relative to that of pure silica coatings of comparable thickness prepared utilizing chemical vapor deposition.

See more of this Session: Nanomaterials for Energy Production and Fuel Cells II
See more of this Group/Topical: Topical 5: Nanomaterials for Energy Applications

Effect of Carbon Nanotube Concentration and Functional Group Length On the Properties of Melt Extruded Polypropylene Nanocomposites

Vinod K. Radhakrishnan, Department of Chemical Engineering, Auburn University, Auburn, AL

Virginia A. Davis, Department of Chemical Engineering, Auburn University, Auburn, AL

The chemistry and concentration of the nanotubes have a significant effect on the thermal stability and mechanical properties of the final melt extruded nanocomposite. We report the influence of functional group length and nanotube concentration on the properties of polypropylene (PP) nanocomposites containing SWNTs, or functionalized SWNTs. Functionalization can improve dispersion by increasing steric hindrance between tubes and improving interfacial interactions between the SWNTs and polymer chains. However, functionalization also deteriorates SWNT properties and introduces impurities that can catalyze polymer degradation. The nanotubes and PP were mixed by spray mixing followed by melt extrusion. The level of nanotube dispersion and nanotube-polymer interaction were evaluated by a combination of microscopy, rheology, Raman spectroscopy, thermal and mechanical analyses.

See more of this Session: Nanoscale Structure in Polymers III: Polymer Nanocomposites
See more of this Group/Topical: Nanoscale Science and Engineering Forum

Optimization of Tailored Multifunctional Nanocomposite Structures

Timothy Shenk, Chemical and Biological Engineering, South Dakota School of Mines and Technology, Rapid City, SD

R. M. Winter, Cbe, South Dakota School of Mines and Technology, Rapid City, SD

Kenneth Benjamin, Cbe, South Dakota School of Mines and Technology, Rapid City, SD

Polymer nanocomposites provide unique solutions to industrial and scientific applications where weight must be minimized and functionality maximized. Researchers are interested in improving the ability to tailor a product to meet specific weight, thermal, optical, mechanical and electrical requirements. Historically functional composite structures have been realized through a top-down approach. With the advent of atomic level measurement tools and experimental techniques a bottom-up approach to the creation of multifunctional structures is receiving intense study. We are developing unique multifunctional structures using such a bottom-up approach with the intent of developing molecular simulations to guide such a process. Properties of polymeric nanocomposite structures are tailored and optimized through a fundamental understanding of intermolecular forces. While macroscopic models of bulk properties of polymer nanocomposites have been characterized, much less is known on the dynamics of their interfacial characteristics, which must be fully developed in order to be able to tailor fabrication of multifunctional nanocomposites using the bottom-up approach. Self assembly, targeted functionalization and spin coating are used to provide consistent means of creating multilayer multifunctional thin film composite strictires allowing for the investigation of multifunctional composites. We investigate the affects of sonication, high shear mixing, and surface modification on the ability to control dispersion to create and predict multifunctional layers of epoxy spin coated nanocomposites and control desired mechanical properties such as conductivity, optical transmission and absorption, loss and storage moduli and coefficient of thermal expansion. The Interfacial Force Microscope (IFM), coupled with the Biaxial Loading Instrument, which allows for refined and uncoupled control of mixed-mode characterization of interface adhesion, will be used to determine interfacial characteristics of these multi-functional composites.

See more of this Session: Nanoscale Structure in Polymers III: Polymer Nanocomposites
See more of this Group/Topical: Nanoscale Science and Engineering Forum

Cellulose Acetate-Laponite® RD Nanocomposites: An Investigation of Model Materials for Advanced Organic-Inorganic Membranes

J.R. Johnson, School of Chemical & Biomolecular Engineering, Georgia Institute of Technology, Atlanta, GA

W.J. Koros, School of Chemical & Biomolecular Engineering, Georgia Institute of Technology, Atlanta, GA

Membranes offer attractive alternatives to traditional thermally-driven separations. Polymeric materials offer attractive, low-cost processing; however, more selective and productive membranes would provide new application opportunities. Multi-component membranes utilizing organic and inorganic materials offer opportunities to achieve these goals.

Hybrid membranes incorporating traditional zeolite materials, with unit aspect ratios, limit options to create thin selective layers. Recent studies have considered high aspect ratio sieving materials (i.e. flakes or platelets) with significantly higher aspect ratio (L/D above 10) rather than traditional zeolites. These higher aspect ratios increase the tortuous path of permeation for rejected species, so a low volume fraction of platelet materials can be used to produce high performance membranes.

Incorporating nanoplatelet materials requires a high degree of exfoliation and complete particle dispersion to achieve optimal transport properties. Unfortunately, traditional melt-blended nanocomposite techniques are not directly applicable because high performance membranes are produced from solution. Our work focuses on addressing the challenges of preparing a composite mixture for membrane formation using non-thermal methods.

We have chosen cellulose acetate as the organic matrix because it is well-studied and industrially relevant as a material for gas separations membranes. Laponite® RD was chosen as the inorganic filler because its geometry ($D < 1\mu m$) is favorable for the transition to the production of asymmetric hollow fiber membranes. In this presentation we will discuss our non-thermal technique for producing composite solutions for membrane/barrier applications. Furthermore, we will discuss the transport properties of the dense-film membranes produced from these solutions and compare these results with common transport models.

See more of this Session: Nanoscale Structure in Polymers III: Polymer Nanocomposites
See more of this Group/Topical: Nanoscale Science and Engineering Forum

Biomimetic Nanocomposites Fabricated by Controlled Assembly of Nanoplatelets

Inkook Jun, Material Science and Engineering, University of Florida, Gainesville, FL

Wei-Han Huang, Cheimcal Engineering, University of Florida, Gainesville, FL

Tzung-Hua Lin, Cheimcal Engineering, University of Florida, Gainesville, FL

Peng Jiang, Chemical Engineering, University of Florida, Gainesville, FL

The nacreous layer of mollusk shells is both strong and stiff. This is attributed to the unique three-dimensional (3D) assembly of platelet-like ceramic building blocks which are embedded in a polymer matrix. Layer-by-layer (LBL) assembly of clay nanoplates and polyelectrolytes has been developed to make polymer nanocomposites that mimic the natural structures and mechanical properties of nacres. However, the LBL approach is tedious in assembling thick coating. Here we report the rapid fabrication of large-area nacreous coatings by utilizing capillary and electrical forces. Hexagonal gibbsite nanoplatelets with ~200 nm diameter and ~10 nm thickness can be assembled into ordered multilayers with controlled thickness. The interstitials between gibbsite particles can be filled up by polymers and metals. The resulting biomimetic nanocomposites exhibit notable stiffness and strength.

See more of this Session: Nanoscale Structure in Polymers III: Polymer Nanocomposites
See more of this Group/Topical: Nanoscale Science and Engineering Forum

Dynamic Relaxation Properties of Polymer Nanocomposites

Anthony C. Comer, Chemical and Materials Engineering, University of Kentucky, Lexington, KY

Alexander L. Heilman, Chemical and Materials Engineering, University of Kentucky, Lexington, KY

Douglass S. Kalika, Chemical and Materials Engineering, University of Kentucky, Lexington, KY

The introduction of nanoscale filler particles into a polymer matrix results in the creation of vast amounts of polymer-particle surface area and the potential for physical confinement of the polymer chains at the nanoscale. Polymer-particle interactions, as well as confinement effects, can substantially alter the local and segmental mobility of the matrix polymer chains, as reflected in the dynamic relaxation properties of the nanocomposite. In this study, we explore the sub-glass and glass-rubber relaxation characteristics of a series of model glassy polymer nanocomposites based on poly(etherimide) [PEI] and poly(methyl methacrylate) [PMMA] in combination with silica (SiO_2) nanoparticles.

Evaluation of the thermomechanical properties of polymer nanocomposites, and their correlation with filler loading, dispersion and the quality of the polymer-particle interface is an area of intense interest. In filled polymers, the inclusion of inorganic particles can lead to an increase in measured glass transition temperature (Tg) owing to interactions and potential adsorption phenomena that limit segmental mobility. Conversely, in systems that exhibit poor wetting characteristics, decreases in Tg have been reported. In the current work, we examine a series of PEI and PMMA nanocomposites formulated with varying levels of native and modified SiO_2 nanoparticles. Dynamic mechanical analysis and broadband dielectric spectroscopy are used to elucidate polymer chain dynamics across the glass-rubber relaxation in relation to nanocomposite morphology, and the nature of the polymer-particle interaction.

The inclusion of native SiO_2 nanoparticles (~10 nm diameter; up to 30 wt% loading) in the PEI matrix had only a very modest influence on the glass transition properties of the resulting materials; a 'bulk' Tg of ~225°C (1 Hz) was observed for all samples, essentially independent of polymer loading. The introduction of various surface-modified (i.e. hydrophobic) fumed silicas led to enhanced particle dispersion, and also to the emergence of a dual-Tg response that encompassed a primary glass transition at 225°C, as well as a second Tg event offset upwards by 60°C to 100°C. This high-temperature transition appeared to correspond to a distinct glass-rubber relaxation originating from regions of reduced mobility in the vicinity of the polymer-particle interface. The relative intensities of the two glass transitions, as well as the position and breadth of the high-temperature relaxation, correlated with the degree of particle loading and the character of the nanoparticle surface. Complementary dielectric spectroscopy measurements were undertaken to further establish the morphological basis for the observed dual-Tg response, and to probe the influence of the nanoparticles on the dynamic relaxation characteristics of the composites across the sub-glass region. Parallel studies on PMMA-SiO_2 composites indicated similar response properties, as related to the underlying polymer-particle compatibility.

See more of this Session: Nanoscale Structure in Polymers III: Polymer Nanocomposites
See more of this Group/Topical: Nanoscale Science and Engineering Forum

The Effect of Polymer Grafting of Polymers On Single-Walled Carbon Nanotubes (SWcNTs) On Glass Transition Behavior in Polymer Composites

Brian P. Grady, School of Chemical, Biological and Materials Engineering, University of Oklahoma, Norman, OK

Warren T. Ford, Department of Chemistry, Oklahoma State University, Stillwater, OK

Abhijit Paul, Chemistry, Oklahoma State University, Stillwater, OK

The addition of single-walled carbon nanotubes to a polymer generally increases the Tg of the polymer slightly; which is certainly not a very unique or surprising phenomena. This shift is accompanied by a decrease in the heat capacity change at the glass transition, due to the formation of a rigid amorphous fraction. However, quite interestingly, a recent publication submitted by our group also indicates an increase in the heat capacity change at the glass transition temperature at high nanotube contents. In this work, these results will be presented along with any changes that are seen when a polymer of lower molecular weight than the matrix polymer is chemically bonded to the carbon nanotubes.

See more of this Session: Nanoscale Structure in Polymers III: Polymer Nanocomposites
See more of this Group/Topical: Nanoscale Science and Engineering Forum

Nanotechnology for Tissue Engineering: A State of the Art Review

Thomas J. Webster, Division of Engineering and Department of Orthopeadics, Brown University, Providence, RI

This review will summarize studies which have demonstrated enhanced in vitro and in vivo tissue growth on nanostructured metals, ceramics, polymers, and composites thereof compared to currently-used implants. Tissue growth to be emphasized will consist of bone, cartilage, vascular, skin, bladder, nervous system, and other tissues. These results strongly imply that nanomaterials may improve tissue regeneration compared to what is being implanted today. This review will also focus on a fundamental explanation why tissue growth is enhanced on nanostructured compared to conventional tissue engineering materials summarizing a wide range of research efforts, including those which have already received FDA approval for implantation. Thoughts on the necessary future studies for the field of nanotechnology and tissue engineering to progress will also be presented.

See more of this Session: Nanostructured Scaffolds for Tissue Engineering
See more of this Group/Topical: Nanoscale Science and Engineering Forum

Nanorods for Preventing Normal and Tumor Cell Adhesion

Jiyeon Lee, Department of Chemical Engineering, University of Florida, Gainesville, FL

Byung Hwan Chu, Department of Chemical Engineering, University of Florida, Gainesville, FL

Fan Ren, Department of Chemical Engineering, University of Florida, Gainesville, FL

Anand Gupte, Division of Gastroenterology, Department of Medicine, University of Florida, Gainesville, FL

Tanmay Lele, Department of Chemical Engineering, University of Florida, Gainesville, FL

Cell interactions with nanostructures are of broad interest because of applications in controlling tissue response to biomedical implants. A key goal is to prevent mammalian cell adhesion and proliferation on implanted devices. In this study, we explored the potential of densely coated nanorod monolayers for controlling normal and tumor cell adhesion. Densely packed silicon dioxide (SiO_2) coated nanorods were fabricated with a solution-based technique. Fibroblasts, vascular endothelial cells and esophageal tumor epithelial cells were unable to adhere on dense nanorods. Cells could not assemble vinculin-marked focal adhesions, and were poorly spread. Cell survival in adherent cells was reduced by more than 100-fold on nanorods. When nanorods were spatially patterned on micron-length scales with intervening flat areas, cells preferred to stay on flat areas, but migrated frequently over nanorods to other flat regions. Our results support a model in which nanorods interfere with integrin clustering at the nanoscale, thereby preventing cell spreading and promoting cell death. We propose that dense monolayers of nanorods are a promising nanotechnology for preventing mammalian cell fouling of biomaterials.

See more of this Session: Nanostructured Scaffolds for Tissue Engineering
See more of this Group/Topical: Nanoscale Science and Engineering Forum

Selenium Nanocluster Coatings for Anti-Cancer, Anti-Bacterial Orthopedic Applications

Phong A.Tran[1], Erik Taylor[2], Love Sarin[2], Robert H. Hurt[2], and Thomas J. Webster[3]
[1] Physics Department, Brown University, Providence, RI 02912
[2] Division of Engineering, Brown University, Providence, RI 02912
[3]Division of Engineering and Department of Orthopeadics, Brown University, Providence, RI 02912

Abstract - Two common causes leading to bone implant failure are: insufficient bonding between the implant and surrounding bone and infection. Patients with implant failure often undergo revision surgeries which are costly and painful. Moreover, for patients who receive orthopedic implants after cancerous tissue is removed, it would be beneficial to implant an anti-cancer material which can impede the return of cancerous tissue growth that may develop from cancer cells not removed during surgery. Therefore, the objective of this in vitro study was to create a coating material that can: (i) promote healthy, normal bone growth, (ii) inhibit bacterial attachment and (iii) impede cancer growth. To achieve that objective, titanium, a conventional orthopedic implant material, was coated with selenium (Se) nanoclusters. Different coating densities were achieved by varying Se concentration in the reaction mixture. Se nanocluster coatings were shown to enhance healthy osteoblast (bone-forming cell) functions and inhibit cancerous osteoblast (osteosarcoma) proliferation in either separate culture experiments or co-culture experiments. Moreover, *S. epidermidis* (one of the leading bacteria that infects implants) functions were inhibited on materials coated with Se-nanoclusters compared to uncoated materials. Therefore, a new orthopedic implant coating material is introduced here that may be ideal for promoting bone growth and inhibiting cancer as well as bacteria growth.

I. INTRODUCTION

Cancer can return to an implant site after surgical resection and implant insertion into bone cancer patients due to the fact that cancer cells are not always completely removed. However, currently used materials for bone implants (such as Ti) do not have any mechanisms to prevent cancer reoccurrence. In addition, infection is one of the key problems all implants today face. Infection occurs when bacteria from the outside environment (such as from the patient's own skin, surgical instrument, etc.) attach to implant surfaces. Therefore, the objective of this in vitro study was to create an implant coating material which can inhibit cancer growth and bacterial functions at the implant surface but still promote healthy bone growth. For this purpose, Ti was coated with nanoclusters of Se which is an anti-cancer element [1,2]. This composite material was shown to improve healthy osteoblast (bone-forming cell) functions, inhibit cancerous osteoblast functions and reduce *S. epidermidis* functions.

II. MATERIALS AND METHODS

Material synthesis

Clean Ti substrates (Alfa Aesar) were used for Se nanocluster coatings. The substrates were placed in a 4:1 molar mixture of glutathione and sodium selenite solutions. Glutathione reduces sodium selenite to form seleno-diglutathione which further disintegrates as the pH is increased above 7 releasing elemental Se into solution. Various concentrations of the reactants (Table 1) were used in the reaction mixture to achieve variable amounts of Se nanocluster density on the Ti surfaces.

Table 1
Reagent volumes used in the colloidal synthesis of Se nano-clusters in the presence of Ti substrates

Reagent	Preparation Method *[Se] = 0.42 mM "Low Dose"	[Se] = 5 mM "Medium Dose"	[Se] = 11.7 mM "High Dose"
Deionized water	14.5 ml	9 ml	1 ml
100 mM GSH	0.25 ml	3 ml	7 ml
25 mM Na_2SeO_3	0.25 ml	3 ml	7 ml
Final Volume	15 ml	15 ml	15 ml

*[Se] = Final concentration of selenium in the colloidal synthesis solution

The substrates were exposed to UV light for 24 hrs on each side to sterilize them. Scanning electron microscopy (SEM) and atomic force microscopy (AFM) were used to characterize the substrates.

Co-culture assays

Since previous results [3] demonstrated increased healthy osteoblast density and decreased cancerous osteblast density with increasing Se coating density, in this study both types of cells were co-cultured on either uncoated Ti or Ti coated with the highest Se density. Healthy primary osteoblasts (HCO, ScienCell Research Laboratories) and cancerous osteoblasts (CLR-2837, ATCC) were pre-stained with fluorescent dyes Vybrant DiO and Vybrant DiD (Molecular Probes), respectively, before seeding. Cells were seeded at 750 cells/cm^2 per type in media consisting of 50% Dulbecco's Modified Eagle Media (DMEM) (supplemented with 10% fetal bovine serum (Hyclone) and 1% Penicillin / Streptomycin (Hyclone)) and 50% osteoblast medium (ObM, ScienCell Research Laboratories) in an incubator under standard cell culture conditions (37°C, 5% CO_2, 95% humidified air). Media was exchanged and cells were rinsed with PBS and observed using fluorescence microscopy at 4, 17, 24, 40, 53 and 65 hrs. Cells were counted under fluorescence microscopy in five random fields and were averaged for each substrate.

Bacterial assays

Bacteria used were *S. epidermidis* #35984 obtained in freeze-dried form (American Type Culture Collection, Manassas, VA). The dry pellet was rehydrated in 6 ml Luria broth (LB) consisting of 10 g tryptone, 5 g yeast extract, 5 g NaCl (all obtained from Sigma, St. Louis, MO) per liter double distilled water with pH adjusted to 7.4. The solution was incubated (37 °C, 5% CO_2, humidified environment) and agitated until the bacteria reached late stationary phase (about 24 hours). The second passage was diluted at a ratio of 1:200 into fresh LB and incubated until late stationary phase whereby it was mixed with equal proportions of 50% glycerol (Sigma) and frozen at −18 °C.

Centrifuge tubes were prepared with 3 ml LB and inoculated with a sterile 10 μL (Sigma) loop followed by agitation at 250 revolutions per minute (rpm) until the bacteria culture reached stationary phase (about 18 hours). At that point, cells were diluted in LB to an optical density of 0.52 at 562 nm using a microplate reader. This value was equivalent to 30% absorbance correlating to a three on the McFarland Scale, or 9×10^8 bacteria per ml. Bacteria were further diluted and inoculated at 3×10^6 cells per well onto substrates of interest. After 3 days of inoculation, substrates were put into 15ml centrifuge tubes followed by vortexing and sonicating to collect cells. 200 μL of each medium containing cells was placed into a well of a 96 well plate and analyzed for cell concentration using a spectrophotometer (Spectra Max, Molecular Devices). Light absorbance was measured at 562nm wavelength.

All experiments were conducted in triplicate and repeated 3 times. Statistical differences were assessed using Student's t test.

III. RESULTS

SEM images demonstrated that Se nanoclusters were clearly present on the substrates when coated at low, medium and high dose Se concentrations (Fig. 1). Se nanoclusters were approximately 80 nm in diameter.

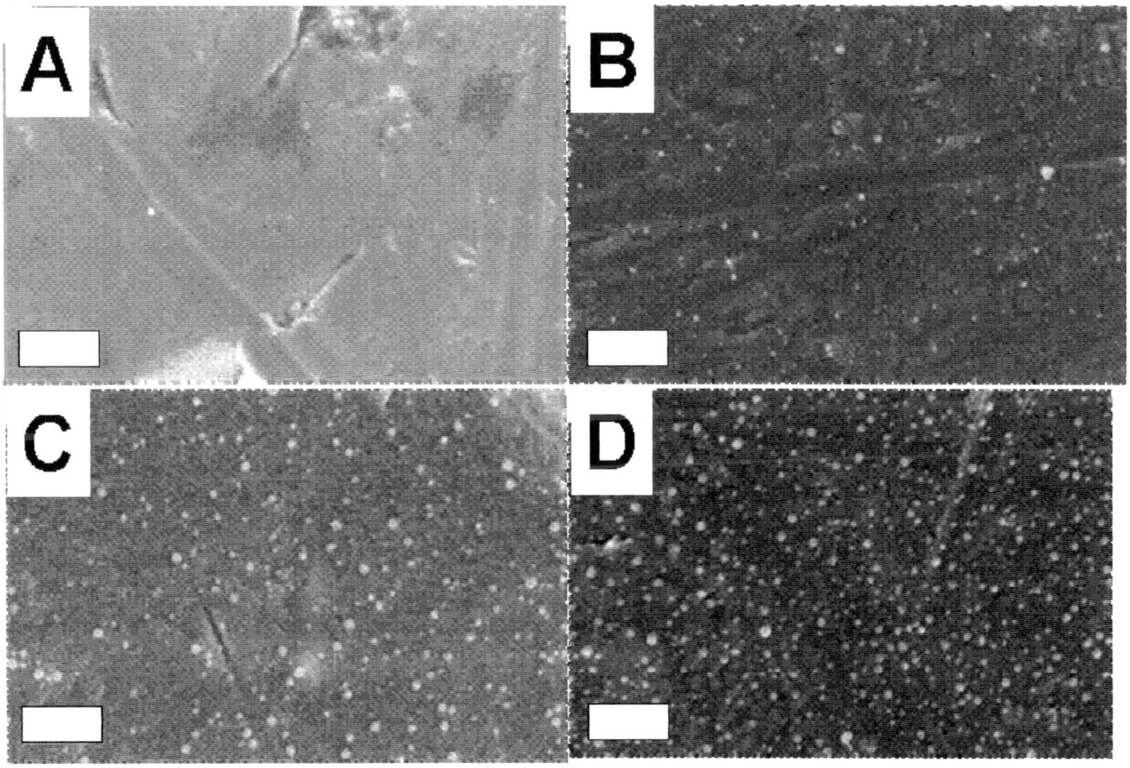

Fig. 1. SEM images of uncoated Ti (A) and Ti coated with low (B), medium (C) and high (D) doses of Se. Bars = 500nm.

The substrates will be referred to as uTi, Low-nSe-Ti, Medium-nSe-Ti and High-nSe-Ti for uncoated Ti, and low, medium and high Se nanocluster density coated Ti, respectively.

AFM analysis (Fig. 2) showed increasing nano-scale surface roughness of the substrates with increased Se density.

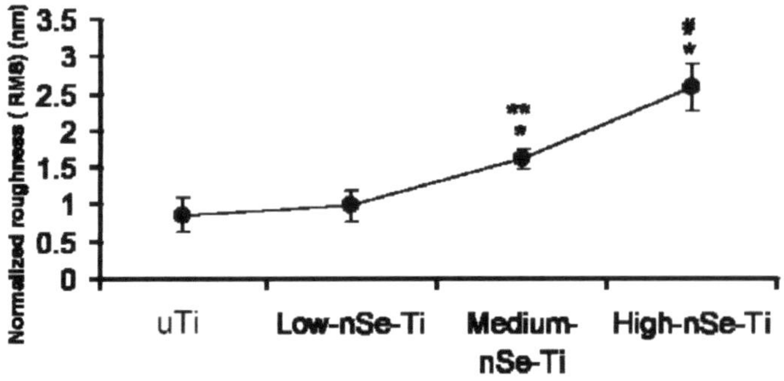

Fig. 2. Normalized RMS of the substrates (normalized to unit scanned area). Data = mean ± SEM; N=3; * $p<0.01$ compared to uTi; ** $p<0.01$ compared to Low-nSe-Ti; # $p<0.01$ compared to Medium-nSe-Ti. AFM scan size was 5μm x 5μm.

When co-cultured on the uncoated Ti substrates, osteoblast cancer cell density increased with time while healthy osteoblast density did not change significantly among the different time points (Fig. 3).

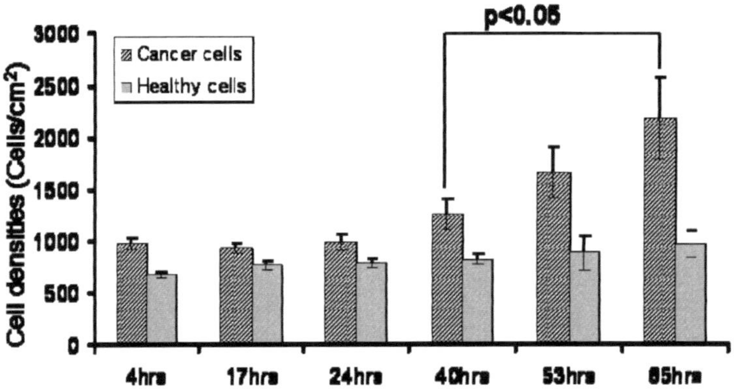

Fig. 3. Increased cancerous osteoblast density after 53 and 65 hrs of co-culturing on uTi. Healthy osteoblast density showed no significant change with time. Data = mean ± SEM; N=3.

In a contrast, significantly increased healthy osteoblast densities after 53 and 65 hrs were observed when co-culturing on High-nSe-Ti substrates compared to 40 hrs timepoint. Importantly, there was no significant change in cancerous cell densities among the different time points (Fig. 4).

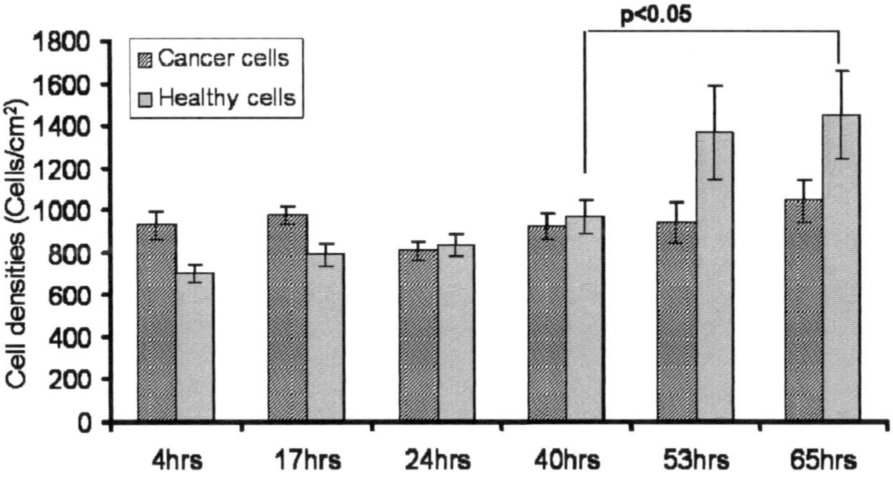

Fig. 4. Increased healthy osteoblast density after 53 and 65 hrs of co-culturing on High-nSe-Ti. Cancerous osteoblasts did not show any significant cell growth.
Data = mean ± SEM; N=3.

Bacterial inoculation results show that, after 3 days, bacterial densities were lower on Ti substrates coated with Se than uncoated Ti substrate (Fig. 5).

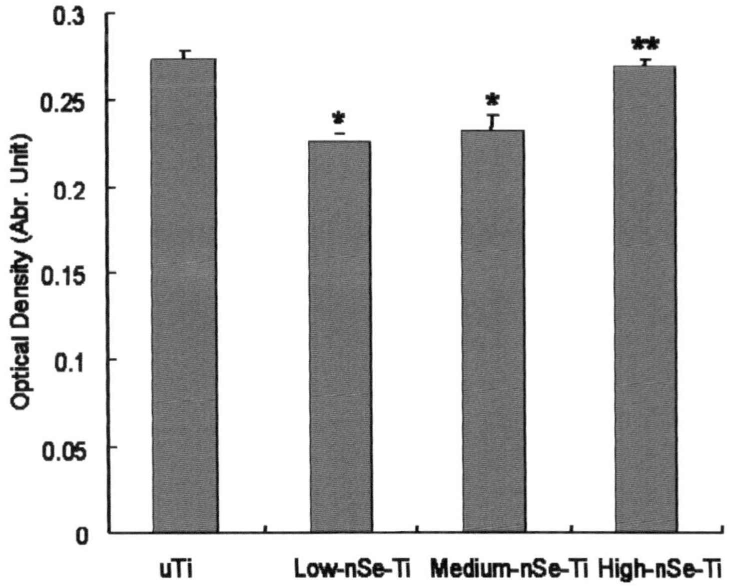

Fig. 5. Decreased *S. Epidermidis* densities after 3 days on substrates coated with Se nanoclusters.
* $p<0.05$ compared to uncoated substrate (uTi). ** $p< 0.05$ compared to Low-nSe-Ti and Medium-nSe-Ti. Bacterial density on High-nSe-Ti was lower (but not significantly) than that on uTi.
Data = mean ± SEM; N=3.

IV. DISCUSSION

The co-culture results on uncoated Ti showed an increasing trend in cancerous osteoblast density. This trend was expected since the uncoated substrates do not posses any properties to stop cancer growth. However, healthy osteoblast densities did not significantly change over time on uncoated Ti. Although this needs more investigation, this could be due to the competition for nutrients and implant space occupied by the cancerous cells that inhibited the proliferation of healthy osteoblasts.

On the other hand, the results from co-culturing cells on High-nSe-Ti (Fig. 4) showed an opposite trend: healthy cell proliferation was enhanced after 53 and 65 hrs while the proliferation of cancerous cell was inhibited during those time periods. More in-depth studies are needed to understand the role of Se nanoclusters for these cancerous cell-inhibitory and healthy cell-promoting effects. However, studies have shown increased healthy osteoblast functions on nano-structured surfaces [4,5]. Therefore, it is hypothesized that the nano-structures on the Ti surfaces brought about by the presence of Se nanoclusters are a factor that promoted healthy osteoblast proliferation.

Results from bacterial assays showed another interesting observation. All Se coatings (i.e., all three coating densities) inhibited bacterial growth. However, there was no dose-dependent effect. Particularly, substrates with the highest density of Se nanoclusters had cell densities significantly higher than substrates coated with lower Se nanoclusters. It is hypothesized that this effect was the result of two contributing factors: the amount of Se and the level of nanoscale surface roughness. More Se will inhibit more bacterial growth. However, more Se nanoclusters will also give rise to surface nanoscale roughness and associated surface area that, in turn, can increase cell adhesion. Therefore, a key future direction will be finding an optimal coating density window in which we can optimize (i) healthy bone cell functions, (ii) inhibit cancerous bone cell growth and (iii) strongly decrease bacterial functions.

V. CONCLUSIONS

This study showed that Se nanocluster coatings on Ti might be an ideal candidate for anti-cancer, anti-bacterial applications where one not only needs to inhibit cancerous growth and bacterial functions but also promote healthy bone growth.

ACKNOWLEDGEMENTS

The authors thank Anthony W. McCormick for help with AFM and the Hermann Foundation for funding.

REFERENCES

[1] L.C. Clark, et al. Effects of selenium supplementation for cancer prevention in patients with carcinoma of the skin. A randomized controlled trial. Nutritional Prevention of Cancer Study Group. *JAMA* 1996;276(24):1957-1963.

[2] L.C. Clark, et al. Decreased incidence of prostate cancer with selenium supplementation: results of a double-blind cancer prevention trial. *Br J Urol* 1998;81(5):730-4.

[3] P.A. Tran, L. Sarin, R.H. Hurt, T.J. Webster. Titanium surfaces with adherent selenium nanoclusters as a novel anti-cancer orthopedic material. *J Biomed Mater Res A.* 2009 in press.

[4] T.J.Webster, C. Ergun, R.H. Doremus, R.W. Siegel, R. Bizios. Enhanced functions of osteoblasts on nanophase ceramics. *Biomaterials* 2000;21(17):1803-10.
[5] K.L. Elias, R.L. Price, T.J. Webster. Enhanced functions of osteoblasts on nanometer diameter carbon fibers. *Biomaterials* 2002;23(15):3279-87.

Assessment of Nanomodified Endotracheal Tubes in a Bench Top Airway Model

Mary C. Machado, *Brown University, Providence RI;*
Daniel Chang, *Brown University, Providence RI;*
Keiko M. Tarquinio, *Rhode Island Hospital; Providence RI;*
Thomas J. Webster, *Brown University, Providence RI*

Abstract

Ventilator associated pneumonia (VAP) is a serious and costly clinical problem. Specifically, receiving mechanical ventilation over 24 hours increases the risk of VAP and is associated with high morbidity, mortality and medical costs. This illness is especially difficult to diagnose in children because of non-specific clinical and radiographic signs as well as a lack of sensitive diagnostic methods. Cost effective endotracheal tubes (ETTs) that are resistant to bacterial infection would help to prevent this problem.

The objective of this study was twofold, first to develop an *in vitro* airway model to simulate the dynamic conditions of the pediatric airway and secondly to assess *in vitro* bacterial adhesion or biofilm formation of clinically used ETTs and nanomodified (with surface features of less than 100 nm in at least one direction) ETTs under these dynamic conditions. In preliminary tests, nanomodified polyvinyl chloride (PVC) ETTs have been shown to be effective at reducing bacterial adhesion. This study also sought to evaluate the bacterial resistance of these ETTs more effectively by creating a bench top airway model, which can create a similar environment to the flow system that ETTs are exposed to *in vivo*.

The airway model designed to test ETTs has two plexiglas chambers representing the oropharynx and the lungs, a tube representing the trachea and finally an intricate pumping system to the oropharynx with bacteria flow and to the lung with simulated compliance and resistance. Endotracheal tubes will be connected to a ventilator and will pass through the oropharynx chamber into the trachea and will be observed under both mechanical ventilation and continuous contamination within the artificial saliva flow system.

In no less than three separate trials in the airway chamber, each ETT will be tested for its effectiveness at the reduction of bacterial growth within the airway by sampling from both lung and oropharynx chambers during continuous operation. Special attention will be given to the long-term effects on the ETT by including a study that lasts longer than five days. Both the bacterial proliferation in the two chambers and on the ETT itself will be carefully analyzed. This specialized testing should yield valuable information on the efficacy of nanomodified ETT in airway conditions and will provide further evidence to determine if nanomodified ETTs are a valid solution to VAP.

Introduction

Ventilator associated pneumonia (VAP) is a serious and costly problem. It is one of the most common causes of hospital associated infection in children and adults. This medical condition results from hospital acquired infection which results in pneumonia. *Pseudomonas aeruginosa* (*P. aeruginosa*) is the most common gram negative organism associated with VAP, but multi-drug resistant strains of bacteria are becoming an increasing problem in the treatment of VAP. Two notable strains, *P. aeruginosa* and Methicillin-resistant *Staphylococcus aureus* (MRSA) are of particular concern to clinicians because of their increasing prevalence within hospitals. Eight to 28% of all patients receiving mechanical ventilation will develop VAP. Depending on the pathogen involved, the patient's underlying condition, and the length of intubation, VAP can have a high mortality rate, ranging between 38% and 76%. VAP can also be very costly for both patients and hospitals. This condition adds an estimated $40,000 dollars to each hospital admission due to the extra intensive care required for treatment. (1)

Diagnosis of VAP is especially challenging in pediatrics. Radiographic and clinical criteria for such diagnoses are often unspecific causing delays in targeted treatment and increasing the use of a broad-spectrum of antibiotics. The diagnosis of VAP in pediatric patients is hindered further by the lack of VAP studies in children and infants.

Endotracheal tubes, one of the main sources of bacterial colonization in the airway, are essential to the process of mechanical ventilation. Like any other device implanted within the body, ETTs interact with an environment that contains not only the airway epithelium, but also potentially harmful microorganisms. Endotracheal tubes present a special concern to clinicians because they are often colonized by oropharyngeal bacteria during long-term mechanical ventilation. These tubes provide a conduit from the outside environment to the more sterile area of the lungs by impairing some of the body's natural defenses like coughing and muscociliary motion of the trachea. Moreover, injury to the epithelial cells of the trachea can result from the movement of the tube in the airway or even from the suction of secretions during nursing care. This causes "opportunistic adherence" of bacteria to the airway. Another problem caused by ETTs is chronic airway inflammation. Inflammation releases proteases that digest fibronectin. Fibronectin bound to the buccal epithelial surface has been shown to inhibit binding of certain gram negative bacteria. Fibronectin is an oposonin, mediating phagocytosis by macrophages. Recently the gel phase of mucus in the trachea has been shown to have high levels of fibronectin and this glycoprotein is thought to aid in the mechanical removal of bacteria. (2) Endotracheal tubes can also increase airway secretions creating a substrate for, *P. aeruginosa*, a bacteria that binds to receptor sites on the mucus. (3)

One of the key challenges to inhibition of bacterial growth on ETTs occurs when certain bacteria exude an exopolysaccharide, which adheres bacteria together. Bacteria in this type of extracellular matrix (known as a biofilm) are especially resistant to both antibiotics and the immune system of the patient. Ventilation through these infected tubes, or even condensation of humidified air can break off pieces of this biofilm bringing bacteria deeper into the lungs and spurring growth on other areas of the tube. (4)

In the past, efforts to reduce VAP have concentrated on decreasing bacterial contamination during intubation by modifying medical procedures. It has become increasingly apparent that the elimination of bacteria also depends on the reduction of bacterial growth on the ETTs.

Cost effective ETTs that are resistant to bacterial infection provide an essential tool to prevent VAP. Nanomodified devices used within the body mimic their natural surroundings and subsequently often lack detrimental immune responses to the foreign object. Research shows a connection between nanophase surfaces (that is, surfaces with features less than 100 nm in at least one direction) and increased bacterial metabolism suggesting the topography of nanostructures influence the biological processes of bacteria. (5) Addition of nanomaterials to these bacteria modifying surfaces could enhance their anti-microbial properties. Preliminary research has uncovered a number of materials that could produce an effective antimicrobial ETT. These materials display antibacterial properties using a number of different methodologies to interfere with bacterial growth and adhesion. Many of these nanomaterials have also been shown to increase fibronectin adsorption thus inhibiting bacterial growth.

However, these preliminary studies were performed under static conditions and do not take into account the dynamic forces that occur in the human airway. Our model seeks to characterize the antimicrobial properties of nanomodified ETTs within a continuously contaminated airway model. This study hypothesizes that airflow and continuous bacterial contamination impact biofilm formation on ETTs and are essential to the determination of the bacterial resistance of these tubes in the pediatric airway.

Materials and Methods

The first type of antibacterial ETT tested in our system was polyvinyl chloride (PVC) ETT with nano-roughened surfaces. These tubes were created using lipases from the bacteria *Candida cilindracea* and *Rhizopus arrhisus* (Sigma Aldrich) which enzymatically degrade the PVC material creating a nano-roughened surface. The procedure included exposing the PVC tubing to a 0.1% mass solution of either *C. cilindracea* or *R. arrhisus* lipase dissolved in a potassium phosphate buffer at 37° C. The samples were then gently agitated for 24 hours at which point they were washed with distilled water and exposed to fresh enzyme solution. Thus, the tubes remained in contact with the lipases for a total of 48 hours. The activity of the *R. arrhisus* used in the experiment was 10.5 U/g and that of *C. cilindracea* was 7.29 U/g/cm^2 where one unit was defined as the amount of enzyme that catalyzed the release of 1 µmol of oleic acid per minute at pH 7.4 and 40°C. The samples were observed under a scanning electron microscope (LEO 1530 SEM, Zeiss) to determine the amount of nano features formed. Each of the samples was coated with a gold palladium mixture to increase conductivity. Endotracheal tubes placed in the system will only be modified on the inner surface.

To test the effectiveness of nanomodified ETT under the conditions present within the airway a bench top model was constructed. Hartmann et al. (1999) created a continuously contaminated airway system to test the effectiveness of the silver coated endotracheal tubes (SCET). The airway model for this study was based upon this general design. Figure 1 contains an illustration of the Hartmann model while Figure 2 depicts the design changes implemented for this study.

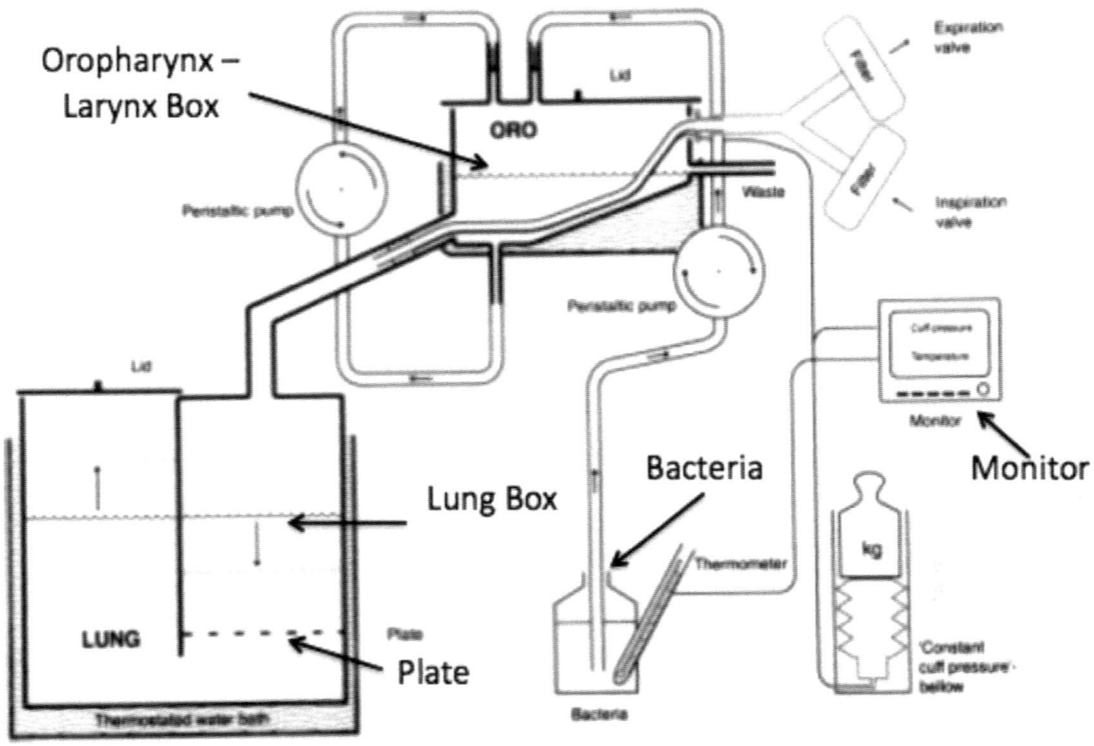

Fig. 1: Endotracheal Tube System (6)

The model in Figure 1 contains two polymethylmethacrylate (Plexiglas) chambers that simulate the structures found in the airway. The first chamber, or oropharynx-larynx box, simulates the conditions found in the oropharynx- larynx and the second chamber, or lung box, simulates the compliance and resistance found within the lung. A Plexiglas tube, representing the trachea, connects the two boxes. Cuffed nanomodified ETTs run through the oropharynx chamber, where they are exposed to continuous bacterial media, into the trachea, where cuff pressure conditions dominate.

Our study will concentrate on the conditions found within the pediatric airway, so the system was reconfigured to better represent conditions in a smaller patient.

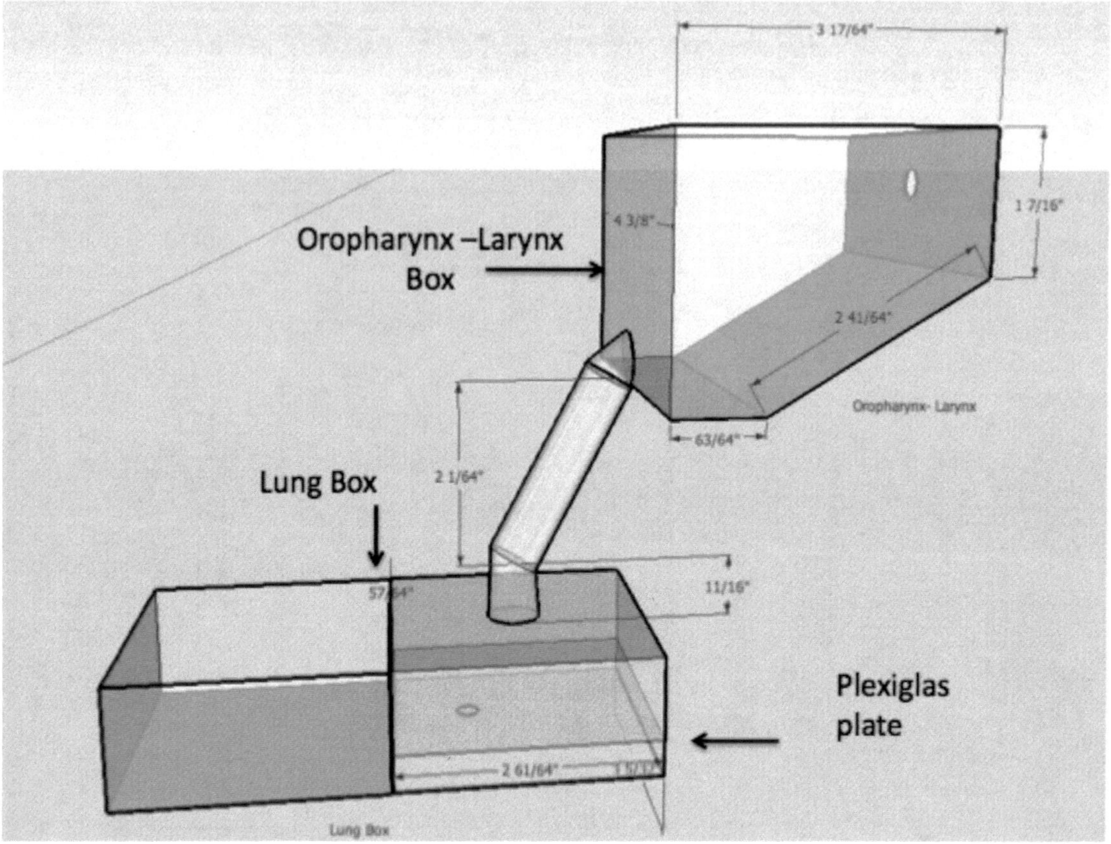

Fig. 2: Pediatric Endotracheal Tube System

The oropharynx-larynx box of the pediatric model is the first chamber where the ETT will be exposed to bacterial media. This chamber, constructed from Plexiglas, will be filled with 100 mL of bacterial media. The length and width of the chamber are based upon the length of the oropharynx and larynx of an 18 month old child.

Bacterial media will be continuously added to the system using a peristaltic pump (Curlin 4000 CMS). This additional media will either drain down the trachea or out of an overflow valve on the side of the chamber. It is important to note that the system contains a method to agitate the chamber before samples are taken using an additional Curlin pump. This allows us to take samples from the system that are homogeneous and that accurately represent the total bacteria content in the oropharynx. Agitation also will prevent any sedimentation of cells within the chamber. This chamber constitutes the first step towards the simulation of the conditions found in the body during mechanical ventilation, because it places the mechanically ventilated ETT in direct contact with bacteria and media.

The second part of our model is a Plexiglas tube that represents the trachea. This tube simulates the pressures on the ETT cuff *in vivo*. The end of this tube is exposed to airflow from the ETT. The diameter of this tube was calculated based on an average of the thoracic inlet tracheal size, aortic arch tracheal size and supracarinal tracheal size (Equations 1, 2, and 3) of an 18 month old child with a height (H) of 80 centimeters.

$$\text{Thoracic inlet} = (0.478 * H) - (0.00446 * H^2) - (0.0000147 * H^3) - 9.80 \pm 1.94 \quad (1)$$
$$\text{Aortic arch} = (0.358 * H) - (0.00324 * H^2) - (0.0000108 * H^3) - 5.88 \pm 1.68 \quad (2)$$
$$\text{Supracarinal} = (0.388 * H) - (0.00346 * H^2) - (0.0000115 * H^3) - 7.02 \pm 1.69 \quad (3)$$

The lung is the last organ modeled within the system. It is represented by a dual chambered Plexiglas box. This simple model simulates both the resistance and compliance of the lung. These two factors are the most important mechanical stresses on the airway system and are important for accuracy using the ventilated model. The total volume of the two chambers is 400 mL and they will be filled with 236 mL of media. The geometry of the lung chamber is based upon the compliance of the lung. The compliance in an 80 cm, 18 month old child was determined to be 34 mL/cmH_20 by equation 4.

$$C_{lung}\ (\text{ml/cm } H_2O) = 0.0017 * \text{length}^{2.26} \quad (4)$$

The resistance of the lungs was determined to be 40 cmH_2O/liters/second and is simulated by a Plexiglas plate with 64 boreholes with a diameter of 6.4 millimeters each using equation 5 where $\Delta p/Q$ is the resistance (dynes/ cm), η is the viscosity, L is the length of the sieve tube (cm), R is the radius of the sieve tube (cm), r is the sieve pore radius (cm), l is the sieve plate thickness (cm), and N is the total number of pores.

$$\Delta p = \frac{8}{\pi} \frac{\eta Q}{R^4} \left[L + \frac{\ell}{N} \left(\frac{R}{r} \right)^4 \right] \quad (5)$$

The model will be connected to an Infant Star 950 ventilator. It will be tested at an inspiratory flow rate of 8.5 mL/sec. In addition the positive end-expiratory pressure (PEEP) will be set at 5 cmH_2O. This will allow for a peak airway pressure of about 11.7 cmH_2O at a tidal volume of 100 mL. (7) The fraction of inspired oxygen (FiO_2) will be maintained at a value of 1 for this trial, to limit flow variability.

Other features of the system that are essential to the experiment are the many controls that make our data reproducible and lessen confounding influences: lids for the exposed areas of the lung and the oropharynx chambers, lower contact with the surrounding environment as well as lessen the chance for potential outside contamination. To better reproduce the conditions of the body, both the oropharynx-larynx chamber and the lung chamber will be immersed in water baths keeping these chambers at a constant 37° C. Humidified air will also be used in the system in order to better replicate clinical conditions. This is an important feature of the airway model because humidified airflow through contaminated ETTs has been shown to cause increased bacterial transport to the lungs. The mechanism of this transport is thought to be the removal of bacteria from biofilms formed on the ETT by the condensation of humidified air. (1)

Cuff pressure and temperature of the bacterial media will be measured by a portable monitor (Datascope Passport 2), while humidity will be recorded in the lung box and the oropharynx using two Varisala HMT337 sensors.

An artificial saliva was chosen as the media for the airway system. The artificial saliva has the following composition: 0.1 L each of 25 mM K_2HPO_4, 24 mM Na_2HPO_4, 150 mM $KHCO_3$,

100 mM NaCl, and 1.5 mM $MgCl_2$. 0.006 L of 25 mM citric acid and 0.1 L of 15 mM $CaCl_2$ will be added to this simple solution. To achieve a pH of 6.7 either NaOH or HCl will then be added to the solution.

Finally to quantify the number of bacteria within the system, 2 mL samples of the fluid in the oropharynx chamber and the lung chamber will be extracted at regular intervals. The system will remain running during sample collection. These samples will then be diluted, plated, and incubated for 48 hours. The number of colonies that are formed will help to determine the effectiveness of the ETT. Biofilms on the ETT will be analyzed using a crystal violet assay in addition to confocal microscopy and a live/dead stain.

Results

This first treatment of PVC tubes cut and submersed in either lipase solution (*C. cilindracea* or *R. arrhisus*) showed visible nanofeatures on the ETT tube surface. The visible differences between the nanomodified tubes were small compared to the untreated PVC (Figures 3-5).

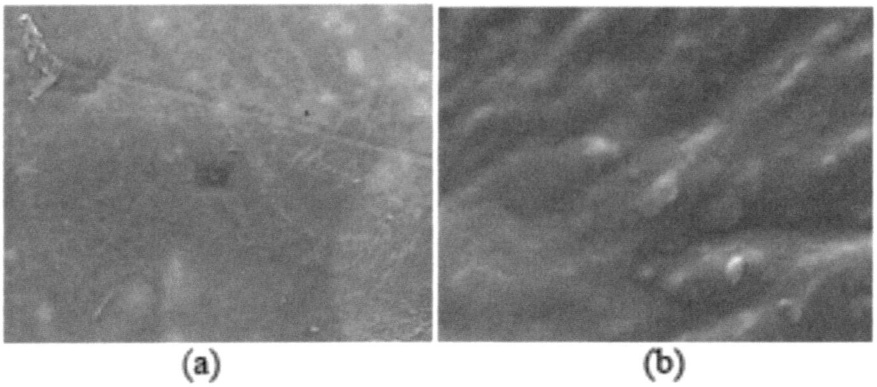

Fig. 3: SEM micrographs of untreated PVC samples at magnifications of (a) 1k and (b) 42k. (8)

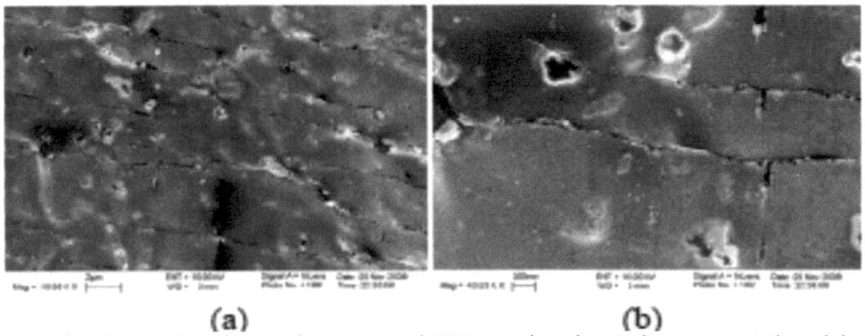

Fig. 4: SEM micrographs of 0.1% *R. arrhisus* solution treated PVC samples after 48 hours at (a)10k and (b) 42k. (8)

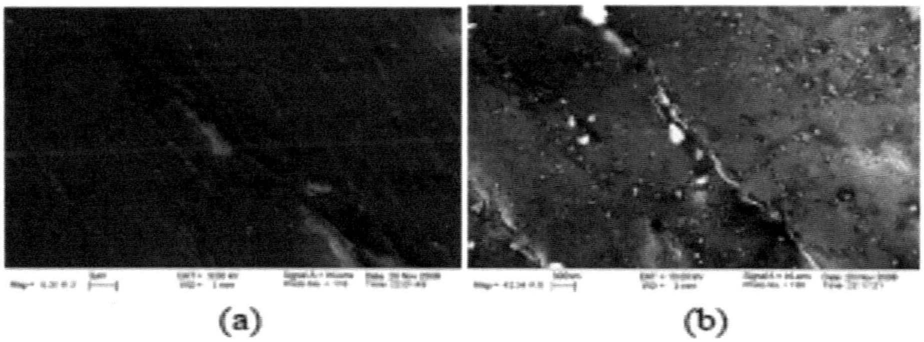

Fig. 5: SEM micrographs of 0.1% *C. cilindracea* solution treated PVC samples after 48 hours at magnifications of (a) 5k and (b) 42k. (2)

Discussion

Past evaluations of pediatric ETT effectiveness have concentrated on either static contamination or dynamic airway simulations. This study aimed to simulate these important attributes of the human airway and to use this knowledge to better quantify the antimicrobial properties of ETTs. A number of unique aspects included within this study are the introduction of humidified air to the system, the media used for the samples, and even the redefined size and shape of the two chambers that simulate the lung and oropharynx. These parameters not only separate this study from its predecessors, but will also yield valuable insights into bacterial colonization and proliferation on ETTs.

Conclusions

The results obtained by the treatment of PVC tubes with bacterial lipase suggest that this is a viable method for nanomodification of ETT. Further tests will determine whether homogenous modification of the PVC surface inside the ETT is possible and additionally will look at the effects of enzymatic modifications on the ETT cuff. Future work includes fluid characterization of the Plexiglas system as well as dual gas analysis within the chambers. Different FiO_2 levels will be analyzed within the system to determine the effects of clinical conditions on nanomodified ETT function. Also planned is the introduction of other nanomodified ETTs including selenium and iron oxide coated tubes. Long term studies of bacterial contamination will provide another area of interest.

Acknowledgements

The authors of this paper would like to thank the Rhode Island Science and Technology Advisory Committee for funding this project.

References

1. Office of Quality and Performance (2007), *FY 2008, Q1 technical manual for the VHA performance measurement system*, Oct, pp. 315.
2. Liu H and *T. J.* Webster (2007), "Nanomedicine for implants: a review of studies and necessary experimental tools." *Biomaterials, 28,* pp. *354-369*.
3. Carson, Stevens (1989), *Fibronectin in Health and Disease* New York: CRC Press, Inc.
4. Marini, J. J. and A. S. Slutsky (1998), *Physiological basis of ventilatory support.* New York: Marcel Dekker.
5. Cardinal, P., Jessamine, P. , Carter-Snell, C., Morrison, S., and G. Jones. (1993), "Contribution of Water Condensation in Endotracheal Tubes to Contamination of the Lungs." *Chest*, pp. 127-129.
6. Hartmann, M., Guttmann, J., Muller, B., Hallmann, T., and K. Geiger (1999), "Reduction of the bacterial load by the silver-coated endotracheal tube (SCET) a laboratory investigation." *Technology and Health Care*, pp. 359-70.
7. National Center for Health Statistics, Center for Disease Control and Prevention (2009), *Clinical Growth Charts*, August, Set 1
8. Cheng, D., and T. J. Webster (2008) "Formation of nano surfaces on endotracheal tubes using bacterial lipase solutions." *Bioengineering Conference, IEEE 35th Annual Northeast*, pp. 1-2.
9. Wilson, W. C., Grande C. M., and D. B. Hoyt (2007), *Trauma Emergency Resuscitation, Perioperative Anesthesia, Surgical Management, Volume I.* New York: Informa Healthcare.

Regulate Embryonic Stem Cell Fates Using Microscale Bioactive Elastic Hydrogel

Michael R. Zonca Jr., The College of Nanoscale Science and Engineering, University at Albany, State University of New York, Albany, NY

Nurazhani Abdul Raof, The College of Nanoscale Science and Engineering, University at Albany, State University of New York, Albany, NY

Magnus Bergkvist, The College of Nanoscale Science and Engineering, University at Albany, State University of New York, Albany, NY

Yubing Xie, The College of Nanoscale Science and Engineering, University at Albany, State University of New York, Albany, NY

Embryonic stem cells and induced pluripotent stem cells have great potential to treat and understand diseases. One of the major challenges in stem cell research is how to maintain these stem cells in an undifferentiated state during expansion and how to effectively differentiate them into a specific lineage. Embryonic stem cell maintenance and differentiation are governed by the unique local 3-D microenvironment, which is composed of soluble factors, elastic extracellular matrices (ECM) and neighboring cells. Cells and tissues in the body are organized into 3-D architectures with nanoscale interactions. We have developed a microscale hydrogel system to mimic the 3-D microenvironment. Hydrogels can be tuned to resemble the elasticity and shape of a specific tissue type where soluble factors controlled release to support stem cell development. Mouse embryonic stem cells were used as a model system to demonstrate how tunable microenvironments affect stem cell maintenance and differentiation. Our results demonstrate that a gel-like microenvironment facilitates stem cell maintenance and a liquid-like microenvironment promotes embryoid body formation. Leukemia inhibitory factor (LIF) and retinoic acid (RA) are encapsulated in the hydrogel to achieve controlled release into embryonic stem cell cultures, respectively. The maintenance and differentiation of stem cells are confirmed by cell morphology and immunocytochemistry of specific cell markers. The combination of controlled release and 3-D hydrogels with tunable elasticity will provide a valuable platform for understanding microenvironment-stem cell interactions and efficiently differentiating embryonic stem cells.

See more of this Session: Nanostructured Scaffolds for Tissue Engineering
See more of this Group/Topical: Nanoscale Science and Engineering Forum

A Novel Scaffold for the Construction of Polyelectrolyte Thin Films for Non-Viral Gene Delivery

Raymond E. Samuel, Chemical Engineering, Massachusetts Institute of Technology, Cambridge, MA

Paula T. Hammond, Chemical Engineering, Massachusetts Institute of Technology, Cambridge, MA

The avian egg shell membrane (ESM) is an abundant and environmentally friendly complex of two semi-permeable membranes, which are composed of highly cross-linked extra-cellular proteins. The ESM has been shown to be biocompatible and a potential biomaterial in tissue engineering applications. We have utilized ESM as scaffold for the delivery of non-viral vectors deposited within polyelectrolyte multilayer (PEM) thin films constructed by the alternating deposition of hydrolytically degradable polycations (poly â-amino ester) and polyanions (plasmid DNA vectors). These ESM-PEM gene delivery devices progressively released the plasmid DNA sequentially deposited on the 3-D fibrous scaffold. The PEM-modified surfaces of the ESM were able to support the adhesion, proliferation and differentiation of MC3T3-E1 pre-osteoblast cells. The engrafted MC3T3-E1 cells were distributed throughout the full-thickness of the ESM and expressed the genes encoded in the plasmid DNA vectors. The increased surface-to-volume ration of the ESM and its relative ease of handling may make ESM a highly versatile 3-D scaffold for tissue engineering applications, such as bone fracture healing.

See more of this Session: Nanostructured Scaffolds for Tissue Engineering
See more of this Group/Topical: Nanoscale Science and Engineering Forum

A Simple Two-Phase Synthesis/Ligand Exchange Reaction for Creating Biocompatible Quantum Dots

Aaron R. Clapp, Dept. of Chemical & Biological Engineering, Iowa State University, Ames, IA

Allison M. Schnoes, Chemical & Biological Engineering, Iowa State University, Ames, IA

Yanjie Zhang, Chemical & Biological Engineering, Iowa State University, Ames, IA

In this talk we describe a method for producing a wide variety of functionalized quantum dot nanocrystals using a simple one-pot synthesis and ligand exchange procedure. As prepared, most highly crystalline quantum dots are organically soluble and not suitable for biological applications. Significant attention has focused on thiol-based ligands that replace entirely the native hydrophobic ligands with customized ligands, many of which contain water-soluble functional functional groups. For a number of reasons, many of the common mono- and dithiol ligands are challenging to synthesize and purify, are difficult to stably exchange with existing ligands, compromise quantum yield, and/or diminish the colloidal stability of the quantum dots. To address these issues, we have utilized a two-phase reaction and exchange procedure to create unique dithiocarbamate ligands at the liquid-liquid interface which subsequently displace the native capping ligands present on the nanocrystal surface. Lacking an undesirable byproduct, the reaction does not require explicit purification of the thiolated ligand which improves the yield. Additionally, the new ligand naturally exchanges with the native ligand which then drives the previously organically-soluble nanocrystals entirely into the aqueous phase. There are several key benefits to this method. First is the ability to perform the reaction and exchange steps in a single vessel which simplifies the process. Second, dithiocarbamates have been shown to better preserve the luminescence properties of the nanocrystals (e.g., maintain high quantum yields in aqueous solutions). Third, the chemistry is general and allows for a wide range of potential ligands opening the possibility for targeted delivery and unique interactions with their surroundings. A diversity of molecules such as amino acids, proteins, glycans, and polymers can be modified in a straightforward manner to serve as capping ligands on quantum dots.

See more of this Session: Nanotechnology for In Vivo and in Vitro Imaging
See more of this Group/Topical: Nanoscale Science and Engineering Forum

Cationic Multifluorescent Quantum Dot Liposomes for Cancer Cell Imaging

Geoffrey D. Bothun, Chemical Engineering, University of Rhode Island, Kingston, RI

Amy E. Rabideau, Chemical Engineering, Biomedical and Pharmaceutical Sciences, University of Rhode Island, Kingston, RI

Matthew A. Stoner, Biomedical and Pharmaceutical Sciences, University of Rhode Island, Kingston, RI

Quantum dot (QD) nanocrystals provide highly efficient and photostable spectral emission for medical imaging and diagnostics.[1-3] However, QDs have also exhibited cytotoxicity[4,5] and are natively hydrophobic, as a result of the synthesis process, which requires that they be surface modified with biocompatible hydrophilic coatings to be dispersed in aqueous phases. An alternative strategy that has been recently explored is to encapsulate QDs within liposomes.[6-8] A potential advantage of this strategy includes incorporating multiple QDs and delivering them to specific cell types using liposome-targeting techniques.

In this work, we describe the design of a liposomal system that can deliver multiple QDs as imaging agents to carcinoma cells. Cationic and zwitterionic multifluorescent quantum dot liposomes (QD-Ls) have been prepared with both green hydrophobic (3 nm diameter; 545 nm emission) and red hydrophilic (11 nm diameter; 620 nm emission) CdSe/ZnS core/shell quantum dots by reverse phase evaporation. Cationic liposomes were composed of dipalmitoylphosphatidylcholine (DPPC)/dipalmitoylethylphosphocholine (ethyl-DPPC)/cholesterol (6:2:2 molar basis) and zwitterionic liposomes were composed of DPPC or DPPC/cholesterol (8:2). The hydrophobic QDs were embedded within the lipid bilayer and the hydrophilic QDs were encapsulated in the aqueous core during liposome assembly at lipid to QD molar ratios of 10,000:1. QD incorporation was confirmed by fluorescence and confocal microscopy, as colocalized green and red QDs produced yellow. Incorporation did not affect QD photoactivity or damage bilayer or liposome structure, as determined from differential scanning calorimetry. Cell uptake was examined in human hepatocellular carcinoma cells (HuH-7) and showed that cationic QD-Ls were stable in vitro, exhibited high uptake, and collected near the nuclear membrane. In contrast, zwitterionic QD-Ls aggregated and exhibited low uptake. We are currently examining cancer cell targeting strategies using headgroup-modified lipids and the effect of serum proteins on QD-L stability and uptake. Given that liposomes are established and versatile platforms for creating cell-targeting therapeutic agents, we envision that multifluorescent QD-Ls may offer advanced techniques for imaging hydrophobic and hydrophilic domains simultaneously. If coupled with an encapsulated drug, QD-Ls could be multifunctional and provide detection and drug delivery in a single assembly.

1. Drbohlavova, J., Adam, V., Kizek, R., Hubalek, J. Quantum Dots - Characterization, Preparation and Usage in Biological Systems, Int. J. Mol. Sci. 10,2 (2009) 656.

2. Gao, X., Dave, S.R. Quantum dots for cancer molecular imaging, Adv. Exp. Med. Biol. 620 (2007) 57.

3. Michalet, X., Pinaud, F.F., Bentolila, L.A., Tsay, J.M., Doose, S., Li, J.J., Sundaresan, G., Wu, A.M., Gambhir, S.S., Weiss, S. Quantum dots for live cells, in vivo imaging, and diagnostics, Science 307,5709 (2005) 538.

4. Kirchner, C., Liedl, T., Kudera, S., Pellegrino, T., Javier, A.M., Gaub, H.E., Stolzle, S., Fertig, N., Parak, W.J. Cytotoxicity of colloidal CdSe and CdSe/ZnS nanoparticles, Nano Lett. 5,2 (2005) 331.

5. Wang, L., Nagesha, D., Selvarasah, S., Dokmeci, M., Carrier, R. Toxicity of CdSe Nanoparticles in Caco-2 Cell Cultures, J. Nanobio. 6,1 (2008) 11.

6. Al-Jamal, W.T., Al-Jamal, K.T., Bomans, P.H., Frederik, P.M., Kostarelos, K. Functionalized-quantum-dot-

liposome hybrids as multimodal nanoparticles for cancer, Small 4,9 (2008) 1406.

7. Al-Jamal, W.T., Al-Jamal, K.T., Tian, B., Lacerda, L., Bomans, P.H., Frederik, P.M., Kostarelos, K. Lipid-quantum dot bilayer vesicles enhance tumor cell uptake and retention in vitro and in vivo, ACS Nano 2,3 (2008) 408.

8. Gopalakrishnan, G., Danelon, C., Izewska, P., Prummer, M., Bolinger, P.Y., Geissbuhler, I., Demurtas, D., Dubochet, J., Vogel, H. Multifunctional lipid/quantum dot hybrid nanocontainers for controlled delivery of live cells, Angew. Chem. Int. Ed. 45 (2006) 5478.

See more of this Session: Nanotechnology for In Vivo and in Vitro Imaging
See more of this Group/Topical: Nanoscale Science and Engineering Forum

DESIGN OF SOLID LIPID PARTICLES WITH IRON OXIDE QUANTUM DOTS FOR THE DELIVERY OF THERAPEUTIC AGENTS

Dr. Ronald Holser, USDA-ARS-RRC, Athens, GA
Dr. Jaya Sundaram, USDA-ARS-NPRL, Dawson, GA
Dr. Samantha Hawkins, USDA-ARS-RRC, Athens, GA
Dr. Chari Kandala, USDA-ARS-NPRL, Dawson, GA
Dr. Bosoon Park, USDA-ARS-RRC, Athens, GA

Introduction

Solid lipid nanoparticles (SLN) were pioneered by Müller with the subsequent development of numerous biomedical and pharmaceutical applications (1, 2). The lipophilic character and small particle size of these systems is particularly advantageous for controlled release applications. Recent work with quantum dots offers a complementary technology to visualize therapeutic agents and evaluate the efficacy of a targeted delivery system. Most of the current research on involves CdS quantum dots that can be easily detected by their strong fluorescence. Combining a targeted therapeutic agent with a quantum dot offers the possibility of tracking the agent to the target site. In vivo applications using this technique require that the system exhibit low toxicity and provide a signal that can be readily mapped as the therapeutic agent moves through the organism to the target. The CdS-based technology presents some complications in both of these areas. There is a level of toxicity associated with Cd which is typical of heavy metals. However, this may be acceptable in certain clinical applications. Also, fluoresence can present a challenge to detect in a tissue or other complex analytical matrix where the signal is quenched by other structures. These issues may be addressed by using iron oxide as the imaging component in place of CdS. Iron oxide exhibits relatively low toxicity (3). It can form particles of 30-40 nm diameters by self assembly and such particles can serve as imaging centers for bioprobes or therapeutic compounds that offer selective binding to target sites. The encapsulation of such structures can provide a controlled release system to facilitate transport and delivery of the bioactive compounds (4). There are several strategies available to encapsulate bioactive compounds. The use of thermal inversion is an attractive low-energy technique that is compatible with thermally sensitive compounds.

Materials and Methods

Iron oxide particles were obtained from Ocean NanoTech, LLC (Springdale, AR). These particles were bonded to oleic acid with a diameter of 15 nm. Triglycerides were used to prepare the solid lipid nanoparticles. Tripalmitin and tristearin were obtained from Nu-Chek Prep (Elysian, MN). Lecithin was used as the surfactant and obtained from Sigma-Aldrich (St. Louis, MO). SLN were prepared following a standard method of heating the lipid slightly above its melting point and dispersing the iron oxide particles by sonication into the melt (5). Iron oxide was added at 0.1, 0.5, and 1.0 wt%.

Results and Discussion

The mean diameters of particles prepared with different amounts of iron oxide are presented in Table 1. The particles were much larger than expected although still submicron. There appeared to be a non-linear relationship between the amount of iron oxide used in the formulation and the mean diameter of the resulting particle. The limit of detection for iron oxide is quite low and would permit a further reduction in wt% to be used. If the projected trend follows the observed behavior then a formulation with 0.05 wt% iron oxide would provide a reasonably useful particle size with a strong signal for imaging.

Table 1. Average Particle Size for Different Levels of Iron Oxide

wt% Iron Oxide	nm
1	759.5
0.5	470.0
0.1	247.7

References

1. Müller, R.H., Schwarz, C., Mehnert, W. and J. S. Lucks, (1993), "Production of solid lipid nanoparticles (SLN) for controlled drug delivery," Proceed. Intern. Symp. Control. Rel. Bioact. Mater., 20:480-481.

2. Müller, R.H., Mehnert, W., Lucks, J.S., and Schwarz, C., zur Mühlen, A., Weyhers, H., Freitas, C. and D. Rühl, (1995), "Solid Lipid Nanoparticles (SLN) -- An alternative colloidal carrier system for controlled drug delivery," Eur. J. Pharm. Biopharm., 41:62-69.

3. Müller, R.H., Maaβen, S., Weyhers, H., Specht, F., and J. S. Lucks, (1996), "Cytotoxicity of magnetite-loaded polylactide, polylactide/glycolide particles and solid lipid nanoparticles,"Int. J. Pharma 138:85-94.

4. Shidhaye, S.S., Vaidya, R., Sutar, S., Patwardhan, A., and V. J. Kadam, (2008), "Solid lipid nanoparticles and nanostructrued lipid carriers - Innovative generations of solid lipid carriers," Current Drug Delivery, 5:324-331.

5. Anton, N., Benoit, J. P., and P. Saulnier, (2008), "Design and production of nanoparticles formulated from nano-emulsion templates—A review," J. Controlled Release 128:185-199.

PLGA-Encapsulated Graphitic Carbon and Iron Oxide Nanoparticles for Simultaneous Fluorescence and MRI Imaging

Dhananjay Thakur, Chemical and Biomolecular Engineering and Biophysics, The Ohio State University, Columbus, OH

Ning Han, Chemical and Biomolecular Engineering, The Ohio State University, Columbus, OH

Jessica Winter, Chemical and Biomolecular Engineering, The Ohio State University, Columbus, OH

There has been considerable interest in recent years in the development of multimodal contrast agents. In particular, the combination of magnetic and fluorescent imaging has attracted attention because of the possibility of combined magnetic resonance imaging (MRI) and fluorescent histopathology or MRI and intrasurgical fluorescence imaging. However a major stumbling block in the implementation of such probes is the toxicity of the fluorescent molecules employed (e.g. quantum dots). We have combined fluorescent carbon and iron oxide nanoparticles in poly(lactic-co-glycolic) acid (PLGA) to produce biodegradable micro/nano-particles that can be used for fluorescent and magnetic imaging, as well as drug delivery. Graphitic carbon nanoparticles of < 5 nm diameter were generated from carbon soot, and exhibited fluorescence originating from nitrogen vacancy centers that give rise to an emission peak at 465 nm. The quantum yield was comparable with that of water soluble cadmium sulfide quantum dots. Iron oxide nanoparticles ~10 nm in diameter were produced through aqueous co-precipitation. Both nanoparticles were embedded in PLGA nanospheres through a double emulsion process. Initial experiments with non biofunctionalized PLGA spheres have shown non-specific endocytosis into human SK-N-SH and glioblastoma cells (neural cell lines). These particles could be used for multimodal imaging. For example, two photon excitation with Ti: Sapphire or Nd:YAG lasers with excitation wavelengths greater than 900 nm would permit fluorescence images from deep tissues. These images could then be correlated to those from obtained from MRI images, enhancing the ability to obtain weighted images.

See more of this Session: Nanotechnology for In Vivo and in Vitro Imaging
See more of this Group/Topical: Nanoscale Science and Engineering Forum

Design and Characterization of Biodegradable Gold Nanoclusters for Optical Imaging

Avinash K. Murthy, Chemical Engineering, University of Texas at Austin, Austin, TX

Jasmine M. Tam, Chemical Engineering, University of Texas at Austin, Austin, TX

Justina O. Tam, Biomedical Engineering, University of Texas at Austin, Austin, TX

Li Leo Ma, Chemical Engineering, University of Texas at Austin, Austin, TX

Davis R. Ingram, Chemical Engineering, University of Texas at Austin, Austin, TX

Konstantin Sokolov, Departments of Biomedical Engineering and Imaging Physics, M.D. Anderson Cancer Center, Houston, TX

Keith P. Johnston, Chemical Engineering, University of Texas at Austin, Austin, TX

Gold nanoparticles which exhibit surface plasmon resonance in the near-infrared region (NIR) of the optical spectrum have been found to be extremely useful for biomedical imaging applications[1]. These imaging agents, however, must eventually be cleared from the body. Clearance has been shown to be highly dependent by the size of imaging contrast agents, and current imaging agents typically exceed the clearance requirement of approximately 5 nm in diameter[2].

We have developed clusters of 4 nm gold nanoparticles which are less than 100 nm in diameter and possess the ability to biodegrade. UV-visible NIR spectroscopy indicates that these nanoclusters present strong absorbance in the 650 to 900 nm region. The nanoclusters biodegrade over the course of approximately a week into particles small enough for efficient clearance. Cluster formation was induced by the addition of a polymer templating agent, and primary particles were capped with different ligands in order to prevent uncontrolled aggregation. Hyperspectral microscopy, darkfield reflectance imaging and transmission electron microscopy (TEM) confirmed that the nanoclusters biodegrade in macrophage cells as well as in solution. ζ-potential measurements have also been conducted on clusters and deaggregated particles in order to approximate surface charge and electrostatic forces. The results can be described with DLVO theory. In conclusion, biodegradable clusters of gold nanoparticles, which show substantial NIR absorbance, were created and shown to degrade at physiological conditions, providing new opportunities for biomedical imaging without accumulation of toxic nanoparticles in the body.

(1) Anker, J.N., et al. Biosensing with plasmonic nanosensors. Nature Materials 7 (2008) 442-453

(2) Choi, H.S.. et al. Renal clearance of quantum dots. Nature Biotechnology 25 (2007) 1165-1170

See more of this Session: Nanotechnology for In Vivo and in Vitro Imaging
See more of this Group/Topical: Nanoscale Science and Engineering Forum

Gold Nanoparticle-Assisted AFM Study of DNA Damage and Repair

Enis Demir, Chemical and Biological Engineering, Koc University, Istanbul, Turkey

Hande Asimgil, Materials Science and Engineering, Koc University, Istanbul, Turkey

Halil Kavakli, Chemical and Biological Engineering, Koc University, Istanbul, Turkey

Seda Kizilel, Chemical and Biological Engineering, Koç University, Istanbul, Turkey

The purpose of this research is to investigate gold nanoparticles as a potential scaffold for the characterization of DNA damage and repair using atomic force microscopy techniques. The procedure consists of functionalizing the surface of gold nanoparticles with DNA, which are then immobilized onto amine modified silicon or glass surfaces. The objective of this study is to examine various lesions in individual DNA molecules and follow their direct reversal by DNA repair enzymes through AFM imaging with the support of UV-vis spectrum results of the functionalized nanoparticles. In order to achieve this goal DNA repair proteins such as photolyase (repairs UV damaged thymine dimers in DNA using blue light energy) is used as damage markers and imaged in complexes with DNA by AFM in order to locate and identify the damage sites. The force spectroscopy measurements then determine the mechanical fingerprints of various types of DNA damage caused by UV and gamma radiation and directly follows damage reversal caused by photolyase repair activity.

See more of this Session: Nanotechnology for In Vivo and in Vitro Imaging
See more of this Group/Topical: Nanoscale Science and Engineering Forum

Facile Synthesis of Nanoparticles for Imaging by Fluorescent Microscopy, CAT Scan, and MRI

Mustafa Akbulut, Chemical Engineering, Princeton University, Princeton, NJ

Robert K. Prudhomme, Princeton University, Princeton, NJ

Marian E. Gindy, Chemical Engineering, Princeton Unversity, Princeton, NJ

There is increased demand for nanoparticles with various imaging modalities that have the desired surface functionalization and particle size to act as nano-probes. Here, we describe a simple, rapid, and robust method, Flash NanoPrecipitation (FNP), to produce such nanoparticles. This process involves encapsulation of hydrophobic or hydrophobically-modified imaging agents with an amphiphilic biocompatible diblock copolymer in a kinetically frozen state. Using this approach, fluorophores, x-ray contrast agents, or magnetic nanoparticles were successfully incorporated into biocompatible polymeric nanoparticles. In addition, imaging agent and drugs were simultaneously incorporated into a single nanoparticle, producing multifunctional nanoparticles having simultaneous drug delivery and biological imaging capabilities. Moreover, functionalization of such nanoparticle surfaces with disease-specific ligands permitted precise cell targeting. These features make FNP-produced nano-probes highly desirable for various biological and medical applications.

See more of this Session: Nanotechnology for In Vivo and in Vitro Imaging
See more of this Group/Topical: Nanoscale Science and Engineering Forum

Self-Assembly of CdTe Nanoparticles with DNA and Proteins

Sudhanshu Srivastava, Chemical Engineering, University of Michigan, Ann Arbor, MI

Elizabeth J. Stewart, Chemical Engineering, University of Michigan, Ann Arbor, MI

Felipe N. Pereira, Chemical Engineering, University of Michigan, Ann Arbor, MI

Kai Sun, Materials Science and Engineering, University of Michigan, Ann Arbor, MI

Nicholas A. Kotov, Chemical Engineering, University of Michigan, Ann Arbor, MI

Nanoparticles (NPs) provide exclusive apparatus for material applications as they feature unique optical, electronic, and magnetic properties. The advanced knowledge of nanoparticle organization can bridge the gap between the synthetic "bottom up", and lithographic "top down" assembly processes. Similarly, protein-protein interaction and DNA engineering is also used for fabricating advanced architectures. As biomolecules and NPs are at same length scale, introducing proteins or DNA directly in the assembly with inorganic NPs is an exceptional way for combining the intrinsic properties of biomolecules with advanced physical properties of inorganic crystals. In this study, the already known CdTe NP assemblies into nanowires and nanosheets were exploited to insert proteins and DNA using charge dipolar interactions. The study provides (1) formation of new morphologies having the properties from both the components; (2) biomolecular structures can be tuned in the assembly for generating dynamic assemblies and (3) the variety of proteins with metallic/semiconductor NPs used can lead to bio-nanocomposites for specific photo-electronic materials, optical sensors and delivery agents.

See more of this Session: Self-Assembled Biomaterials II
See more of this Group/Topical: Nanoscale Science and Engineering Forum

Spotted Vesicles and Striped Worms: Ligand-Induced Phase Separation

David A. Christian, Chemical & Biomolecular Engineering, University of Pennsylvania, Philadelphia, PA

Aiwei Tian, Chemical & Biomolecular Engineering, University of Pennsylvania, Philadelphia, PA

Ilya Levental, Bioengineering, University of Pennsylvania, Philadelphia, PA

Karthikan Rajagopal, Chemical & Biomolecular Engineering, University of Pennsylvania, Philadelphia, PA

Paul A. Janmey, Department of Physiology, University of Pennsylvania, Philadelphia, PA

Tobias Baumgart, Department of Chemistry, University of Pennsylvania, Philadelphia, PA

Dennis E. Discher, Chemical & Biomolecular Engineering, University of Pennsylvania, Philadelphia, PA

Mixtures of amphiphiles are ubiquitous and assemble into various morphologies, including vesicles and cylinder micelles that raise the possibility of mesoscopic segregation within the assemblies - perhaps even in response to binding of small ligands. Here, anionic and neutral polymer amphiphiles are mixed within vesicle and cylinder morphologies, and divalent cations are shown to induce meso-scale domains, generating 'responsive Janus structures'. Calcium crossbridging of the anionic amphiphiles rigidifies the charged structures and leads to lateral phase separation. Domains in vesicles and stripes in micelles have tunable sizes, shapes, and/or spacing that can last for years, but domain formation is not a simple consequence of negative charge. A systematic phase diagram for these robust assemblies shows that domains occur only within a narrow regime near the polyanion's pK's for protonation and cation association. The phase behavior appears well described by a relatively simple model in which - among electrostatic and entropic contributions - counterion entropy outcompetes attractive crossbridging to drive remixing of the highly charged polyacid at high pH. Initial observations extend from polymers to lipid assemblies containing the polyanionic lipid phosphatidylinositol (4,5)-bisphosphate, thus highlighting the generality of ligand-induced domain formation. The results portend a new means to control patterning of self-assemblies.

See more of this Session: Self-Assembled Biomaterials II
See more of this Group/Topical: Nanoscale Science and Engineering Forum

Wrapping of a Single Live Bacterium by Biochemically Modified Graphene (BMG) Sheets: Avenues to Build Nano-Biomachines

Nihar Mohanty, Chemical Engineering, Kansas State University, Manhattan, KS

Vikas Berry, Chemical Engineering, Kansas State University, Manhattan, KS

In this talk, we will discuss the structural and interfacial properties of single bacterium completely wrapped with biochemically modified graphene (BMG). We will demonstrate that tethering Concanavalin-A (ligand) on graphene is highly effective in producing strong and specific interfacing of the BMG with the teichoic acid molecular-groups on bacterial surface. We will also show that the bacteria can remain alive for 12 hours after wrapping and in absence of external nutrients, thereby conferring a substantial window for study and development of bio-driven applications. We will show that the as-produced BMG-bacteria biohybrids have the maximum theoretically possible interfacial area with concomitant resilient interface strength. This research will enable development of biological-interfacing of the sensitive graphene sheets for applications in nano-biomachines, bio-batteries and smart electrochemical circuitry.

See more of this Session: Self-Assembled Biomaterials II
See more of this Group/Topical: Nanoscale Science and Engineering Forum

Thermo-Reversible Nanoparticle Formation Using a Novel Elastin-Like Three-Armed Star Polypeptide

Ali Ghoorchian, Chemical and Biomedical Engineering, Cleveland State University, Cleveland, OH

James Cole, Chemical and Biomedical Engineering, Cleveland State University, Cleveland, OH

Nolan Holland, Chemical and Biomedical Engineering, Cleveland State University, Cleveland, OH

Elastin like polypeptides are responsive materials that have been studied for the past three decades because of their promising characteristics which can be exploited in many different areas including targeted drug delivery, tissue engineering, surface modification, protein purification and more. These materials are a member of a class of materials that show lower critical solution temperature (LCST) and if crosslinked, they produce thermally responsive hydrogels. They can be designed in such a way to respond to different environmental stimuli including temperature, pH, ionic strength and light. The primary sequence of these polypeptides is GVGVP in which the two valines can be substituted with most of the naturally occurring amino acids to change the polypeptide hydrophobicity and consequently its transition temperature. In recent years and thanks to the technical advances in molecular biology, it has become possible to design and construct these materials at the gene level with total certainty regarding their sequence and length. As a result, the transition temperature of these materials can be controlled.

In this study, we successfully designed and expressed a new three-armed star polypeptide, based on pentapeptide repeats of GVGVP followed by a trimer-forming oligomerization domain at the C-terminus of the ELP. These new polypeptides were expressed in E.coli and purified by thermal transition cycling. By characterizing the purified protein using SDS-PAGE and circular dichroism (CD) spectroscopy, we confirmed the formation of the three-armed star polypeptide.

By heating up this star polypeptide above its transition temperature we managed to get stable small micellar particles with the hydrophobic ELP tails being stabilized by the charged oligomerization domains as the head groups. These micelles showed thermal stability without any measurable aggregation up to the temperatures well above the ELP transition temperature and even higher than the oligomerization domain stability temperature. This is a very interesting and different response in comparison to all the typical ELP molecules. Normally for an ELP at the transition temperature, the chains would aggregate to very big random particles which is the start of a phase separation process which eventually leads to soluble and coacervate phases if the solution is kept above transition temperature for a longer time. But the newly designed molecules are stable in their micellar form independent of the time they are kept above their transition temperature without forming bigger aggregates or phase separation.

The transition temperature of the these novel three-armed star polypeptides can be tuned during the gene synthesis by changing the length of ELP molecule or by changing the hydrophobicity of the molecule using different amino acids.

Since these nano-particles can be designed to respond to temperature and pH changes and considering their very small controllable size and their perfectly controllable transition temperature they have a very good potential of being used especially in drug delivery of poorly soluble drugs and modified for targeted drug delivery.

See more of this Session: Self-Assembled Biomaterials II
See more of this Group/Topical: Nanoscale Science and Engineering Forum

Self-Organization and Interactions Between Nanoparticles and β-Amyloid

Elizabeth J. Stewart, Chemical Engineering, University of Michigan, Ann Arbor, MI

Nicholas A. Kotov, Chemical Engineering, University of Michigan, Ann Arbor, MI

Alzheimer's disease in the most common type of dementia; currently, there are 5.3 million people in the United States living with Alzheimer's disease. One of the characteristic abnormalities in Alzheimer's patients is a large deposit of β-amyloid plaques in the brain. There is no treatment available to slow or stop the deterioration of brain cells in Alzheimer's disease[1]. Nanoparticles and β-amyloid peptides are on similar size scales and structural parallelism may lead to the creation of nanoparticle-β-amyloid superstructures that can be manipulated to disrupt the formation of β-amyloid plaques or prevent the aggregation of β-amyloid peptides entirely. This study considers the interactions and assemblies of nanoparticle- β-amyloid systems.

The key challenges of nanotechnology and materials science are: constructing and controlling the specific organization of nanoparticles, determining the unique physical properties of materials at the nanoscale, and developing technologies from nanoscale architectures[2, 3]. In addition to the potential applications to disrupting plaques of Alzheimer's patients, this work develops insight into a fundamental understanding of the interactions between biological materials and inorganic nanoparticles, in turn contributing to solving the key challenges of nanotechnology.

The overall objective of this work is to investigate the interactions between nanoparticles and β-amyloid to determine if nanoparticle- β-amyloid interactions can disrupt the formation of β-amyloid plaques within the brains of Alzheimer's patients through static or dynamic assemblies. The initial stages of this work seek: (a) to create stable organizations of nanoparticles and β-amyloid, (b) characterize these interactions, and (c) determine if a dynamic manipulation of the nanoparticle- β-amyloid structures is needed to disrupt the formation of β-amyloid plaques.

1. 2009 Alzheimer's Disease Facts and Figures, in Alzheimer's and Dementia. 2009, Alzheimer's Association.

2. Mirkin, C.A., Programming the Assembly of Two- and Three-Dimensional Architectures with DNA and Nanoscale Inorganic Building Blocks. Inorganic Chemistry, 2000. 39(11): p. 2258-2272.

3. Whitesides, G.M. and B. Grzybowski, Self-assembly at all scales. Science, 2002. 295(5564): p. 2418-2421.

See more of this Session: Self-Assembled Biomaterials II
See more of this Group/Topical: Nanoscale Science and Engineering Forum

Self-Assembled Peptide-Conjugated Polymer Nanoparticles for Tumor Targeting

Angel Mercado, Chemical Engineering, University of South Carolina, Columbia, SC

Esmaiel Jabbari, University of South Carolina, Columbia, SC

Short sequences of amino acids are especially attractive for self-assembly because nanostructures with varying size, shape, morphology, and surface functional group can be assembled by arranging the 24 naturally accruing amino acids in different sequences or by changing the sequence length. It has been demonstrated that the peptide CV6K2 spontaneously self-assembles in aqueous environment to form nanospheres (NPs) with relatively uniform size distribution. Poly(lactide fumarate) (PLAF) macromer alone does not produce NPs but blends of PLAF and amphiphilic poly(lactide-co-ethylene oxide fumarate) (PLEOF) macromers produce NPs with relatively wide distribution. The objective of this work was to investigate self-assembly and release characteristics of chemotherpauetic drugs from CV6K2-conjugated PLAF nanoparticles. CV6K2 peptides self-assembled into NPs with 70 nm average size and narrow distribution of 30-110 nm. PLAF macromers alone did not produce particles but blends of PLAF and PLEOF macromers produced NPs with 300 nm average size and relatively wide distribution of 160-430 nm. Interestingly, when CV6K2 was conjugated to PLAF and self-assembled in the absence of PLEOF, NPs with 110 nm average size and relatively narrow distribution of 50-170 nm were produced. Furthermore, When a mutant C(V2K)2V2 peptide (a peptide sequence with the same composition as CV6K2 that does not self-assemble into NPs) was conjugated to PLAF, the conjugate self-assembled into NPs with average and distribution similar to PLAF-PLEOF blend. These results demonstrate that the CV6K2 peptide induced self-assembly when conjugated to a non self-assembling PLAF macromer. Furthermoer, the self-assembled CV6K2-PLAF NPs degraded linearly by matrix erosion and released Paclitaxel or Doxorubicin at a constant rate concurrent with matrix erosion. The narrow size distribution peptide conjuagted NPs are potentially attractive for targeted tumor drug delivery.-->

See more of this Session: Self-Assembled Biomaterials II
See more of this Group/Topical: Nanoscale Science and Engineering Forum

Biomimetic Self-Assembly of Triskelion Peptide Analogues Into Supramolecular Polyhedra

Weiping Wang, Chemical and Biomolecular Engineering and Graduate Program of Nano Science and Technology, The Hong Kong University of Science and Technology, Hong Kong, Hong Kong

Ying Chau, Chemical and Biomolecular Engineering, The Hong Kong University of Science and Technology, Hong Kong, Hong Kong

We designed a triskelion molecule using the self-assembling peptide analogue, Fmoc- Phenylalanine-Tyrosine (FPT). The triskelion peptide analogue was obtained by conjugating FPT to Tris(2-aminoethyl)amine (TAEA). The resulting molecule, (FPT_3-TAEA), resembles the three-fold symmetry of the cage-forming protein clathrin. Interestingly, FPT_3-TAEA molecules are found to self-assemble into nanocages not in the pure form, but when mixed with the linear FPT molecules. Dynamic light scattering shows that the diameter of nanocages is around 17 nm. High-resolution TEM image of the samples indicate similar dimension and a hollow structure. We propose that the structure of the nanocage follows the architecture of a Buckyball that has 60 vertices. One FPT_3-TAEA molecule is located at each vertex. Two FPT molecules form anti-parallel β-sheet, which results in two clingy ends linking two legs of different triskelion molecules to form each polygon edge. This model is supported by the measurement of dimension. To validate this hypothesis, we study the change in the formation of nanocages when altering the ratio of FPT_3-TAEA to FPT. The study provides some criteria to construct polyhedra structures from peptide self-assembly and insights to control the pore size and cage dimension.

See more of this Session: Self-Assembled Biomaterials II
See more of this Group/Topical: Nanoscale Science and Engineering Forum

AUTHOR INDEX

AUTHOR INDEX

AUTHOR INDEX

AUTHOR INDEX

AUTHOR INDEX

AUTHOR INDEX

AUTHOR INDEX

AUTHOR INDEX